Food Supply Chains in Cities

Emel Aktas • Michael Bourlakis
Editors

Food Supply Chains in Cities

Modern Tools for Circularity and Sustainability

palgrave
macmillan

Editors
Emel Aktas
School of Management
Cranfield University
Cranfield, Bedfordshire, UK

Michael Bourlakis
School of Management
Cranfield University
Cranfield, Bedfordshire, UK

ISBN 978-3-030-34064-3 ISBN 978-3-030-34065-0 (eBook)
https://doi.org/10.1007/978-3-030-34065-0

This Palgrave Macmillan imprint is published by the registered company Springer Nature Switzerland AG.
The registered company address is: Gewerbestrasse 11, 6330 Cham, Switzerland

Foreword

Keeping the world's city dwellers well nourished becomes an ever-greater logistical challenge. The proportion of the global population living in urban areas is expected to rise from just over a half to two-thirds over the next 30 years. Meanwhile the chains supplying our bourgeoning towns and cities continue to lengthen and diversify as consumers are offered, and conditioned to expect, an expanding range of fresh produce all year round. The pervasive adoption of just-in-time replenishment across the food distribution networks that supply our cities has driven down inventory levels, making them more vulnerable and less resilient. Within cities, food distribution is being gradually transformed by consumers switching to online grocery retailing and the home delivery of ready-cooked meals. As their diets change, and will have to change more radically to meet greenhouse gas emission targets, food supply chains in and around cities will need to be reconfigured. Their management will also require re-engineering to minimise the shocking levels of food waste, food poverty and malnutrition that one finds side-by-side in many major cities. It is not hyperbole to say that the quality of billions of people's health and well-being depends on food supply chains at the urban level being effectively planned and managed. This in turn requires a substantial research effort to find new and improved ways of delivering vast quantities of food efficiently and reliably to urban populations.

This book successfully showcases some of the recent studies that have been done on this vitally important subject. It illustrates the thematic, methodological and geographical breadth of the research currently underway. The chapter authors address many different aspects of the subject including food security, supply chain collaboration, the digitalisation of food logistics, the optimised routing of food deliveries, the creation of food hubs, the development of new reverse logistics networks to maximise the recycling of food waste and the costs and benefits of 'vertical farming' within urban areas. Many of the studies are linked by a common interest in promoting the long-term sustainability of food distribution. In the course of their research the authors have deployed many different approaches and techniques and drawn data from a variety of sources, thereby setting a good example to future researchers in this field. The have also provided an invaluable service in reviewing the extensive literature that has already emerged at the interface between food distribution, supply chain management and city logistics.

The book's geographical coverage is particularly impressive as it focuses on cities in five continents and in countries at very different levels of economic development. The cities themselves, such as Athens, Istanbul, New York, Sao Paulo, Beijing and Beirut, vary enormously in their size, structure, wealth and food supply systems, giving the reader a sense of the extent to which business and policy innovations need to be adapted to local circumstances. There will clearly be no-one-size-fits-all in the upgrading of food supply chains to meet the culinary needs of the next generation of urban dwellers.

Kühne Logistics University
Hamburg, Germany

Alan McKinnon

Preface

We are excited to present you this edited volume on food supply chains in cities. Our interest in food supply chains dates several decades, including projects on agricultural supply chains, food retail, and food distribution. We have decided to edit this volume on food supply chain management with specific focus on urban environments owing to the increasing urbanisation and the permanent need to feed the populations that reside in our cities. At the time of completing this project, the population projections suggest that we will have 8.5 billion people living on Earth by 2030 and 9.7 billion by 2050 (United Nations, 2019). Since urban areas are anticipated to grow with increasing population, we face the challenge of making our cities sustainable and food supply chains play a key role in addressing this challenge. The Sustainable Development Goal 11 is dedicated to making cities and human settlements inclusive, safe, resilient, and sustainable and the reader will find several papers in the edited volume that are aligned with this goal.

We open the book with a chapter entitled 'Towards inclusive food supply chains' authored by Virva Toumala. This chapter addresses the growing concern around food security, highlighting the links between food supply chain management and urban food security. Synthesising the literature, Virva concludes five themes that affect urban food security: Rural bias, urban political ecology, urban poverty, food supply chain management, and food retail. Food security indicators are grouped under dietary diversity, consumption behaviour, experiential measures, and self-assessment measures. Conclusions pave the way for future research on reducing urban food insecurity, resilient food supply chains, and micro retailers specifically in developing countries.

The second chapter is entitled 'An Economic and Environmental Comparison of Conventional and Controlled Environment Agriculture (CEA) Supply Chains for Leaf Lettuce to US Cities' and it is authored by Charles F Nicholson, Kale Harbick, Miguel I Gómez, Neil S Mattson. The chapter addresses an increasingly important area of vertical farming unravelling implications towards reducing food miles and food waste, depleting lower resources, and increasing accessibility to fresh produce in cities. The authors build a simulation model that integrates economic and environmental considerations into a comparison of lettuce supply chains: field-grown, greenhouse-grown, and plant factory-grown. Key conclusions suggest that just because the produce is grown in or near cities, it does not necessarily mean that it is more environmentally-friendly. Total economic and environmental costs should be considered when making investment decisions.

The third chapter, authored by Markus Rabe, Moritz Poeting, Astrid Klueter, is entitled 'Evaluating the Benefits of Collaborative Distribution with Supply Chain Simulation'. Investigating the urban food distribution networks, the authors use simulation to identify the location of an urban consolidation centre that will be used by multiple manufacturers collaboratively to ship goods to retailers in cities. Using primary data from Athens, Greece, the authors evaluate the impact of collaborative transport practices on the number of deliveries, establishing the reduction in environmental and social indicators such as traffic congestion, noise, and emissions. Future work suggests investigating different types of transport vehicles such as electric vans or e-cargo bikes to fulfil the last mile.

The fourth chapter investigates a typical urban distribution problem: vehicle routing with simultaneous pick-up and deliveries. Authored by Mustafa Cimen, Cagri Sel, Mehmet Soysal, the chapter is entitled 'An Approximate Dynamic Programming Approach for A Routing Problem with Simultaneous Pick-Ups and Deliveries in Urban Areas' and incorporates a reality of urban environments, time-dependent vehicle speeds, into modelling. A state-of-the-art literature review informs a real-life case study from one of the most crowded cities in Europe: Istanbul, with a population of 15 million residents as of 2017. The authors provide a succinct synthesis of the literature around sustainability and green supply chain management applicable to planning and scheduling problems, inventory management, lot-sizing, and production-distribution problems. The authors propose an approximate dynamic programming-based algorithm for vehicle routing of a soda manufacturer to reduce the distance travelled and the corresponding carbon emissions.

The fifth chapter is entitled 'The Role of Informal and Semi-Formal Waste Recycling Activities in a Reverse Logistics Model of Alternative Food Networks' and it is authored by Luis Kluwe de Aguiar and Louise Manning. Introducing a hierarchy of the waste sector, the authors discuss the barriers to participating in urban waste management. Coupling the literature with a case study from Brazil, the authors develop alternative business models for municipal solid waste management considering not only economic but also social impact of the model. As a case study, they investigate the reverse logistics network for cooking oil and sow the economic gains to stakeholders involved.

The sixth chapter is entitled 'Shortening the supply chain for local organic food in Chinese cities' and, herewith, we turn our attention to local organic food provision with a case from China. The authors Pingyang Liu and Neil Ravenscroft investigate the elements of alternative food networks and find out how local food supply chains operate under immature markets. An interesting aspect of this work is the impact of new digital technologies on enabling farmers access new markets. The chapter provides comparisons with Western food systems and draws lessons for circularity and sustainability of food supply chains. Future research should investigate new instances of internet-enabled tools and how they intermediate the relationship between farmers and consumers.

Luciana Marques Vieira and Daniele Eckert Matzembacher investigate the role of digital business platforms in reducing food waste in the seventh chapter entitled 'How digital business platforms can reduce food losses and waste?'. Underpinning one of the key reasons for food waste, the lack of coordination among the members of the food supply chain, the authors discuss the potential of digital business platforms in reducing food losses and waste. Interviews with 28 stakeholders including producers, cooperatives, and governmental agencies suggested that these platforms can connect production and consumption and create a new market for food that would otherwise be discarded. One of the key conclusions is related to consumer awareness around food waste and digital platforms that can help reduce waste so that sufficient demand is generated for these emerging actors to survive in the food chain.

We then shift our focus to feeding students in Chap. 8, entitled 'The Role of Food Hubs in Enabling Local Sourcing for School Canteens', authored by Laura Palacios-Argüello; Iván Sanchéz-Díaz; Jesús González-Feliu; Natacha Gondran. In this chapter, the authors examine food hubs to serve school canteens within the broad discipline of institutional catering and propose design suggestions for these food hubs to increase local and organic products offered to schools. Combining qualitative and quantitative data, the authors classify food hubs in line with attributes identified from the literature: stakeholder focus, structure, and ownership, logistics functions, and commercial services. Then they solved vehicle routing problems to distribute food from food hubs to school canteens in the Auvergne-Rhône-Alpes region in France. Their proposed food hub allocation resulted in 25% reduction in distance travelled.

In a similar theme, João Roberto Maiellaro, João Gilberto Mendes dos Reis, Laura-Vanessa Palacios-Arguello, Fernando Juabre Muçouçah, Oduvaldo Vendrametto investigate food distribution from local farms to schools and, subsequently, Chap. 9 is entitled 'Food Distribution in School Feeding Programmes in Brazil'. Using Geographical Information System data on Mogi das Cruzes, Brazil, the authors established the current operating conditions and proposed new operating principles that minimised cost of transportation. Their proposed solution meant that 74% of the schools served in Mogi das Cruzes were within 5 km of a farm.

In the final chapter entitled 'A Descriptive Analysis of Food Retailing in Lebanon: Evidence from a Cross-Sectional Survey of Food Retailers', Rachel A. Bahn and Gumataw K. Abebe draw our attention to supermarkets and their competition with traditional food retailers from a country with limited evidence on the extent and the process of supermarket penetration. Reporting information collected from 49 food retailers in Lebanon, the authors conclude that there is no clear evidence of a pattern in expansion (i.e. socio-economic, geographical, or food product-based expansion). Retailers in Lebanon tend to maintain long-term written agreements with suppliers, which are chosen on the basis of quality and food safety, followed by price. Their findings are expected to inform policy makers on sustainable food system development, which also include monitoring consolidations in the food chain, availability of diverse and nutritious food in the market for urban and rural consumers, and participation of specific actors in the upstream supply chain. Based on the breadth of the works included, we are confident that we have brought together a plethora of fascinating areas of research in food supply chain management such as food security, food distribution, alternative food networks, food waste, school feeding programmes, and food retail. We are cognisant of possible limitations of this venture and, therefore, we acknowledge that it is not possible to include every possible research stream in a volume like this one. However, we do anticipate that the works included in this volume will generate and stimulate future research on food supply chains and will pave the way for more sustainable food supply chain practices.

Bedfordshire, UK
2019

Emel Aktas
Michael Bourlakis

Reference

United Nations. (2019). *World population prospects*. Retrieved from https://population.un.org/wpp/Publications/Files/WPP2019_Highlights.pdf.

Contents

Towards Inclusive Urban Food Supply Chains 1
Virva Tuomala

An Economic and Environmental Comparison of Conventional and Controlled Environment Agriculture (CEA) Supply Chains for Leaf Lettuce to US Cities 33
Charles F. Nicholson, Kale Harbick, Miguel I. Gómez, and Neil S. Mattson

Evaluating the Benefits of Collaborative Distribution with Supply Chain Simulation 69
M. Rabe, M. Poeting, and A. Klueter

An Approximate Dynamic Programming Approach for a Routing Problem with Simultaneous Pick-Ups and Deliveries in Urban Areas 101
Mustafa Çimen, Çağrı Sel, and Mehmet Soysal

The Role of Informal and Semi-Formal Waste Recycling Activities in a Reverse Logistics Model of Alternative Food Networks 145
Luis Kluwe de Aguiar and Louise Manning

Shortening the Supply Chain for Local Organic Food in Chinese Cities 171
Pingyang Liu and Neil Ravenscroft

How Digital Business Platforms Can Reduce Food Losses and Waste? 201
Luciana Marques Vieira and Daniele Eckert Matzembacher

The Role of Food Hubs in Enabling Local Sourcing for School Canteens 233
Laura Palacios-Argüello, Ivan Sanchez-Diaz, Jesus Gonzalez-Feliu, and Natacha Gondran

Food Distribution in School Feeding Programmes in Brazil 265
João Roberto Maiellaro, João Gilberto Mendes dos Reis, Laura Palacios-Argüello, Fernando Juabre Muçouçah, and Oduvaldo Vendrametto

A Descriptive Analysis of Food Retailing in Lebanon: Evidence from a Cross-Sectional Survey of Food Retailers 289
Rachel A. Bahn and Gumataw K. Abebe

Index 347

Notes on Contributors

Gumataw K. Abebe, PhD (Wageningen University) is Assistant Professor of Agribusiness Marketing and Management at the Faculty of Agricultural and Food Sciences, American University of Beirut, Lebanon. His research focuses on exploring the efficiency and effectiveness of agri-food supply chain governance in sub-Saharan Africa and Middle East and Northern Africa in response to recent trends for food safety and food quality; the extent of food retail expansion, its drivers and impact on food security; and the economics of agricultural technology adoption in arid and semi-arid regions.

Luis Kluwe de Aguiar, is the Course Manager for BSc (Hons) Food Degree Programmes and Senior Lecturer on Food Security and Sustainability as well as Food Marketing at Harper Adams University, Newport, Shropshire, UK. Prior to becoming a full time academic, he worked in different agribusiness-related roles, namely as Agricultural Attaché at the Foreign and Commonwealth Office British Embassy (Brazil), the Ministry of Foreign Relations of Japan; Commercial Manager of agricultural storage systems; Food Technologist for barley malting, and Sales Promoter for an importer of horticultural seeds. He has undertaken international consultancy work and has an extensive publication portfolio in the field of food marketing and sustainability.

Rachel A. Bahn is Coordinator of the Food Security Program and Instructor of Agribusiness at the Faculty of Agricultural and Food Sciences, American University of Beirut. Her research focuses on issues related to food security including food retail transformation, sustainability of food systems, and the intersection of agriculture and conflict. Previously she served as an economist with the United States Department of Treasury and the United States Agency for International Development. She holds degrees from the Johns Hopkins University School of Advanced International Studies (SAIS) and Saint Joseph's University.

Mustafa Çimen is an Assistant Professor of Management Science section of the Hacettepe University Business Administration Department, in Turkey. He holds BSc degree in Business Administration in Hacettepe University (Turkey), MSc degree in Production Management and Quantitative Techniques from Hacettepe University (Turkey), and a PhD in Management Science from Lancaster University (United Kingdom). Dr. Çimen's research is mainly focused in the field of approximate dynamic programming, particularly applied in green logistics and inventory optimization problems. He is author and co-author to a number of research papers. He also serves as a reviewer in several international journals of operational research, transportation and logistics.

Miguel I. Gómez is Associate Professor in the Charles H. Dyson School of Applied Economics and Management, Cornell University. He concentrates his research and outreach program on two interrelated areas under the umbrella of food marketing and distribution. One is Food Value Chains Competitiveness and Sustainability, which involves multidisciplinary collaborations to assess supply chain performance in multiple dimensions. The second is Food and Agricultural Markets, combines theory and outreach methods, emphasizing key concepts such as price transmission, demand response, buyer-seller negotiations, and market power. In addition, his research extends to economic development, examining the incentives smallholder farmer participation in food chains.

Natacha Gondran is an assistant professor at the École des Mines de Saint-Étienne, Environment City and Society. She defended in 2015 her Habilitation Degree about the evaluation and representation of environmental issues with a focus on ecological and energy transition. She holds a PhD (2001) and an energy and environmental engineering degree from INSA Lyon. She is currently involved in the social and environmental responsibility delegation of Mines Saint-Etienne. She has co-authored 31 articles in peer-reviewed journals (Ecological Indicators, International Journal of Sustainable Development, Journal of Cleaner Production, among others). She co-authored 3 books, among which one which is often quoted as a reference for ecological footprint, and 9 chapters of books.

Jesus Gonzalez-Feliu is associate professor at École des Mines de Saint-Étienne, Environment, City and Society. He holds a MSc. in Civil Engineering (INSA Lyon, France), a PhD. in Computers and Systems Engineering (Politecnico di Torino, Italy) and is Dr. Hab. in Transport Economics (Université Paris Est). He is also member of the Institute of Urban Transport and Mobility (Lyon, France), and has led various projects on sustainable logistics systems. His main research interests are sustainable urban logistics actions and policies, sustainability evaluation and sustainable dashboards, group decision support, consensus search and social evaluation of transport systems and logistics organizations. He is a recognized expert in the field of sustainable urban logistics. He is co-editor more than 10 special issues of peer-reviewed journals and participates actively on other editorial activities.

Kale Harbick studied computer science and robotics at the University of Southern California, earning a PhD in 2008. He researched crater detection for Mars landers and autonomous helicopters at NASA-JPL, before shifting his career focus to energy. Kale taught many courses in physics and energy management for over 10 years. He managed a program at Oregon Department of Energy which implemented energy efficiency measures in over 800 K-12 schools. He was twice asked to testify for the Oregon State Legislature on residential energy efficiency programs. His current research in the CEA group at Cornell focuses on environmental controls and modeling of energy and light.

A. Klueter is a full-time Ph.D. Student at the department of IT in Production and Logistics at the TU Dortmund University. She holds two B.Sc. from the faculty of mechanical engineering and the faculty of

Business, Economics and Social Sciences. For her M.Sc. she focused on the field of logistics. She graduated with a master thesis on a concept for the implementation of a collaborative route planning approach using a data-driven supply chain simulation model.

Pingyang Liu is an Associate Professor in the Department of Environmental Science and Engineering, and Director of the Center for Land and Resource Economics Studies, at Fudan University, China. He holds a PhD in economics, and his current research interests include rural environmental management and economic transition, collective action and natural resource management, and regional sustainable development under the context of rapid urbanization. As Principal Investigator, Dr Liu has undertaken over 20 research projects sponsored by NSFC, WWF China, FORHEAD, Central and local governments of China.

João Roberto Maiellaro has a Ph.D. (in progress) in Production Engineering at Universidade Paulista, Master's degree in Production Engineering at Methodist University of Piracicaba (2004), Post-graduation in Production Engineering at the Polytechnic School of the University of São Paulo, and Bsc. in Production Management at Universidade São Judas Tadeu. Coordinator of Logistics Technology course and Full Professor of Operational Research and Simulation Modelling, Quality Management and Supply Chain Management in Centro Estadual de Educação Tecnológica Paula Souza. Large management experience in production engineering, with emphasis on planning, designing, and control of manufacturing and quality systems.

Louise Manning, Professor in Agri-food and Supply Chain Security at the Royal Agricultural University, Cirencester has worked for over 30 years in the agri-food supply chain in a range of roles. She lives on a mixed-enterprise family farm in Herefordshire. Her expertise is in the area of food integrity including food safety, food quality, food crime, and governance, social and corporate responsibility. She advises both government and corporate businesses on supply chain risk assessment techniques and associated mitigation strategies. Louise has published over 65 publications in a range of peer-reviewed journals and regularly presents at conferences and writes books, book chapters and for trade press and electronic media.

Neil S. Mattson joined Cornell University in 2007, having previously earned a PhD from the University of California Davis and MSc at the University of Minnesota. He is currently an Associate Professor and greenhouse extension specialist. His research focuses on the physiology of horticultural crops in controlled environments especially strategies to reduce energy use, LED lighting, nutrient management, and abiotic stress physiology. Mattson seeks to translate research into industry adoption through several outreach mechanisms (bulletins, webinars, and online tools). Mattson is the director of Cornell's Controlled Environment Agriculture program.

Daniele Eckert Matzembacher is a Ph.D. student in Administration at the Federal University of Rio Grande do Sul (UFRGS), in the area of Innovation, Technology, and Sustainability. Her research interests are in the impact of food waste reduction through entrepreneurship and sustainability in food supply chains.

Fernando Juabre Muçouçah is a graduate in Agronomic Engineering at São Paulo State University, master in Agronomy (horticulture) from São Paulo State University and Ph.D. in Agronomy (irrigation and drainage) by the State University of São Paulo. MBA in project management at FGV. Experience in the field of agronomy, with an emphasis on horticulture and fruit growing. Consultant in ornamental plants and flower production. Professor in the course of Technology in Agribusiness at Faculty of Technology (Fatec) Mogi das Cruzes and Dean of the Faculty of Technology (Fatec) Ferraz de Vasconcelos.

Charles F. Nicholson is an Adjunct Associate Professor in the Charles H. Dyson School of Applied Economics and Management and an International Professor of Applied Economics and Management in the College of Agriculture and Life Sciences at Cornell University. His work focuses on the economic and environmental impacts of changes to food systems in both high- and low-income country settings. He has PhD and MSc degrees from Cornell University and a BSc in Economics from the University of California, Davis.

Laura Palacios-Argüello has a Ph.D. in Industrial Engineering from École des Mines de Saint-Étienne, France. Master's Degree in Information Systems Management, speciality Supply Chain Management, from I.A.E., Grenoble, France. Master's Degree in Industrial Engineering from the National University of Colombia. Researcher in ELUD project—(Efficacité Logistique Urbaine alimentaire et Durable) funded by University of Lyon-LabEx IMU, working on providing a scenario simulation framework for evaluation and analysis of the sustainable performance of new urban food delivery systems. Previously, research assistant at the Department of Industrial Engineering and member of research group SEPRO (Society, Economy and Productivity), at the National University of Colombia. Research interests include food supply and distribution, city logistics, urban food distribution systems, sustainable food supply chains and eco-responsible demand.

M. Poeting is full-time PhD Student and researcher at the Institute of Transport Logistics at TU Dortmund University. He holds a B.Sc. and M.Sc. in Logistics from TU Dortmund University and member of the "Simulation in Production and Logistics" section of the simulation society ASIM. His research focus is on parcel logistics, the simulation of logistics facilities, and sustainable urban transport.

M. Rabe is Full Professor for IT in Production and Logistics at the TU Dortmund University. Until 2010 he had been with Fraunhofer IPK in Berlin as head of the corporate logistics and processes department, head of the central IT department, and a member of the institute direction circle. His research focus is on information systems for supply chains, production planning, and simulation. Markus Rabe is vice spokesman of the "Simulation in Production and Logistics" section of the simulation society ASIM, member of the editorial board of the Journal of Simulation, member of several conference program committees, has chaired the ASIM SPL conference in 1998, 2000, 2004, 2008, and 2015, Local Chair of the WSC'2012 in Berlin and Proceedings Chair of the WSC'18. More than 190 publications and editions report from his work.

Neil Ravenscroft is Professor and Head of the School of Real Estate and Land Management at the Royal Agricultural University, UK. He holds a PhD in land management from the University of Reading and is a

Professional Member of the Royal Institution of Chartered Surveyors. His research interests include sustainable farm and rural businesses, land tenure and access to land for land-poor farmers, ecological economics and community supported agriculture. Dr Ravenscroft has undertaken a wide range of research projects, funded by the UK Research Councils, EU Framework and Joint Initiative Programmes, charities, NGOs and the UN FAO.

João Gilberto Mendes dos Reis has graduated in Logistics at Faculty of Technology (Fatec) Zona Leste, master in Production Engineering at Universidade Paulista and Ph.D. in Production Engineering from Universidade Paulista. Postdoctoral by Universidade Paulista and the University of Porto. Experience in production and logistics management. He is a member of International Federation for Information Processing (IFIP) in the Work Group 5.7—Advances in Production Management Systems and the scientific committee of International Conference on Information Systems, Logistics and Supply Chain Management (ILS). Currently is Full Professor at the Universidade Paulista (UNIP) in the Graduate Studies Program in Production Engineering and Administration. Besides, is Professor of Graduate Studies in Agribusiness at Universidade Federal da Grande Dourados.

Ivan Sanchez-Diaz is a Senior Lecturer at the Division of Service Management and Logistics at Chalmers University of Technology (Sweden), where he is part of the Urban Freight Platform funded by the VREF. Dr Sánchez-Díaz received his BS in Civil Engineering from the Universidad del Norte (Colombia); and his MSc and PhD in Transportation Engineering from the Rensselaer Polytechnic Institute (New York). Dr Sánchez-Díaz has been involved in multiple projects related to freight demand modelling, freight policy and innovations in the United States and in Sweden. He is an expert in urban freight transportation policy and modelling; his research involves econometric analyses, behavioural modelling and analytical modelling.

Çağrı Sel is an Assistant Professor of the Industrial Engineering Department at Karabük University in Turkey. He completed his undergraduate education in 2007 and received his MSc in 2010 from Industrial

Engineering department at Kırıkkale University (Turkey). He received his PhD in 2015 from Industrial Engineering at Dokuz Eylül University (Turkey). He was also a Guest Researcher of the Operations Research and Logistics group at Wageningen University (The Netherlands). His research areas are supply chain management, production scheduling and distribution planning, mathematical programming, constraint programming, heuristic and metaheuristic methods.

Mehmet Soysal is an Associate Professor of the Operations Management section of the Hacettepe University Business Administration Department, in Turkey. He has a B.Sc. degree in Management and a M.Sc. degree in Operations Research from Hacettepe University, in Turkey. He conducted his Ph.D. research regarding decision support modeling for sustainable food logistics management in the Operations Research and Logistics Group at Wageningen University, The Netherlands. Here, he participated in several European Commission funded projects as a researcher. His main research interest is on developing decision support tools for green logistics and supply chain problems. Dr Soysal's research is published in international peer reviewed journals, book chapters, technical reports and at international conferences. He also serves as a reviewer in several international journals of operational research, transportation and logistics.

Virva Tuomala is a doctoral candidate in supply chain management and social responsibility at Hanken School of Economics in Helsinki. The topic of her thesis is inclusive supply chain management for urban food security, looking particularly at societal issues that prevent poor urban dwellers from sufficient access to nutrition. The research is interdisciplinary, drawing from geography and development studies in addition to SCM. Empirical work for the thesis has been conducted in a township in the Western Cape, South Africa and a new study is planned to be completed in Bangkok, Thailand. Virva has BSc and MSc degrees in logistics from University of Turku and an MSc in environment and development from University College London. Before embarking into academia, she worked for an IT consulting company in Manila, Philippines.

Oduvaldo Vendrametto has BSc. in Physics at Universidade de São Paulo, master in Physics at Universidade de São Paulo and Ph.D. in Production Engineering at Universidade de São Paulo. Has experience in production engineering projects, focusing on industrial and agricultural areas. Currently is Full Professor of Universidade Paulista and Coordinator of Graduate Studies Program in Production Engineering at Universidade Paulista.

Luciana Marques Vieira holds a PhD from the University of Reading and is currently a Professor in Operations Management and Sustainability at Fundacao Getulio Vargas (FGV) in Sao Paulo, Brazil. Her main research interests are related to the governance of agri-food chains in an emerging country context. She has published in journals such as Journal of Cleaner Production, International Journal of Emerging Markets, British Food Journal, Industrial Management & Data Systems, among others.

List of Figures

Towards Inclusive Urban Food Supply Chains

Fig. 1 Review methodology. (Source: Adapted from Colicchia & Strozzi, 2012) 3
Fig. 2 Division among disciplines 7
Fig. 3 Timeline of publications included 7
Fig. 4 Factors in urban food security. (Source: Author) 22

An Economic and Environmental Comparison of Conventional and Controlled Environment Agriculture (CEA) Supply Chains for Leaf Lettuce to US Cities

Fig. 1 System boundary for analysis of costs and environmental impacts of three lettuce supply chains. (Source: Authors) 38

Evaluating the Benefits of Collaborative Distribution with Supply Chain Simulation

Fig. 1 The 2015 share of transport in the European Union. (Source: EEA, 2017) 70
Fig. 2 Schematic operating principle of a UCC. (Source: Authors) 75
Fig. 3 First UCC scenario to be evaluated. (Source: Rabe et al., 2018) 86
Fig. 4 Second UCC scenario (selected areas) to be evaluated. (Source: Rabe et al., 2018) 88

An Approximate Dynamic Programming Approach for a Routing Problem with Simultaneous Pick-Ups and Deliveries in Urban Areas

Fig. 1 The interactions between the research fields reviewed. (Source: Author) 105
Fig. 2 Number of research papers per year. (Source: Author) 106
Fig. 3 The soda distribution network of the carbonated soft-drink company. (Source: Google Maps website) 128
Fig. 4 The resulting two vehicle routes for non-urban travels with pickup and deliveries. (Source: Google Maps website) 130
Fig. 5 The resulting three vehicle routes for urban travels with pickup and deliveries. (Source: Google Maps website) 132

The Role of Informal and Semi-Formal Waste Recycling Activities in a Reverse Logistics Model of Alternative Food Networks

Fig. 1 Informal waste sector hierarchy. (Source: Adapted from Masood & Barlow, 2013; Sandhu, Burton, & Dedekorkut-Howes, 2017) 147
Fig. 2 Map of Rio Grande do Sul. (Source: Google Maps, 2019) 153
Fig. 3 AFN Model for Municipal Waste Collection Service (DMLU) of Porto Alegre via pig production. (Source: Authors' own elaboration) 156
Fig. 4 Map of Brazil—Campo Grande. (Source: Google Maps, 2019) 158
Fig. 5 AFN Model for waste oil collection and reuse. (Source: Authors' own elaboration) 160

Shortening the Supply Chain for Local Organic Food in Chinese Cities

Fig. 1 Location of Shared Harvest Farm (Shunyi Base and Tongzhou base). (Source: Drawn by authors) 175
Fig. 2 Weibo (L) and WeChat (R) of Shared Harvest Farm online platform. (Source: Mobile phone screen shot of the official Weibo site and the online wechat store of Shared Harvest, by authors) 177
Fig. 3 Supply chains for fresh produce in Shanghai. (Source: Drawn by authors) 180
Fig. 4 Sources of information. (Source: Drawn by authors) 184
Fig. 5 Timeline of Shared Harvest Farm. (Source: Drawn by authors) 185

Fig. 6 Online products from Shared Harvest Farm (left) and other farms (right, usually marked by fair trade 公平贸易). (Source: Mobile phone screen shot of the online wechat store of Shared Harvest, by authors) 188
Fig. 7 Integrated organization structure at Shared Harvest Farm. (Source: Liu, Ravenscroft, Ding, & Li, 2019) 189

How Digital Business Platforms Can Reduce Food Losses and Waste?

Fig. 1 DBP as promoters of coordination mechanisms to reduce food waste. (Source: Authors' own elaboration) 210
Fig. 2 Retail-rejected food returned to producer for failing to meet aesthetic standards. (Source: Authors' own elaboration) 215
Fig. 3 Potatoes reject by retail. (Source: Authors' own elaboration) 215
Fig. 4 Disposal of food on land due to the lack of marketing channels. (Source: Authors' own elaboration) 216

The Role of Food Hubs in Enabling Local Sourcing for School Canteens

Fig. 1 The flow of information among institutional catering stakeholders. (Adapted from Lessirard et al., 2017). (a) The consumer could be the parent or the student who eats at the school canteen; the patient in the hospital; or the employee in the enterprise, among others. (b) The customer or purchaser could be a territorial authority, public institution, or private enterprise 236
Fig. 2 Food Hub Stakeholders. (Adapted from Palacios-Argüello & Gonzalez-Feliu, 2016) 239
Fig. 3 FH distribution scenarios. (Source: Authors' own elaboration) 253
Fig. 4 Producers' allocations based on administrative subdivision. (Source: Authors' own elaboration) 254
Fig. 5 Producers' allocations based on geographical proximity, using QGIS. (Source: Authors' own elaboration) 255
Fig. 6 Producers' allocations based on geographical proximity using VRP. (Source: Authors' own elaboration) 256

Food Distribution in School Feeding Programmes in Brazil

Fig. 1 Current distribution point. (Source: Authors) 274
Fig. 2 Producers, current distribution point and schools. (Source: Adapted from QGIS 2.18 (2018)) 275

Fig. 3 New distribution centre at School Feeding Department. (Source: Authors) 276
Fig. 4 The proposed distribution point. (Source: Adapted from QGIS 2.18 (2018)) 278
Fig. 5 Kernal hot spot map analysis. (Source: Adapted from QGIS 2.18 (2018)) 279
Fig. 6 Buffer method analysis. (Source: Adapted from QGIS 2.18 (2018)) 280

A Descriptive Analysis of Food Retailing in Lebanon: Evidence from a Cross-Sectional Survey of Food Retailers

Fig. 1 Lebanese population living in urban, peri-urban, and rural areas. (Source: World Bank (2019) and authors' calculations based on World Bank data) 295
Fig. 2 Total and per capital value of retail sales by grocery retailers in selected countries, 2017. (Source: Euromonitor International, 2018) 299
Fig. 3 Retail value of traditional and modern grocery retailers' sales, 2003–2017. (Source: Euromonitor International, 2018) 300
Fig. 4 Frequency of change of suppliers, by food category (Among retailers selling the food product, excluding "do not know" responses. The numbers of responses per food product category are as follows: fruits (n = 22), vegetables (n = 24), dairy (n = 31), fresh meat (n = 11), fresh fish (n = 2), baked goods (n = 30), frozen goods (n = 22), canned goods (n = 35), dry goods (n = 35), and snacks (n = 31).). (Source: Authors) 319
Fig. 5 Selected economic statistics for Lebanon. (Source: World Bank, 2018) 326

List of Tables

Towards Inclusive Urban Food Supply Chains

Table 1 List of journals included 6
Table 2 Categories of food security indicators 19

An Economic and Environmental Comparison of Conventional and Controlled Environment Agriculture (CEA) Supply Chains for Leaf Lettuce to US Cities

Table 1 Selected characteristics of field, CEA GH and CEA PF operations analysed 40
Table 2 Total landed cost for delivery of 1 kg lettuce to wholesale produce markets in New York City and Chicago from field-based production, a CEA greenhouse (GH) and a CEA plant factory (PF) 47
Table 3 Environmental impacts for the delivery of 1 kg lettuce to wholesale produce markets in New York City and Chicago from field-based production, a CEA greenhouse and a CEA plant factory 48
Table 4 Comparison of environmental outcomes from current and previous studies of lettuce production and supply chains 50
Table 5 Comparisons of the baseline and best case total landed cost of 1 kg lettuce delivered to wholesale markets in New York and Chicago metropolitan areas 51

Table 6 Characteristics and assumptions for energy modeling of GH and PF production systems 55
Table 7 Detailed calculations for field-based lettuce production operating costs per acre 56
Table 8 Detailed calculations for field-based lettuce production structure and equipment costs per acre 57
Table 9 Detailed calculations for field-based lettuce transportation costs 58
Table 10 Annual operations costs for greenhouse and plant factory operations 58
Table 11 Total investment costs for structures, land and equipment for greenhouse and plant factory operations 60
Table 12 Detailed calculations for CEA lettuce transportation costs 61
Table 13 Energy use and cost calculations for New York metropolitan area greenhouse and plant factory operations 61
Table 14 Energy use and cost calculations for Chicago metropolitan area greenhouse and plant factory operations 63
Table 15 Detailed calculations of CO_2 equivalent emissions, field, greenhouse and plant factory operations 65

Evaluating the Benefits of Collaborative Distribution with Supply Chain Simulation

Table 1 Simulation results for the second UCC scenario (selected areas) 90
Table 2 Post-calculated KPIs for the second UCC scenario (selected areas) 91

An Approximate Dynamic Programming Approach for a Routing Problem with Simultaneous Pick-Ups and Deliveries in Urban Areas

Table 1 Journals publishing the research papers 106
Table 2 Research areas in the field 108
Table 3 The most cited research papers 109
Table 4 The keywords used in the research papers 110
Table 5 The research papers classified in each category 116
Table 6 The demand and the return amounts per replenishment cycle and service times for each crossdocking point 129
Table 7 The demand and the return amounts per replenishment cycle and service times for each dealer in İstanbul 129

Table 8	Distances in kms between the soda-filling facility (node 0) and the crossdocking points (nodes 1–10)	135
Table 9	Distances between the crossdocking point in İstanbul (node 0) and the dealers in İstanbul (nodes 1–18)	136

How Digital Business Platforms Can Reduce Food Losses and Waste?

Table 1	Framework to analyze inter-organisational relationships providing solutions to food waste	207
Table 2	Interviews in data collection	213
Table 3	Summary of the results in phases 1 and 2	222

The Role of Food Hubs in Enabling Local Sourcing for School Canteens

Table 1	Considerations of FH	240
Table 2	Information about interviewees	242
Table 3	FHs characteristics	245
Table 4	Stakeholder focus of regional FH	247
Table 5	Structure of regional FH	248
Table 6	Functions of regional FH	248
Table 7	Economic impacts of regional FH	250
Table 8	Social impacts of regional FH	250
Table 9	Environmental impacts of regional FH	250
Table 10	Producers per department	252
Table 11	FH locations, product origins and customer locations	252
Table 12	Distance results for producers' allocations based on administrative subdivision	254
Table 13	Distance results for producers' allocations based on geographic proximity	257

Food Distribution in School Feeding Programmes in Brazil

Table 1	Schools distances from the distribution points	279

A Descriptive Analysis of Food Retailing in Lebanon: Evidence from a Cross-Sectional Survey of Food Retailers

Table 1	Retailer categories and market orientation: primary location of customers' residence	300
Table 2	Retailer classifications and market orientation: *p*-value resulting from means testing (Chi-square)	301

Table 3 Retailers' operations: mean values and *p*-values resulting from means testing 303
Table 4 Monthly business expenditures by line item per store 306
Table 5 Average total monthly business expenditures per store, area, and employee 307
Table 6 Retailers' marketing and customer demand/profile: *p*-value resulting from means testing 308
Table 7 Retailers' perceived changes in demand across food groups 310
Table 8 Retailers' perception of customer income category 311
Table 9 Retailers' perceptions of determinants of customer food outlet selection 312
Table 10 Retailers offering additional customer services 312
Table 11 Retailers' supply chain structures: mean value and *p*-value resulting from means testing 314
Table 12 Average number of suppliers, by food product, across retailer types 316
Table 13 Retailers' perceptions of important features in selecting suppliers 320
Table 14 Population and survey responses across registered retailers 327
Table 15 Complete statistical analysis results: *p*-values resulting from means testing 328

Towards Inclusive Urban Food Supply Chains

Virva Tuomala

1 Introduction

Urban food insecurity is a complex issue with multiple levels and contexts. The FAO (2009) defines food security as "a situation that exists when all people, at all times, have physical and economic access to sufficient safe and nutritious food that meets their dietary needs and food preferences for an active and healthy life". The concept is further divided into four pillars: availability, access, utilisation and stability. Urban food security is particularly concerned with access to food rather than its fundamental availability. As urbanites are generally dependent on the market for food, the stability of the market in terms of e.g. prices is also an important consideration. As such, food insecurity affects the urban poor disproportionally, creating a challenge for food supply chains.

This chapter uses an interdisciplinary perspective to explore food security in an urban context. Through a literature review of three fields in the

V. Tuomala (✉)
Hanken School of Economics, Helsinki, Finland
e-mail: virva.tuomala@hanken.fi

E. Aktas, M. Bourlakis (eds.), *Food Supply Chains in Cities*,
https://doi.org/10.1007/978-3-030-34065-0_1

social sciences—supply chain management (SCM), urban geography, and development studies—a holistic point of view of the particularities of urban contexts and how some of them result in food insecurity is presented. A majority of the previous research in urban food insecurity has been done in the fields of urban geography and development studies. SCM provides an appropriate additional lens into the flows of food that are brought into the city, and the powers that govern them, even if the articles do not explicitly discuss food security per se. Sustainable SCM (SSCM) brings insights into global supply chains functioning in different societies and contexts. An inclusive supply chain serves all strata of society, including those disproportionally affected by food insecurity. In this chapter, marginalized neighbourhoods and households, urban poverty, and informality are analysed as underlying factors behind urban food insecurity, highlighting stable access to food over availability concerns. The lens of urban political ecology (UPE) is used to tie the social, environmental, and economic flows together into a coherent urban reality.

The chapter is organized in the following manner: the next section briefly presents the review method, followed by the main themes that emerge from the analysis. It then moves on to a discussion section and finishes with a conclusion section that draws the chapter together and suggests future empirical research avenues in urban food insecurity.

2 Review Method

Literature reviews are an essential part of academic research and constitute what empirical work is derived from to locate its relevance in the discipline. However, they can also be a self-standing academic contribution that produces an overview of a discipline and synthesis of a potentially scattered collection of literature (Seuring & Gold, 2012). Tranfield, Denyer, and Smart (2003) present two aims for literature reviews: first, they map the discipline being researched through existing literature to gain a holistic picture of it. Second, they identify gaps in the body of knowledge that need to be addressed for the discipline to develop.

The methodology for our paper draws mainly from the principles of a systematic literature review (SLR) (Tranfield et al., 2003) and content analysis (Seuring & Gold, 2012), forming a critical review of the extant literature on supply chain management (SCM), urban/human geography, and development/food studies. The SLR approach is an efficient choice in performing initial searches within existing research and selecting the most relevant articles to be included (Colicchia & Strozzi, 2012). The content analysis method for literature reviews then provides a more robust technique of categorising and evaluating the collected material through interpretation of the underlying arguments presented in the literature (Seuring & Gold, 2012). This combination was inspired by the methodology presented by Colicchia and Strozzi (2012), which used an SLR with citation network analysis. Content analysis was chosen for this review as the research areas are interdisciplinary and as such cross-disciplinary citations would be lacking. The review methodology is presented in Fig. 1.

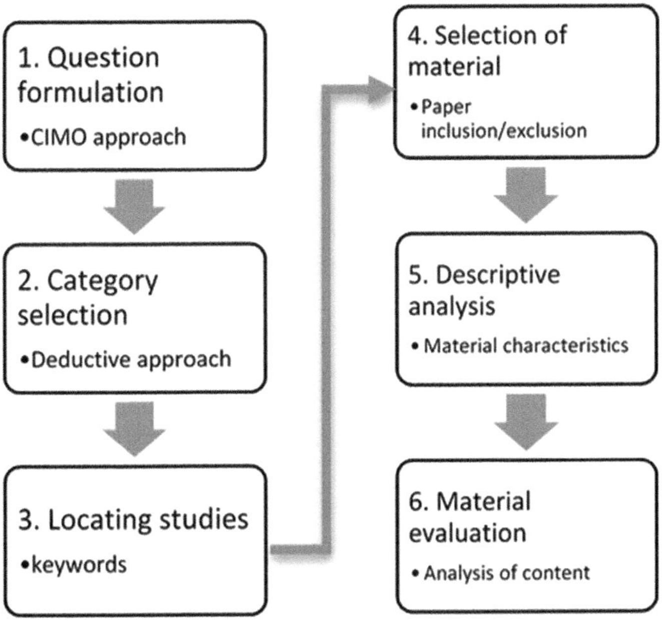

Fig. 1 Review methodology. (Source: Adapted from Colicchia & Strozzi, 2012)

An SLR requires certain criteria to be set prior to any literature search to determine what is included in the review (Denyer & Tranfield, 2009). This is in reference to transparency, which is the first of four core principles for SLRs in management and organisation studies proposed by Denyer and Tranfield (2009). First, the author must be clear and concise about the processes used in the review, as well as about any prior knowledge regarding the literature in question. The second principle, *inclusivity* refers to the range of studies in the review, while the *explanatory* principle synthesises the literature into a coherent whole. The *heuristic* nature of the review points to practical solutions that derive from the SLR. Together these four pillars make the SLR an objective overview of extant research and eliminate bias and error (Colicchia & Strozzi, 2012; Denyer & Tranfield, 2009). Content analysis on the other hand delivers two layers of interpretation: first through a statistical and superficial analysis of the literature followed by delving thoroughly into the underlying ideas of the research (Seuring & Gold, 2012).

Applying the methodology presented in Fig. 1 to the context of this review, the research objectives of this chapter are first formulated using the Context, Intervention, Mechanisms, and Outcome (CIMO) approach (Denyer & Tranfield, 2009). The *context* is the relationship between urban food security and modern food supply chains and their management (Battersby & Peyton, 2014; Lorentz, Kittipanya-Ngam, & Singh Srai, 2013; Silvestre, 2015). Increased urbanization and consequently urban poverty are the underlying phenomena that frame the context of the study (*intervention*). Urban dwellers are generally completely dependent on the market for food, so the food supply chain becomes an important *mechanism* in the lives of urban citizens. Issues such as the location of grocery retailers and the informal food sector are examined in this research as well. The *outcome* of the research is comprehensive insight into how supply chain management can be harnessed into providing solutions for urban food insecurity.

Materials were collected through a keyword search as well as through a snowballing technique of following up on references. Keywords included "food security", "urban", "supply chain" and "sustainable" and combinations thereof. Searches were carried out in the Scopus database. A deductive approach was used when determining the categories for the literature, i.e. the disciplines of SCM, urban geography and development studies were selected before the search was conducted. Delimiting the search area was an imperative step to take prior to any searches, as food security can be approached from a variety of different perspectives, such as nutritional science or public health. The unit of analysis was defined as one English language article, report or book chapter. Reports from organizations such as the United Nations contain relevant data for this type of subject, therefore the review is not limited to academic publications.

All in all, 61 pieces of literature were included in the review: 52 academic articles and 9 from other sources such as book chapters and reports. Publications were read carefully to surmise their relevance to the review. Division between the three main disciplines was fairly equal, although development studies had both the most articles in total (21) and the most cited journal (*World Development*, with five articles). Table 1 presents the journals included in the SLR. The division among the disciplines, presented in Fig. 2, was made according to the source journal.

The material was limited to the twenty-first century, save two articles from the latter half of the 1990s. A majority of the material is from the past five years, with a peak in 2013–2014. Descriptive statistics are the first layer of content analysis. The types of journals and the timeline of the articles provide important insight on the materials, and the trends that emerge. Figure 3 depicts the timeline of the selected publications.

In content analysis, material evaluation is the process of reviewing the literature collected, which begins in the next section with the main themes of urban food insecurity that emerged from the literature. The underlying layers of the material were always analysed with the category in mind, i.e. whether this article appeared in an SCM journal or development studies. It was, however, noted that development studies and urban geography overlapped considerably, and SCM stood out with different themes, such as urban logistics (Lagorio, Pinto, & Golini, 2016) and challenging markets (Amine & Tanfous, 2012). This chapter brings new approaches into SCM, the *recipient* discipline, from the *referent* fields of development studies and urban geography. Five subthemes were recognised from the literature that fed into the overarching theme of urban food insecurity:

Table 1 List of journals included

Journal title	#	CiteScore	SNIP
World Development	5	3.92	2.542
Geoforum	4	3.09	1.543
International Journal of Retail & Distribution Management	4	2.52	1.263
Food Security	3	2.66	1.569
Global Food Security	3	4.72	2.151
International Journal of Physical Distribution & Logistics Management	3	4.15	1.5
International Journal of Production Economics	3	5.42	2.386
Urban Forum	3	1.51	1
Applied Geography	2	3.75	1.569
Food Policy	2	4.53	2.453
Supply Chain Management: An International Journal	2	5.45	1.793
Antipode	1	3.88	2.412
Area	1	2.09	0.949
British Journal of Management	1	3.23	1.818
Capitalism Nature Socialism	1	0.49	0.274
Development Policy Review	1	1.47	1.177
Development Southern Africa	1	0.8	0.931
Environment and Urbanization	1	2.54	1.582
Geography compass	1	2.74	1.336
Habitat International	1	3.37	1.704
International Business Review	1	3.2	1.61
Journal of Business Ethics	1	2.91	1.639
Journal of Economic Geography	1	4.14	2.634
Journal of Industrial Ecology	1	3.93	1.439
Journal of Supply Chain Management	1	7.04	2.934
Science as Culture	1	1.38	0.841
Sociological Quarterly	1	1.68	1.194
Theory, Culture & Society	1	2.17	1.73
Urban Studies	1	2.91	1.826
Total number of articles	52		

- Rural bias
- Urban political ecology
- Urban poverty
- SCM in diverse contexts
- Urban food retail

The following section discusses the findings from the literature, with the subsections further indicating a subtheme and how it relates to urban food insecurity.

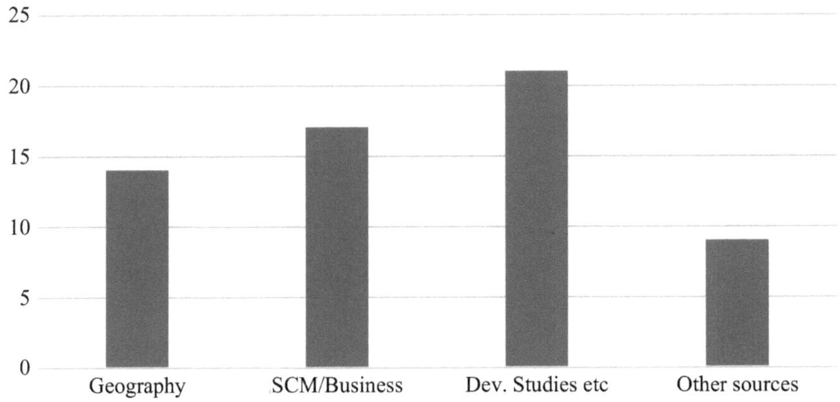

Fig. 2 Division among disciplines

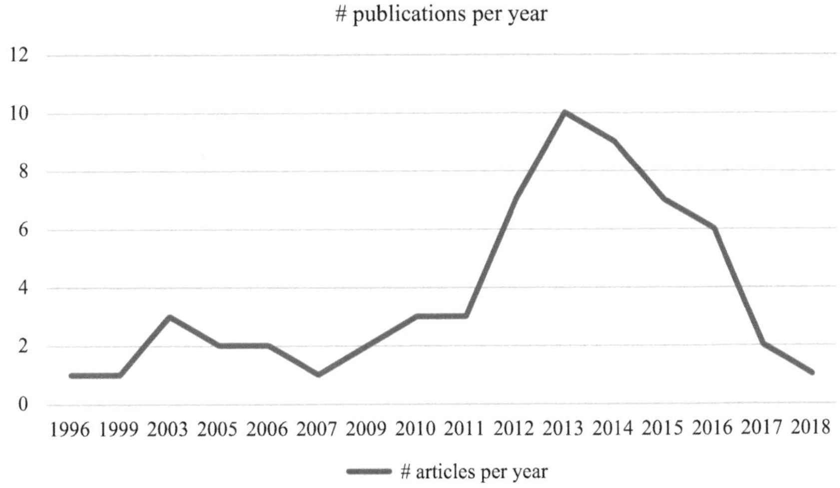

Fig. 3 Timeline of publications included

3 Urban Food Insecurity

As an outcome of the content analysis, several main themes emerge from the literature. These are a rural bias, which has dominated the food security conversation in the past; urban political ecology, acting as a theoretical framework for the analysis; urban poverty and food supply chain management, as well as urban food retail.

3.1 Rural Bias

Food security is defined to include all people having physical, social and economic access to sufficient, safe and nutritious food to maintain a healthy and active life (FAO, 2009). According to the United Nations (2015), over half of the global population resides in urban areas and the number is estimated to reach 66% by 2050. Considering these statistics, when referring to "all people" as in the FAO definition of food security, the urban population should be prominently featured. Nevertheless, policy and planning still consider food security issues to affect mainly rural communities. In an exhaustive overview of both international and specifically African food security agendas, Crush and Frayne (2011) found very little reference to the urban context of food security or the different dynamics of malnutrition among the urban poor as opposed to the rural poor.

Cities worldwide face the challenge of planning and managing their resources to meet the needs of their inhabitants (Lynch, Maconachie, Binns, Tengbe, & Bangura, 2013). As opportunities for livelihoods become scarce in rural areas, people flock to urban centres in search of employment and increased quality of life (Zezza, Carletto, Davis, & Winters, 2011) Often urbanisation happens at such a pace that it is impossible for planning to keep up, placing poverty and urbanization among the biggest development challenges of the twenty-first century (Frayne, Crush, & McLachlan, 2014). Urban environments cannot be considered as merely physical spaces, but rather they are a nexus of social, environmental and economic activities brought together by a myriad of people from all strata of society (Battersby & Marshak, 2013). The contexts explored here are consequences of the various factors at play in an urban setting that lead to food insecurity and other manifestations of inequality.

While the widely accepted FAO (2009) definition of food security mentions *access* to food as an explicit factor, it is ensuring availability to food through increased agricultural production that garners attention in food security discourse (Battersby & Crush, 2014). Maxwell (1999) credits this focus to a rural bias, wherein urban dwellers' nutritional challenges are downplayed, as urban markets seemingly have ample food on offer. However, fundamental availability is only one side of the coin. Access to any food that is physically available may be constrained either spatially or financially for many urbanites (Battersby & Crush, 2014). These constraints manifest on a small scale, i.e. neighbourhood or even household scale. Urban food insecurity is thus often overshadowed by other, more visible urban challenges, such as unemployment, infrastructure, and overcrowding (Maxwell, 1999). In addition to availability and access, the definition includes the pillars of utilization and stability of food. Utilization includes e.g. sanitation and the adequate nutritional attributes of the food being consumed. Stability concerns continued access to food through shocks and volatility, which is particularly relevant in an urban context (Carletto, Zezza, & Banerjee, 2013). The four pillars overlap and influence the discourse on food security as a whole.

3.2 Urban Political Ecology

Political ecology (PE) studies the relationship of humans to their physical environment as well as the effect matters of the environment have in terms of the relationships between humans (Brown & Purcell, 2005). This involves for example interactions between people and the places where they live, the elements of power surrounding the division of natural resources and placing power debates into historical context in the political economy (Page, 2003). Power relations are a central notion in an urban context, which is why PE lends itself well to examining the social, economic, and environmental flows that constitute a city (Swyngedouw & Heynen, 2003). It is further emphasized by Swyngedouw and Heynen (2003) that urban political ecology (UPE) considers the city to be a cumulative result of socio-environmental processes mediating natural resources, hence it is counterintuitive to say that the urban is somehow disconnected from the rest of nature.

There are several reasons why UPE is an appropriate lens for examining urban food insecurity (Agyeman & McEntee, 2014). UPE is capable of examining hybrid relationships due to the socio-ecological production of the urban environment at its centre. The analysis of food deeply intertwines physical and social factors. Food is also distinctly rooted in power relations, as neoliberal market forces ensure that food insecurity does not affect everyone in a given city (Agyeman & McEntee, 2014). UPE highlights the material conditions and relationships that determine which communities are more prone to food insecurity and why. This stems from the Marxist history of PE, where the interest of the elite was served at the expense of the marginalised in society (Swyngedouw & Heynen, 2003). UPE places socio-ecological processes into a physical location, such as a city, which is also fitting for the investigation of urban food insecurity as the physical environment is one of the core constructs in this analysis (Agyeman & McEntee, 2014).

In one of the early writings on UPE, Swyngedouw (1996) uses water as an entry point into juxtaposing nature and the city. Water, according to him, has biological properties alongside symbolic, cultural, and socio-economic meanings. Its distribution is wrought with notions of power, affecting who has access to water and why (Swyngedouw, 1996). Similarly, food as a necessity for human life, suffers from similar power debates leading to repression of some and dominion of others. As stated by Engels (quoted in Heynen, 2006, p. 129) "*power lies in the hands of those who own, directly or indirectly, foodstuff and the means of production*". Indirect ownership is especially relevant in the case of food supply chains. It can refer to for example food distribution and the infrastructure these activities require. Food retail outlets, namely supermarket chains, control every part of the supply chain without actually producing anything (Reardon, Henson, & Berdegué, 2007). The modern food industry has made nourishment into a tradeable commodity, but it also acts as an expression of power, encompassing race, gender, and class (Agyeman & McEntee, 2014). Those that do not belong to the "elite" group of owners become more vulnerable and insecure based on their position in the political economy, entitlements, and even physical location. The commodification of food further exposes communities to food insecurity through globalization, complex infrastructure, and indirect access, i.e. the dependence on the market and the power of the food retailers, that define the contemporary food system (Heynen, 2006).

3.3 Urban Poverty

The United Nations (2015) predicts that over two-thirds of the global population will reside in cities by 2050 and most of this urbanization will occur among the poorer members of society. However, official statistics referencing generally agreed upon indicators for poverty (e.g. daily income) commonly show that very few urban dwellers are counted as poor in financial terms (Floro & Swain, 2013). A lack of reliable data on urban poverty is both a cause and an effect of these biased indicators and consequently influences research on issues such as urban food insecurity and informality, both direct consequences of urban poverty (Fox, 2014; Weatherspoon, Oehmke, Dembele, & Weatherspoon, 2015). Many cities in the Global South, e.g. Bangkok and Manila, have seen substantial economic growth in terms of per capita consumption, but the growth does not portray the widening gap between the wealthy and poor cohabiting the same urban space (Floro & Swain, 2013).

Statistically, urbanites may not be considered poor, but there is an inherent contradiction when it comes to the number of people living in informal settlements. The United Nations (2015) estimates that 870 million people reside in urban informal settlements, or slums, which amounts to approximately one third of all urban residents in developing nations. Informal settlements are characterized by precarious dwelling structures, lack of access to services, overcrowding and inadequate water and sanitation facilities (Fox, 2014). As infrastructure is one of the most visible ways to 'measure' development, the prevalence of slums leaves many cities in the Global South wanting in this respect (Frayne & McCordic, 2015). The shortcomings of urban infrastructure and its inability to meet rapid urbanization patterns of many developing countries contributes to the environmental context of urban poverty (Tolossa, 2010), which goes further than measuring mere income.

Informality is at the heart of urban poverty, encompassing habitat, employment, and nourishment. Informal settlements have been traditionally treated as by-products of modernization, where rural migrants are forced to initially dwell before being integrated into urban society as their wealth increases. However, this view is heavily criticized due to the questionable practical and historical evidence of wealth trickling down as societies develop (Fox, 2014). Frayne and McCordic (2015) further highlight that whatever scarce services and infrastructure are available to the urban poor are usually much more expensive than what is provided in more formalized areas. This further exacerbates inequalities experienced by the urban poor.

While slums represent the physical manifestation of informality and urban poverty, the informal economy plays just as relevant a role in the urban landscape. Study of the informal economy encompasses several fields of research and focuses on its size and composition, drivers, causes, and linkages between several areas of development (Chen, 2012). There are many definitions of the informal economy, but most of them include functioning outside of government regulation due to inefficient enforcement as well as not paying taxes (Brown & McGranahan, 2016; Elbahnasawy, Ellis, & Adom, 2016).

The prevalence of urban poverty and informality indicates that social sustainability is still lacking in global supply chains (Yawar & Seuring, 2017). In order to be inclusive of e.g. the urban poor, supply chains must consider the needs of people before profits (Järvensivu et al., 2018; Touboulic & Ejodame, 2016). For example. Järvensivu et al. (2018) suggest that the aim of the food industry should be to provide people with food rather than make profits for stakeholders. This approach would not exclude people based on their financial/societal status. For more inclusive business practices and economies in general, Brown and McGranahan (2016) emphasize the importance of understanding the informal economy for several reasons. First, employment in the informal sector is constantly growing and to build more inclusive economies and supply chains, it must be taken into serious consideration. Second, inclusive supply chains cannot be achieved if the relationship between authorities and the informal sector is not improved. Without sufficient understanding of the dynamics of the sector, this relationship will remain strained and achieving inclusiveness difficult. Third, due to its heterogeneous nature, activities in the informal economy vary tremendously in terms of efficiency and profitability. With sufficient knowledge on the functioning of the sector it is possible to recognize the valuable innovations and harness them appropriately. Finally, the importance of the informal sector for people at the bottom of the pyramid, i.e. the poorest members of society, has to be considered whenever the informal sector is concerned (Brown & McGranahan, 2016).

Whereas Brown and McGranahan (2016) and Chen (2012) treat the informal economy as an opportunity for innovation and research, some view informality as a hindrance to development that should be contained (Elbahnasawy et al., 2016). Proponents of this latter approach have a similar aim to understand the informal economy, but rather than learn from it they wish to develop policies to hinder its spread and adverse effects. Characteristics of the informal economy they wish to forestall are its distorting macroeconomic effects, inefficient resource use, and small-scale, low-return production (Elbahnasawy et al., 2016). This school of thought subscribes to the *voluntarist* view of the informal economy, which maintains that actors choose to remain informal for example to avoid paying taxes or dealing with bureaucracy for registering a company.

In addition to the voluntarist view, there are three other schools of thought described by Chen (2012). *Dualists* consider informality as completely separate from the formal sector, existing as a side effect of insufficient economic growth. The *legalists* consider stringent state regulations and persistent entrepreneurs to be the main drivers of the informal economy. Unlike the voluntarists, legalists view entrepreneurs as being forced to come up with alternative ways to function due to e.g. anti-poor legislation and policies. The final school are the *structuralists*, who consider informality as being driven by capitalist growth via formal companies exploiting cheap informal labour to maximize their profits (Chen, 2012). None of these schools provides definitive answers, as the informal economy derives from a multifaceted set of circumstances, but help in understanding it. In a developing world urban context, informality cannot be bypassed. Brown and McGranahan (2016) even claim that no endeavour seeking social or environmental transformation of the global economy can ignore informality, a view that is echoed in this paper. The dynamics of the informal food sector are extremely relevant in urban food insecurity research and its potential solutions. In line with this view, Brown and McGranahan (2016) suggest adding a fifth element to the four schools of thought on informality suggested by Chen (2012): the *inclusionist* school. This approach emphasizes initiatives whereby marginalized communities can partake in decisions that concern their everyday lives and rights.

3.4 SCM in Diverse Contexts

Urban dwellers, poor and wealthy, are predominately dependent on the market for their nourishment, hence the global food supply chain is a relevant concept in the study of urban food insecurity. The market for food has transformed in the last two decades in several significant ways (Maertens, Minten, & Swinnen, 2012). The most relevant in terms of urban populations are the global level shifts in retail dynamics towards multinational corporations (MNCs) and the increasing integration of supply chains that comes with it. MNCs are usually based in developed or advanced markets, and see developing countries as new business opportunities especially in the wake of globalisation-friendly policies put in place in the 1990s (Reardon et al., 2007). There are however substantial challenges in the performance of operations as well as network designs when entering markets in the Global South.

Lorentz et al. (2013) utilize the framework of geography, resources, and institutions (GRI) (not to be confused with Global Reporting Initiative) to review how market characteristics affect supply chain functions and overall business success. The geographical attributes of a country may bring significant difficulties to logistics operations due to e.g. topography, landlocked status, and population density. Isolation from major trade routes and especially from sea ports have been acknowledged as causes for food insecurity (Brown, Silver, & Rajagopalan, 2013). The levels of infrastructure in a country or city are also a major player in isolating an area from the global market, often resulting from restricted availability or allocation of resources (Lorentz et al., 2013). Urban logistics brings about a particular set of challenges. Cities are not just a physical location, but contain an endless number of stakeholders and 'systems within systems' (Lagorio et al., 2016). Navigating these systems is not just a technical matter but requires careful analysis of underlying factors, which according to Lagorio et al. (2016) is still widely under researched in urban logistics literature. Frameworks like the GRI (Lorentz et al., 2013) are helpful in establishing a more comprehensive research paradigm that will translate into practice as well.

Other resource related challenges suggested by Lorentz et al. (2013) in their framework are the level of competence in the local labour force, issues with currency and interest rates, and a general lack of capital for SCM-related activities. The institutional characteristics also contribute to the potential difficulties of entering a new market or in the isolation of a market from global trade. Economic and political barriers are often more potent than physical ones in hampering participation in global trade (Brown et al., 2013). The institutions that uphold these barriers often control many SCM related activities such as international trade and transport infrastructure, as well as contribute to the overall business and national culture (Lorentz et al., 2013).

Lorentz et al.'s (2013) GRI framework is a useful tool for discussing the role SCM plays in urban food insecurity, because it considers underlying SCM factors that have considerable effects on activities in developing and otherwise challenging markets. Additionally, the food supply chain is extremely complex and dynamic, relevant to all and under constant scrutiny from consumers and industry alike (Beske, Land, & Seuring, 2014). Considerations for the sustainability of the food supply chain have emerged from the scrutiny as negative environmental and social impacts of the food industry come to the attention of the public.

3.5 Urban Food Retail

Urban populations are net food buyers and spend a majority of their income on food (Maitra, 2016) so it is prudent to analyse the existing retail network in conjunction with urban food insecurity. Food security as such is not discussed much in SCM literature, but there is an active stream of literature on supermarkets spreading as the dominant form of food retail in developing or emerging economies. Due to its multifaceted effects on society, this stream is not limited to the retail or SCM literature, but discussed in multiple disciplines across the social sciences and while the research in these different disciplines often overlaps, an interdisciplinary discourse on the subject has not been established (Nguyen, Wood, & Wrigley, 2013).

Studies conducted in North Africa explored consumer reactions to Western supermarket chains entering the market (Amine & Lazzaoui, 2011; Amine & Tanfous, 2012). Research conducted in Morocco (Amine & Lazzaoui, 2011) concludes that the entrance of a major supermarket chain into the market contributes to largely social, even classist consequences. If the retailer fails to sufficiently adapt to local customs and traditions, in terms of store location and functioning, the lower income population generally does not have physical or financial access to these chains. Additionally, if the chains do not cater to local traditions and customs, this can lead to outright boycotting of the new retailers (Amine & Tanfous, 2012). In a North African context this might mean having a wide array of alcohol or pork products for sale, or not offering halal products. Amine and Lazzaoui (2011) further discuss the relative success of a smaller, local chain that eschews suburban locations that are impossible to reach by public transport and upholds local service standards.

In analysing activities of global large retailers (GLR) in Africa, Nandonde and Kuada (2016) echo the above findings. They further specify that micro retailers are one of the keys to operating a food retail business in Africa and the developing context in general. Low-income consumers rely on micro retailers in part due to the social nature of the transactions which gives them the opportunity to e.g. buy with credit. The informal food sector is a relevant part of the micro retailing scene in developing countries. GLRs have been able to penetrate developing countries due largely to the liberalization of trade in these countries as well as the acceleration of foreign direct investment (FDI) into these markets (Nguyen et al., 2013).

Research focusing mainly on price as an access factor sees supermarket expansion into developing countries as positive in terms of food security, as supermarkets can price their products at lower rates thanks to economies of scale (Battersby & Peyton, 2014). However, Minten, Reardon, and Sutradhar (2010) point to several factors that indicate that the urban poor are not fully embracing the supermarket trend. The aforementioned micro retailer domination of the traditional retail scene is one, as the underlying social dynamics permit consumers to negotiate and utilise credit systems when shopping food. Foreign supermarkets are also likely to offer products that are branded, pricing the urban poor out as a result.

Finally, the urban poor prefer to buy products in smaller quantities and more often, which is usually not possible in formal supermarkets (Minten et al., 2010).

Applying parts of the GRI framework of Lorentz et al. (2013) to a specific city could pinpoint some of the individual SCM as well as broader issues faced by that city and what leads its poorer citizens to be food insecure. Examining the geography of supermarkets in Cape Town, Battersby and Peyton (2014) point out that there are eight times as many supermarkets in wealthier areas than in areas where the lowest income residents reside. Supermarkets located near transport hubs had considerably lower stocks of fresh produce than those located in richer neighbourhoods, providing commuters with less nutritious options. While it is very common in Cape Town and other similar cities to run errands while commuting, the physical locations of the supermarkets are still extremely relevant. The supermarkets with the best selections are located in areas inaccessible to the poor, due to poor public transport links. The legacy of apartheid also maintains the wealthy areas far away from the poor areas (Battersby & Peyton, 2014). These flows can lead to poorer areas becoming 'food deserts', which are areas where fresh and healthy food at affordable prices is either difficult or impossible to obtain (Myers & Sbicca, 2015). Battersby and Crush (2014) argue that a food desert is a useful concept for spatializing food insecurity, but it fails to grasp the complexity of the foodscapes many urban poor, especially in a developing context, face. It is not uncommon for a poor urbanite to work far from their place of residence, meaning that a lot of their household expenditure is allocated to transit, as food retail outlets are much more likely to be located around transport hubs and business districts than in poorer communities (Battersby & Crush, 2014).

This legacy highlights the importance of historical and political flows in the context of food security. Many cities in both developing and developed countries experience classist or racist divisions similar to apartheid, which decades later still influence daily activities. This goes to show that geography and institutions are intertwined in the urban landscape, making the GRI framework particularly apt.

The next section introduces several indicators and frameworks for measuring food insecurity from several different perspectives.

4 Measuring Urban Food Insecurity

The complicated and multidimensional nature of food security makes establishing reliable measuring tools challenging. Carletto et al. (2013) recommend using the level of analysis as a starting point. For example, global or regional level food security needs to be approached from a more production-oriented perspective, whereas for urban individuals or households, accessibility is the key factor. Headey and Ecker (2013) elaborate on this view by emphasizing the different types of data that need to be derived from any food insecurity indicators, which they have divided into three broad categories. First, the indicator needs to be comparable among a multitude of different groups, such as regions and social factors. The rural bias is an example of the indicator favouring one type of community over another and providing skewed data. Second, they suggest combining several inter-temporal approaches, i.e. long-term data, seasonality as well as shocks such as disasters and spikes in prices and incomes. Last, any indicators need to address the distinct nutritional needs of different demographics. The needs of infants or the elderly differ greatly from those of an adult male doing physical work for example. This variety does not only refer to physical needs, but differences in cultures and other societal influences (Headey & Ecker, 2013).

Maxwell, Vaitla, and Coates (2014) explore different food security indicators and their primary uses. They have divided them into four categories, depicted in Table 2.

Many of these indicators are being used by organizations, such as the World Food Programme, which work on food security issues. On their own, all of these indicators do capture at least one dimension of food insecurity, but when relying on a single indicator there is a risk of misrepresentation (Coates, 2013). The African Food Security Urban Network (AFSUN) has conducted extensive research projects both in Southern Africa as well as other regions around the world through affiliations and partnered research projects (McCordic & Frayne, 2018). In these projects, they have generally administered the HDDS, MAHFP, and HFIAS indicators in their surveys, but as the surveys were carried out in extremely diverse environments, a lot of the challenges started with the basic

Table 2 Categories of food security indicators

Category	Indicators	Use
Dietary diversity and food frequency	Food consumption score (FCS), household dietary diversity score (HDDS)	What people eat and how often, the diversity of intake
Consumption behaviour	Coping strategies index (CSI), months of adequate household food provisioning (MAHFP)	Behaviour and strategies engaged when food and/or money runs out
Experiential measures	Food insecurity experience scale (FIES), household food insecurity access scale (HFIAS), household hunger scale (HHS)	Household behaviour following insufficient food access
Self-assessment measures	e.g. Gallup polls	Subjective assessment of food security status

Source: Adapted from Maitra (2016), Maxwell et al. (2014), McCordic and Frayne (2018)

definitions and linguistic obstacles. Defining basic concepts such as 'food' and 'household' to both the respondents as well as the research teams required considerable efforts, as well as then figuring out who the best person to survey would be in each household.

The HDDS scale measures dietary diversity by asking respondents whether anyone in the household has consumed any of 12 particular food groups[1] in the past 24 hours. HDDS has been found to correlate with other food security measures used (Maxwell et al., 2014). The HFIAS scale, measuring the social, physical, and economic experiences of limited access to food with a set of questions, has also been found to correlate with other indicators, such as childhood under-nutrition. The questions, such as 'in the past four weeks, did you worry your household would not have enough to eat?', are answered with a never-often scale.[2] The MAHFP measures the months in the past year when the household has gone without proper food provisioning. Information on the reliability of this indicator is limited, but McCordic and Frayne (2018) predict it will become more relevant as research in this field advances. Subjective assessments done through Gallup polls could provide relevant information, but are easily influenced by for example dietary reference points, cultural issues and feelings of shame and are thus easily corrupted (Headey & Ecker, 2013; Maxwell et al., 2014).

The lack of data in urban food insecurity research is partly to blame for the negative bias its research is experiencing (Frayne et al., 2014). The more data that is gathered within similar delimitations and scopes, the more reliable the indicators become. It is however also acknowledged generally that measuring food security will always be difficult and the best method is to conduct more research using a combination of different techniques and indicators (Maxwell et al., 2014). A consensus that rises from the literature is that focusing solely on undernutrition and food availability is no longer an option, and especially in urban environments the experiential indicators as well as those researching coping strategies need to be emphasized (e.g. Headey & Ecker, 2013; Maxwell et al., 2014; McCordic & Frayne, 2018). The next section discusses the multifaceted nature of urban food insecurity in detail, focusing on the factors and metabolic flows that underlie it.

5 Managing Urban Metabolic Flows for Food Security

The classic trifecta of economic, social and environmental factors in sustainability research is almost perfectly encompassed in food (Grant, Trautrims, & Wong, 2017). There is substantial literature that explores the environmental effects of food distribution but social aspects of sustainability are investigated less both in food SCM as well as SSCM literature in general (Yawar & Seuring, 2017). While urban food insecurity has elements of all three aspects of sustainability, its social effects are the most far reaching and complex, encompassing health as well as questions of race and class. Touboulic and Walker (2015) in their literature review of SSCM theories note there appears to be an increasingly holistic approach to SSCM research, focusing on all three sustainability pillars rather than focusing on just one. The incorporation of external stakeholder pressures, as opposed to incentives based purely on profitability, also indicates that SSCM literature is widening its focus (Touboulic & Walker, 2015).

The exclusion of marginalized populations from supply chains is identified by Yawar and Seuring (2017) as an under researched social issue in SCM literature. Marginalization can derive from a number of different contexts, but this paper focuses on poverty as the marginalizing factor. Poverty alleviation remains a top priority in development policy and research, but supply chain studies rarely approach it. Hall and Matos (2010) however investigated the inclusion of poverty stricken communities in supply chains through a study in the Brazilian biofuel industry. They base their research on the ideas laid out by Prahalad (2005) regarding innovation and skills found at the 'bottom of the pyramid' and integrating them into mainstream supply chains. Gold, Hahn, and Seuring (2013) also base their research on this approach, concluding that working at the bottom of the pyramid level can provide new avenues in business logic, which is needed as supply chains become increasingly globalized and function in diverse societies.

Silvestre (2015) suggests that supply chains evolve through a learning process, and sustainability in a supply chain starts to make business sense towards the end of the learning process. The external environment as well as institutional presence are key factors in the learning process, tying Silvestre's (2015) argument in with Lorentz et al.'s (2013) GRI framework. The challenging environments of emerging economies and their lack of institutional strength has contributed to the difficulties in establishing sustainable supply chains in such markets (Silvestre, 2015).

Authors in the field of SCM call for sustainable supply chain management practices to become more holistic, including social issues along with the environmental and the economic (Grant et al., 2017; Touboulic & Walker, 2015; Yawar & Seuring, 2017). As well, researchers note the predominantly Western and MNC perspective on sustainability (Pagell & Shevchenko, 2009; Touboulic & Ejodame, 2016) which, especially in the case of urban food insecurity, ignores the many nuances of cities in different geographical and cultural contexts. The challenges presented in the external environment and the lack of institutional strength suggested by Silvestre (2015) are likewise observed from a Western perspective.

Some factors leading to urban food insecurity have been summarized in Fig. 4.

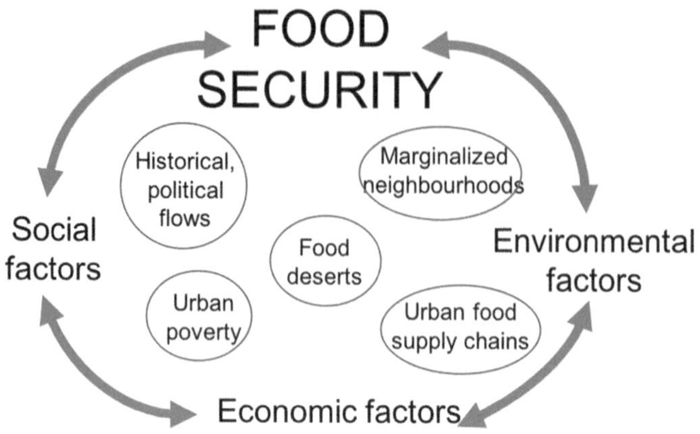

Fig. 4 Factors in urban food security. (Source: Author)

Figure 4 shows the linkages between different SSCM factors as well as some of the socio-environmental elements that underlie urban food security. The previously under-researched social issues are given an equal weight with environmental and economic factors, and all the factors are connected to each other. Historical and political flows are an example of underlying factors for why some neighbourhoods are marginalized and food supply chains therefore may not function in those parts of town as it is not deemed profitable enough.

Focusing on the material and immaterial flows that make up a city and its inequalities, the concept of urban metabolism emerges. The concept entails the material and immaterial flows within a city that produce its socio-economic and biophysical essence (Castán Broto, Allen, & Rapoport, 2012). There are many approaches to urban metabolism, and it is inherently an interdisciplinary notion. Drawing heavily from UPE, it is concerned with the interaction of processes and people on the surrounding environment, and how these processes reproduce certain assumptions and circumstances. Urban metabolism however offers a less abstract approach, focusing on physical, material and financial flows in addition to the social and political ones at the heart of UPE. It is therefore a prudent lens for the analysis of food supply chains and their influence on food security.

Urban food insecurity is produced through a number of social, economic, and environmental contradictions (Heynen, 2006). Using perspectives emerging from urban metabolism and UPE, it is possible to develop more functional cities using inspiration from nature, optimize material flows, address urban inequality, and resignify socio-ecological processes that have the potential to reconfigure existing flows (Castán Broto et al., 2012). SSCM in its quest to become more holistic could incorporate some notions from the urban metabolism. For example in urban logistics, considering the city as an ecosystem could help optimize material flows into closed loop supply chains (Lagorio et al., 2016; Wachsmuth, 2012). Different subsystems feeding into each other, such as food supply chains producing waste, create metabolic flows (Biel, 2014). However, these flows do not function on their own, as they would in nature (Swyngedouw, 2006). Social flows govern the supply chains, leading to reproduction of inequalities in terms of distribution to marginalized neighbourhoods and full-blown food deserts.

Using the urban food desert concept as a lens sheds some light on the complicated circumstances that can lead to food insecurity in certain areas. The neighbourhoods classified as food deserts, especially in developed countries such as the US or the UK, are usually low income and inhabited by ethnic minorities (Weatherspoon et al., 2015). This speaks of classist and racist discrimination in the locations of food retail outlets. The flows that govern food supply chains reproduce spatial and financial inequalities through acts of power. The exodus of supermarkets in favour of larger locations in the suburbs, unreachable without a private vehicle, has been justified with claims of profitability, leaving inner city low-income residents with no access to fresh produce (Battersby & Crush, 2014; Myers & Sbicca, 2015). Racial and cultural aspects play a large role in the formation of food deserts. Anguelovski (2015) investigated a case in Boston where affordable ethnic markets, stocking products from the Latin American countries many of the neighbourhood's residents originated from, were closed down in favour of a gentrified and significantly more expensive supermarket. In this case the residents technically had physical access to produce, but it was unaffordable in the long run and culturally inappropriate. The supply chain cannot therefore be considered inclusive. These types of issues are present in cities all across the world (Anguelovski, 2015; Battersby & Crush, 2014) and would benefit from additional empirical research.

Spatializing food security through food deserts is useful but does not capture the complexity of food security; experiential scales such as the HFIAS and FEIS provide some insight into its intricacies. In urban contexts, one can live in a neighbourhood where food is readily available, but simply be financially constrained from accessing it (Battersby & Crush, 2014). For this reason, experiential data is vital as there can be numerous different flows that affect how and why households or individuals become food insecure. McCordic and Frayne (2018) predict that indicators measuring consumption behaviour, such as the MAHFP, will increase in relevance once research in the field advances. Data on consumption and especially coping strategies of urbanites is valuable to both GLRs trying to break into new markets as well as organizations and researchers looking into urban food insecurity. As GLRs are spreading through the Global South, and often not succeeding as well as they had hoped (Amine & Tanfous, 2012; Nandonde & Kuada, 2016), their SCM strategy needs to be better adapted to the environment, rather than relying on a one-size-fits-all mode of operation. Applying aspects of the urban metabolism perspective to the evolution of urban supply chains in the Global South could result in a more refined version of the current GLR spearheaded grocery retail strategy. An inclusive supply chain can still remain a top-down approach if elements of holistic SSCM are used as the principles. Being aware of the consumption habits and the dynamics of urbanites in diverse contexts, factors such as location decisions and which products to sell would give the GLRs' supply chains the opportunity to have a positive effect on urban food insecurity rather than exacerbating the issue.

Juxtaposing coping strategies with consumption habits, e.g. the informal food sector with GLRs, a more bottom-up approach emerges, where the dynamics of the urban poor are harnessed to alleviate urban food insecurity. The *inclusionist* school suggested by Brown and McGranahan (2016) is in line with this approach. In terms of food security this would give micro retailers, who now function largely uncontrolled within the IFS, a better framework within which to do business. Castán Broto et al. (2012) refer to such structures as 'parallel metabolisms', which function alongside the formal flows. It has been widely discussed in the literature that the IFS is a significant source of food for the urban poor, yet city authorities have a largely unfavourable attitude towards it, often subjecting vendors to stringent checks and even violence (Battersby & Crush, 2014; Patel, Guenther, Wiebe, & Seburn, 2014). Nandonde and Kuada (2016) maintain that the strategies used by micro retailers are the most relevant operational guidelines to be used when entering markets such as Africa. A more open approach to urban planning, where currently marginalized citizens are included in the process, could significantly improve the landscape of micro retailers in the food sector. The current socioecological flows do not harness sustainability, and will continue to produce urban inequalities, but using tools such as UPE and urban metabolism, it is possible to imagine and realise an alternate urban reality (Castán Broto et al., 2012). In this reality parallel metabolisms, such as the IFS, would cease to be marginalized by formal flows and be included in urban supply chains as viable options for the dynamic and diverse needs of urban residents.

6 Conclusions

The world is urbanizing at an unprecedented pace, with over half the global population residing in cities for the first time in human history. Urban areas have a distinctly different dynamic from rural contexts. Approaches, such as increasing food production, which are relevant in the countryside do not apply in an urban context and may lead to urban food policy to be swept aside as a non-issue, as food is deemed readily available. Urban food security suffers from a rural bias, where similar solutions are offered regardless of context. To provide an appropriate course of action for urban food security, its causes and effects must be better understood. Food security in any context is a multifaceted subject, and especially in cities the specific dynamics of the environment make its analysis difficult. While it is being acknowledged as a major challenge in development discourse, urban food security suffers from a lack of interdisciplinary research and data. Other urban issues, such as unemployment and overcrowding, take precedence over food insecurity even though they are all inherently connected. Much focus is also on ensuring food availability through increased production, whereas financial and spatial access constraints to existing resources receive less attention.

As a literature review, this chapter opens avenues for further empirical study. Emphasizing the concept of an inclusive food supply chain would give social issues equal weight, while not disregarding other factors in SSCM. Supply chains evolve through a learning process, and a part of that process is adapting to the environment in which the supply chain functions steadily. In many developing countries, this means not going in with large supermarkets in the suburbs but looking at the dynamics of micro retailers and learning from their strategies. This includes the informal sector, which is a primary source of food and employment for many poor urban residents. In that regard, SCM researchers need to put more emphasis on social and learning structures in supply chain evaluation, design, and optimization as part of a holistic perspective. Additional research should emphasize the informal sector as a goal to harness the current coping mechanisms used by food insecure citizens. This could result in a reliable source of food that respects the dynamics of urbanites in different contexts and improves the opportunities for marginalized populations.

These gaps pertain to both academics and practitioners in terms of the evolution process of supply chains. Micro-retailers and the informal sector appear to be the key to food security on a smaller scale, such as neighbourhood, as well as tapping into the business opportunities emerging from the increasing urban population. This literature review maps these gaps for future empirical studies to tackle in a more practical manner. The following research questions are proposed:

1. How can food supply chains consider specific dynamics of urban poor neighbourhoods to avoid a 'one-size-fits-all' mentality while operating in urban areas?
2. In addition to specific dynamics, how can food supply chains adapt to shocks and disturbances that may disproportionately affect marginalized populations in terms of food access?
3. What is the role and relevance of micro-retailers and the informal food sector for food security, especially in a developing country context?
4. How can the learning process of supply chains include sustainability from the outset rather than waiting for the supply chain to 'evolve' to a stage that is mature enough for sustainable approaches?

The lack of data and negative bias has had its effect on urban food security research. Food supply chains, along with other urban metabolic flows, play an extensive role in urban nutrition. Including marginalized populations and their parallel supply chains into a holistic urban reality would pave the way for sustainability in a traditionally profit-oriented stream. However, more research is necessary both from an academic as well as a practitioner point of view before steps can be taken in policy and planning.

As with all research, this paper has limitations as it only comprises a literature review to investigate linkages of food supply chain management to availability and accessibility of food for urban communities, particularly in poor urban contexts. The review conducted is both rigorous and robust based on the framework used, however this paper offers no empirical research to validate that investigation. Nonetheless, the avenues suggested for further research are an important start to enhancing the interdisciplinary discourse in the subject and bringing SCM more into the debate such that it becomes a referent discipline in its own right.

The chapter contributes to SSCM literature by integrating disciplines which clearly contribute to the discourse in a relevant way, forming a comprehensive view on urban food security. Urban food security as a phenomenon is not tied to any geographical context, but needs to be approached as a global issue. With more studies and data from cities around the world, businesses, governments and NGOs can begin building a comprehensive picture of the world's food insecure people, the underlying reasons behind food insecurity and what can be done to mitigate it.

Notes

1. Cereals; roots & tubers; vegetables; fruit; meat, poultry, offal; eggs; fish & seafood; pulses/legumes/nuts; dairy products; oil/fats; sugar/honey; miscellaneous.
2. Never; Rarely (once or twice in the past four weeks); Sometimes (three to ten times in the past four weeks); Often (more than ten times in the past four weeks).

References

Agyeman, J., & McEntee, J. (2014). Moving the field of food justice forward through the Lens of urban political ecology. *Geography Compass, 8*(3), 211–220.

Amine, A., & Lazzaoui, N. (2011). Shoppers' reactions to modern food retailing systems in an emerging country. *International Journal of Retail & Distribution Management, 39*(8), 562–581.

Amine, A., & Tanfous, F. H. B. (2012). Exploring consumers' opposition motives to the modern retailing format in the Tunisian market. *International Journal of Retail & Distribution Management, 40*(7), 510–527.

Anguelovski, I. (2015). Alternative food provision conflicts in cities: Contesting food privilege, injustice, and whiteness in Jamaica plain, Boston. *Geoforum, 58*, 184–194.

Battersby, J., & Crush, J. (2014). Africa's urban food deserts. *Urban Forum, 25*, 143–151.

Battersby, J., & Marshak, M. (2013). Growing communities: Integrating the social and economic benefits of urban agriculture in Cape Town. *Urban Forum, 24*, 447–461.

Battersby, J., & Peyton, S. (2014). The geography of supermarkets in Cape Town: Supermarket expansion and food access. *Urban Forum, 25*, 153–164.

Beske, P., Land, A., & Seuring, S. (2014). Sustainable supply chain management practices and dynamic capabilities in the food industry: A critical analysis of the literature. *International Journal of Production Economics, 152*, 131–143.

Biel, R. (2014). Visioning a sustainable energy future: The case of urban food-growing. *Theory, Culture & Society, 31*(5), 183–202.

Brown, D., & McGranahan, G. (2016). The urban informal economy, local inclusion and achieving a global green transformation. *Habitat International, 53*, 97–105.

Brown, J. C., & Purcell, M. (2005). There's nothing inherent about scale: Political ecology, the local trap, and the politics of development in the Brazilian Amazon. *Geoforum, 36*(5), 607–624.

Brown, M. E., Silver, K. C., & Rajagopalan, K. (2013). A city and national metric measuring isolation from the global market for food security assessment. *Applied Geography, 38*(1), 119–128.

Carletto, C., Zezza, A., & Banerjee, R. (2013). Towards better measurement of household food security: Harmonizing indicators and the role of household surveys. *Global Food Security, 2*(1), 30–40.

Castán Broto, V., Allen, A., & Rapoport, E. (2012). Interdisciplinary perspectives on urban metabolism. *Journal of Industrial Ecology, 16*(6), 851–861.

Chen, M. A. (2012). *The informal economy: Definitions, theories and policies.* WIEGO Working Paper, 1 (August), 26.

Coates, J. (2013). Build it back better: Deconstructing food security for improved measurement and action. *Global Food Security, 2*(3), 188–194.

Colicchia, C., & Strozzi, F. (2012). Supply chain risk management: A new methodology for a systematic literature review. *Supply Chain Management: An International Journal, 17*(4), 403–418.

Crush, J., & Frayne, B. (2011). Urban food insecurity and the new international food security agenda. *Development Southern Africa, 28*(4), 527–544.

Denyer, D., & Tranfield, D. (2009). Producing a systematic review. In D. Buchanan & A. Bryman (Eds.), *The SAGE handbook of organizational research methods* (pp. 671–689). London: SAGE.

Elbahnasawy, N. G., Ellis, M. A., & Adom, A. D. (2016). Political instability and the informal economy. *World Development, 85*, 31–42.

FAO. (2009). *Declaration of the world summit on food security.* World Summit on Food Security.

Floro, M. S., & Swain, R. B. (2013). Food security, gender, and occupational choice among urban low-income households. *World Development, 42*(1), 89–99.

Fox, S. (2014). The political economy of slums: Theory and evidence from sub-Saharan Africa. *World Development, 54*, 191–203.

Frayne, B., Crush, J., & McLachlan, M. (2014). Urbanization, nutrition and development in southern African cities. *Food Security*, 1–12.

Frayne, B., & McCordic, C. (2015). Planning for food secure cities: Measuring the influence of infrastructure and income on household food security in southern African cities. *Geoforum, 65*, 1–11.

Gold, S., Hahn, R., & Seuring, S. (2013). Sustainable supply chain management in 'base of the pyramid' food projects-a path to triple bottom line approaches for multinationals? *International Business Review, 22*(5), 784–799.

Grant, D. B., Trautrims, A., & Wong, C. Y. (2017). *Sustainable logistics and supply chain management* (2nd ed.). Kogan Page.

Hall, J., & Matos, S. (2010). Incorporating impoverished communities in sustainable supply chains. *International Journal of Physical Distribution & Logistics Management, 40*(1/2), 124–147.

Headey, D., & Ecker, O. (2013). Rethinking the measurement of food security: From first principles to best practice. *Food Security, 5*(3), 327–343.

Heynen, N. (2006). Justice of eating in the city: The political ecology of urban hunger. In N. Heynen, M. Kaika, & E. Swyngedouw (Eds.), *In the nature of cities: Urban political ecology and the politics of urban metabolism* (pp. 124–136). London and New York: Routledge.

Järvensivu, P., Toivanen, T., Vaden, T., Lähde, V., Majava, A., & Eronen, J. T. (2018). *Governance of economic transition*. Retrieved from https://bios.fi/bios-governance_of_economic_transition.pdf

Lagorio, A., Pinto, R., & Golini, R. (2016). Research in urban logistics: A systematic literature review. *International Journal of Physical Distribution & Logistics Management, 46*(10), 908–931.

Lorentz, H., Kittipanya-Ngam, P., & Singh Srai, J. (2013). Emerging market characteristics and supply network adjustments in internationalising food supply chains. *International Journal of Production Economics, 145*(1), 220–232.

Lynch, K., Maconachie, R., Binns, T., Tengbe, P., & Bangura, K. (2013). Meeting the urban challenge? Urban agriculture and food security in post-conflict Freetown, Sierra Leone. *Applied Geography, 36*, 31–39.

Maertens, M., Minten, B., & Swinnen, J. (2012). Modern food supply chains and development: Evidence from horticulture export sectors in sub-Saharan Africa. *Development Policy Review, 30*(4), 473–497.

Maitra, C. (2016). Adapting an experiential scale to measure food insecurity in urban slum households of India. *Global Food Security, 15*(April), 1–12.

Maxwell, D. (1999). The political economy of urban food security in sub-Saharan Africa. *World Development, 27*(11), 1939–1953.

Maxwell, D., Vaitla, B., & Coates, J. (2014). How do indicators of household food insecurity measure up? An empirical comparison from Ethiopia. *Food Policy, 47*, 107–116.

McCordic, C., & Frayne, B. (2018). Measuring urban food security. In B. Frayne, J. Crush, & C. McCordic (Eds.), *Food and nutrition security in southern African cities*. Routledge.

Minten, B., Reardon, T., & Sutradhar, R. (2010). Food prices and modern retail: The case of Delhi. *World Development, 38*(12), 1775–1787.

Myers, J. S., & Sbicca, J. (2015). Bridging good food and good jobs: From secession to confrontation within alternative food movement politics. *Geoforum, 61*, 17–26.

Nandonde, F. A., & Kuada, J. (2016). International firms in Africa's food retail business-emerging issues and research agenda. *International Journal of Retail & Distribution Management, 44*(4), 448–464.

Nguyen, H. T. H., Wood, S., & Wrigley, N. (2013). The emerging food retail structure of Vietnam. *International Journal of Retail & Distribution Management, 41*(8), 596–626.

Page, B. (2003). The political ecology of Prunus africana in Cameroon. *Area, 35*(4), 357–370.

Pagell, M., & Shevchenko, A. (2009). Why research in sustainable supply chain management should have no future. *Journal of Supply Chain Management, 50*(1), 44–55.

Patel, K., Guenther, D., Wiebe, K., & Seburn, R.-A. (2014). Promoting food security and livelihoods for urban poor through the informal sector: A case study of street food vendors in Madurai, Tamil Nadu, India. *Food Security, 6*(6), 861–878.

Prahalad, C. K. (2005). *The fortune at the bottom of the pyramid: Eradicating poverty through profits.* Wharton School Publishing.

Reardon, T., Henson, S., & Berdegué, J. (2007). 'Proactive fast-tracking' diffusion of supermarkets in developing countries: Implications for market institutions and trade. *Journal of Economic Geography, 7*(4), 399–431.

Seuring, S., & Gold, S. (2012). Conducting content-analysis based literature reviews in supply chain management. *Supply Chain Management: An International Journal, 17*(5), 544–555.

Silvestre, B. S. (2015). Sustainable supply chain management in emerging economies: Environmental turbulence, institutional voids and sustainability trajectories. *International Journal of Production Economics, 167*, 156–169.

Swyngedouw, E. (1996). The city as a hybrid: On nature, society and cyborg urbanization. *Capitalism Nature Socialism, 7*(2), 65–80.

Swyngedouw, E. (2006). Circulations and metabolisms: (Hybrid) natures and (cyborg) cities. *Science as Culture, 15*(2), 105–121.

Swyngedouw, E., & Heynen, N. C. (2003). Urban political ecology, justice and the politics of scale. *Antipode, 35*, 898–918.

Tolossa, D. (2010). Some realities of the urban poor and their food security situations: A case study of Berta Gibi and Gemechu Safar in the city of Addis Ababa, Ethiopia. *Environment and Urbanization, 22*(1), 179–198.

Touboulic, A., & Ejodame, E. (2016). Are we really doing the 'right thing'? From sustainability imperialism in global supply chains to an inclusive emerging economy perspective. In L. Bals & W. Tate (Eds.), *Implementing triple bottom line sustainability into global supply chains* (pp. 14–33). Routledge.

Touboulic, A., & Walker, H. (2015). Theories in sustainable supply chain management: A structured literature review. *International Journal of Physical Distribution & Logistics Management, 45*, 16–42.

Tranfield, D., Denyer, D., & Smart, P. (2003). Towards a methodology for developing evidence-informed management knowledge by means of systematic review. *British Journal of Management, 14*(3), 207–222.

United Nations. (2015). *World urbanization prospects: The 2014 revision*. New York: United Nations Department of Economic and Social Affairs, Population Division.

Wachsmuth, D. (2012). Three ecologies: Urban metabolism and the society-nature opposition. *Sociological Quarterly, 53*(4), 506–523.

Weatherspoon, D., Oehmke, J., Dembele, A., & Weatherspoon, L. (2015). Fresh vegetable demand behaviour in an urban food desert. *Urban Studies, 52*(5), 960–979.

Yawar, S. A., & Seuring, S. (2017). Management of social issues in supply chains: A literature review exploring social issues, actions and performance outcomes. *Journal of Business Ethics, 141*(3), 621–643.

Zezza, A., Carletto, C., Davis, B., & Winters, P. (2011). Assessing the impact of migration on food and nutrition security. *Food Policy, 36*(1), 1–6.

An Economic and Environmental Comparison of Conventional and Controlled Environment Agriculture (CEA) Supply Chains for Leaf Lettuce to US Cities

Charles F. Nicholson, Kale Harbick, Miguel I. Gómez, and Neil S. Mattson

1 Introduction

Metropolitan agriculture, the production of food in urban and peri-urban areas, has captured the attention and excitement of municipalities and entrepreneurs as a means to improve fresh food access while contributing to environmental sustainability (Mougeot, 2000). What began as a community gardening movement has been transformed over the last five years with the emergence of larger-scale commercial Controlled

C. F. Nicholson (✉) • M. I. Gómez
Charles H. Dyson School of Applied Economics and Management, Cornell University, Ithaca, NY, USA
e-mail: cfn1@cornell.edu; mig7@cornell.edu

K. Harbick • N. S. Mattson
School of Integrative Plant Science, Cornell University, Ithaca, NY, USA
e-mail: kh256@cornell.edu; nsm47@cornell.edu

E. Aktas, M. Bourlakis (eds.), *Food Supply Chains in Cities*,
https://doi.org/10.1007/978-3-030-34065-0_2

Environment Agriculture (CEA) operations in metropolitan areas. These greenhouses and plant factories enable year-round intensive production of vegetables by creating controlled environments that supply an optimal balance of light, heat, CO_2 and water to optimise plant growth (Harbick & Albright, 2016). These systems have the potential to alter metropolitan food supply chains by decentralising vegetable production, reducing food waste and food miles, using less water than soil-based production, and creating new opportunities for entrepreneurs and workforce development.

A wide-range of CEA growing systems are being considered (Newbean Capital, 2015), but the three most commonly proposed for metropolitan areas are temperature-controlled greenhouses with supplemental lighting (GH-SL), plant factories (PF) with sole source lighting (SSL, i.e., no sunlight) and vertical farms (VF; multi-level buildings with windows and supplemental light (SL)[1]; Kozai, Niu, & Takagaki, 2015). CEA as an urban food production method, contributor to local food systems, and municipal investment strategy, however, is yet to be proven. Examples exist of commercially viable soil-based metro farms and apparently-successful metro-based GH operations, but neither the financial feasibility nor the scalability of metro-based CEA, particularly for plant factories, has been systematically addressed by previous research. The extent to which a city's demand for vegetables can be produced within its boundaries using CEA systems is yet unanswered. To fulfill the potential of metro CEA, a systems approach to analysis of the economic, social, and ecological footprint plus empirical information about the potential outcomes of its implementation is needed. Such efforts will deliver critical analysis and decision support tools to facilitate strategic investments in metro CEA as a key component of urban food supply chains. The potential benefits of metro CEA include lower transportation costs, reduced product waste, and job creation but must be assessed and also weighed against potentially higher land, labour, water, and energy costs and compared with field-based production methods. A supply-chain approach is useful to compare the economics and greenhouse gas emissions, energy use, and water use of representative conventional and metro-based CEA supply chains.

Our principal objective is to compare the economic and environmental performance for representative conventional (field-based) and CEA supply chains for leaf lettuce, which is a major vegetable crop with a production value of more than $850 million in 2015 (USDA, 2017) to two metropolitan areas in the US: New York City and Chicago. Although many CEA operations produce greens targeted to specialty markets, the comparison to conventional leaf lettuce is relevant because some CEA operations aspire to compete with supply chains using conventional production (Johnny Bowman, Edenworks, personal communication). We document and integrate information about production, processing, transportation and other marketing costs and input use for delivery to the ultimate consumers in these two metropolitan areas for both the conventional field production and representative configurations of two types of metropolitan-based CEA supply chains, greenhouses (GH) and plant factories (PF). This supply-chain analysis includes fixed costs, land, transportation, labour, energy, and other inputs required in production, processing, transportation and distribution. For the CEA production component, the analysis builds on energy-modeling analyses that incorporate relevant biological, lighting and other parameters for the specific locations (Harbick & Albright, 2016).

Although a number of previous studies have examined the environmental impacts of lettuce supply chains (e.g., Emery & Brown, 2016; Rothwell, Ridoutt, Page, & Bellotti, 2016), we are not aware of any previous study that has compared both landed costs and environmental outcomes of lettuce supply chains to major US urban areas. Thus, our analysis provides a much-needed comparative assessment of conventional and metropolitan-based agricultural supply chains for a key vegetable crop and provide a framework and example for future assessments of other food products. This information can lead to more informed decisions by potential investors, consumers and metropolitan policy makers with regard to the future configuration of urban food supply chains. We analyze "baseline" CEA supply chain performance based on current industry average performance and then assess a "best case" scenario with improved productivity and lower costs.

2 Literature Review

Few studies have evaluated the landed cost of lettuce from alternative supply chains to US metropolitan areas. Eaves and Eaves (2018) compared the profitability of producing lettuce in a greenhouse (GH) versus what they describe as a vertical farm (VF) but which operates similar to a multi-level plant factory (PF) to supply product to Québec City. They found that despite large differences in the composition of the investment (higher for the VF) and operating costs (higher for the GH), the overall production cost difference was small. Production of 1 kg of lettuce cost $4.66 and $4.51 (US dollars) in the GH and VF, respectively, a difference of about 3%. The authors did not examine the landed cost because they assumed that delivery processes and costs would be the same given their assumptions about the production location. This study utilised methods other than ASHRAE standard calculation methodologies for modeling energy consumption (a simplified spreadsheet model), which only approximates the complex energy flow dynamics.

A large number of studies have evaluated the environmental impacts of alternative lettuce production techniques and supply chain configurations. Most of these studies have used Life Cycle Assessment (LCA) methods that are commonly used to examine the environmental impacts of food supply chains (e.g., Notarnicola et al., 2017; Stoessel, Juraske, Pfister, & Hellweg, 2012). LCA methods typically account for resource use and outputs from production, but also those "embodied" in production inputs, equipment and structures. Although there are international standards for such studies (ISO 14040 and ISO 14044) their empirical implementation varies—often considerably—in terms of system boundaries, data inputs, computational methods and results.

A number of LCA studies exist for lettuce products with differing assumptions about the nature of production and the supply chain. Emery and Brown (2016) compared the production and delivery of lettuce from a commercial California-based field growing operation with a community garden approach to supply the Seattle, Washington market. They concluded that CO_2 emissions per unit for production and delivery were significantly lower (in fact, negative) for the community garden, although they did not consider the "embodied" costs of inputs. Hospido et al. (2009) examined field lettuce and greenhouse (GH) lettuce production systems supplying a retail distribution centre in the UK, using production locations in the UK and Spain. Emissions of CO_2 and cumulative energy demand from non-renewable sources were as much as 10 times larger for GH production systems, although water use per unit of production was only 40% of that for field production in the GH. These results suggest trade-offs between different production and distribution systems depending on the environmental indicators assessed. Rothwell et al. (2016) examined the CO_2 emissions and water use of lettuce production and distribution systems to supply the Sydney, Australia market. They compared three field production techniques and locations with two greenhouse production systems within 60 km of the central produce market. Large-scale field production located more than 900 km from Sydney had the lowest CO_2 emissions per unit for the *production* of 1 kg lettuce, but local lettuce had only 50% of *total* CO_2 emissions per unit including transportation for lettuce delivered to the central market. This was because emissions due to transportation were 2.5 times those for production for the large-scale production system. The large-scale field system also had the largest water use per kg lettuce. A GH production system located at 39 km from the central market had the lowest water use per kg lettuce produced, but the highest delivered CO_2 emissions.

Because the technology is newer and there are fewer commercial operations, analyses of the environmental impacts of plant factories (PF) are limited. Shiina et al. (2011) reported high levels of CO_2 emissions per unit product for two PF configurations producing lettuce and spinach. Graamans et al. (2018) undertook detailed energy and water modeling of greenhouses and potential PF configurations located in the Netherlands, Sweden and the United Arab Emirates, noting that although there are regional differences, PF production required significantly more purchased energy per unit of product than production in greenhouses.

The limited coverage of previous studies is due in part to the challenges associated with compilation of data for the assessment of what are typically many possible production and supply chain configurations. In addition, we noted no previous studies have simultaneously evaluated the environmental and cost components, which will be key information to make informed judgments about which configurations are appropriate and(or) how existing configurations might be transitioned to reduce environmental impacts while maintaining profitability.

3 Methods

Our overall objective is to compare the Cumulative Energy Demand (CED), Global Warming Potential (GWP), Water Use (WU) and Total Landed Cost (TLC) of 1 kg of saleable leaf lettuce delivered to a representative wholesale market location in both New York City and Chicago from a conventional (field-based production) supply chain and two types of CEA-based supply chains. Using the terminology employed in Life Cycle Assessment studies, we adopt a cradle-to-wholesale system boundary (Fig. 1). We omit analysis of the processes after delivery to the wholesale market under the assumption that differences in costs and environmental outcomes for the different production systems would be small after wholesale delivery. The functional unit for comparison is 1 kg of saleable lettuce delivered to a major wholesale produce market in each of the two cities.

For the purposes of assessing cost and environmental impacts, we adopt a simplified version of a Life Cycle Inventory (LCI) approach, that is, the "detailed tracking of all the flows in and out of the product system,

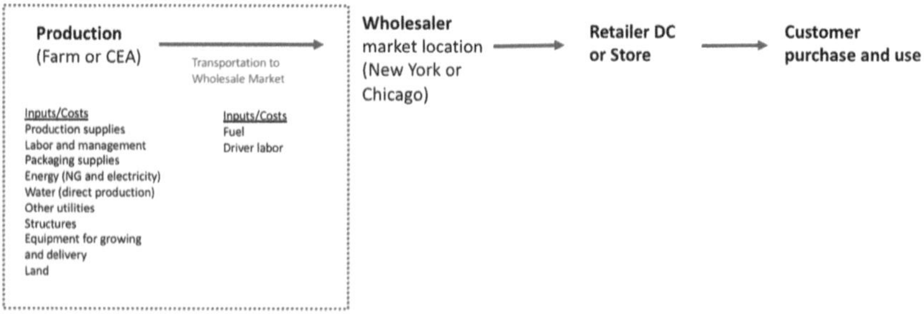

Fig. 1 System boundary for analysis of costs and environmental impacts of three lettuce supply chains. (Source: Authors)

including raw resources or materials, energy by type, water, and emissions to air, water and land by specific substance" (Athena Sustainable Materials Institute, 2017). Specifically, we account for inputs used in direct production but not those resources and impacts embodied in the production of the inputs. We define the lettuce production systems as follows. Conventional field-based leaf lettuce production is assumed to occur on 101 ha (250 acres) of an approximately 600-ha (1500-acre) farm in the Salinas Valley of California, which is the major lettuce-producing location for the US. The assumed yield is approximately 10,600 kg per acre per cropping cycle, about 900 cartons of lettuce packed for shipment. (Typically, there are two cropping cycles per year in the Salinas Valley.) We assume that 30% of the shipped production is not saleable upon arrival, based on industry estimates of shrink. The total production per acre per year is about 1800 cartons, considerably less than the quantities produced by GH and PF, which operate throughout the year. The GH production system assumes the use of a Nutrient Film Technique (NFT) growing system, a freestanding gable greenhouse with a total area of approximately 4460 m^2 (48,000 ft^2) and a net production area of 4000 m^2 (43,200 ft^2) with glass glazing material and artificial lights operated 2575 hours per year for NY, or 2856 hours per year for Chicago, based on 418 HPS luminaires of 1000 W. The PF system assumes an insulated warehouse-type structure with 10 production levels that result in the same total yield as the greenhouse, with a total area of 803 m^2 (8640 ft^2) based on one-tenth of the net production area of the greenhouse and 50% of total required space used for production. The GH and PF operations are non-automated systems assumed to be located in the relevant metro area at a location used by an existing CEA operation, although a peri-urban location is more consonant with the land requirements for GH and PF of the assumed production area. We choose a location for the New York CEA operations very close to the wholesale market and a location for the Chicago CEA operations farther away from the wholesale market to highlight the trade-offs in land versus transportation costs for urban and peri-urban production locations. Additional description of the production systems is provided in Table 1.

Table 1 Selected characteristics of field, CEA GH and CEA PF operations analysed

Production system	Field	GH	PF
Land area for production, ha	101.00	0.45	0.08
Land area for non-production,[a] ha	0.00	0.24	0.24
Total land area, ha	101.00	0.69	0.32
Cropping frequency analysed	1 crop (summer)	Continuous	Continuous
Production amount analysed,[b] kg	7144	454,685	454,685
Location of facility serving New York City	Salinas, CA	Bronx, New York	Bronx, New York
Location of facility serving Chicago	Salinas, CA	Northern Indiana	Northern Indiana
Distance from New York City wholesale market, km	4825	3.5	3.5
Distance from Chicago wholesale market, km	3570	75	75
Land value in New York area, $/ha	–	5,868,748	5,868,748
Land value in Northern Indiana, $/ha	–	753,282	753,282
Land rental cost in Salinas Valley, $/ha	3336	–	–

Source: Authors' own calculations and assumptions
[a]Non-production area is used for cooling, packing, office facilities and parking
[b]For Field production, this is calculated as harvested yield of 10,206 kg less 30% shrink in transit

3.1 Landed Costs

Production costs for field-based lettuce are based on Tourte, Smith, Murdock, and Sumner (2017), which provides detailed cost information for production supplies, packaging, labour, structures and equipment. Transportation cost calculations are based on diesel fuel costs for a tractor-trailer rig loaded with 900 cartons of lettuce achieving a fuel efficiency of 3 km/litre (7 miles per gallon) and requiring 33 hours of driver labour to travel 3570 km (2218 miles) from Salinas to the Chicago International Produce Market and 44 hours of driver labour to travel 4825 km (3000 miles) from Salinas to the Hunt's Point Produce Market in the New York City metropolitan area. Transportation costs assume that a backhaul to California is available for 75% of trips delivering lettuce from the field production operation. Transportation costs also include an estimate of overhead costs in addition to fuel and driver labour. Water use is reported as 1440 m^3 for production of 10,600 kg, or about 135 litres per kg of lettuce produced. (Additional details are provided in Tables 7, 8, and 9.)

Production costs for the CEA GH and PF are derived from information in the Lettuce Interactive Business Tool (Gómez, Mattson, & Nishi, 2017) and Eaves and Eaves (2018), both of which also include costs of production supplies, packaging, labour, structures and equipment. Production supplies include seeds, propagation cubes, beneficial insects, fertilisers, and sanitisers used in direct production. For the GH operation, costs for bio-based fungicides and pesticides and biological control of insects are also included, although they are not in the case of the PF because this system operates without direct access to the outside. Production labour includes that for seeding, transplanting, harvesting and packaging. Additional labour is required for delivery to markets. Production management includes a production manager and administrative support, and a single sales manager is responsible for marketing. A salaried executive position is assumed to oversee all operations. Packaging costs are assumed to be similar for the three systems, using wax cardboard cartons with a capacity of approximately 11 kg. Utilities other than energy and water include sewer, landline telephones and cell phones. Miscellaneous costs include those for advertising and promotions, office supplies laboratory testing, postage, software, professional services (legal and accounting) and participation in trade shows. Water use is assumed to be about 21 litres per kg for the operation of the growing systems (Harbick & Albright, 2016), which does not include the additional water required for evaporative cooling. (Additional details are provided in Table 10.)

Energy costs often account for more than one-third of the total costs for a CEA operation, and likely constitute a main cost difference between field, GH and PF operations (Eaves & Eaves, 2018). In contrast to many previous studies, we used detailed energy modeling simulations specific to the assumed GH and PF structures to determine energy use and related costs for the both operations. EnergyPlus (Crawley et al., 2001) is an energy modeling simulation engine that implements the American Society of Heating, Refrigerating and Air-Conditioning Engineers (ASHRAE) heat balance method (ASHRAE, 2017). It is commonly used for buildings in the commercial sector but was modified to facilitate the modeling of CEA buildings (Harbick & Albright, 2016). To estimate the annual energy use, EnergyPlus calculates loads and system response on sub-hourly time steps using building parameters and Typical Meteorological Year (TMY3; Wilcox & Marion, 2008) hourly weather data. Warehouse building parameters reflect the warehouse type of Department of Energy (DOE) commercial reference buildings, which follow ASHRAE 90.1–2004 standards (ASHRAE, 2004). Heating is assumed to use a natural gas boiler. Cooling the GH is assumed to require evaporative pads, whereas a chiller unit is required for the PF. The required amounts of natural gas (m^3) and electricity (kWh) are calculated for each month based on changes in climate during the course of a year. Electricity use includes lighting and ventilation but not water pumping. The GH CED and GWP values assume DLI control using LASSI (Albright, Both, & Chiu, 2010), which is close to day-wise optimal. Threshold or timer-based lighting control, such as available from greenhouse controls companies, would incur higher CED/GWP for the same yield. The same value of efficacy of supplemental lighting was used for both the GH and PF. Costs for natural gas are calculated based on the reported unit costs per m^3 of natural gas for commercial use from the US Energy Information Administration. Electricity costs are based on a per state-specific average industrial rates (for New York and Indiana) per kWh used, plus a "demand charge" based on the peak number of kW used across all months in the year. Additional assumptions related to energy modeling are presented in Table 6.

The transportation costs for the metro-based GH and PF assume production in locations currently used by CEA operations in both Chicago and New York City, located at distances of 75 km and 3.5 km from the wholesale markets of these two cities, respectively. The difference in distances traveled within the metropolitan area can provide insights regarding the intra-metropolitan-area location decision for CEA operations. We assume 10 round-trip deliveries per week to the wholesale market with a refrigerated reefer truck. (Additional details are provided in Table 12.)

Individual cost components for CEA structures and equipment are difficult to obtain and extant information shows considerable variation among these components. We use a simplified industry rule of thumb of \$538.3/m^2 (\$50/ft^2) for the production-related area of the GH and PF to calculate structure and equipment costs. Our assumed value is a commonly-used average unit cost, although this this can vary based on location and GH or PF configuration. We subsequently evaluate the impact of this assumption with scenario analysis using an industry-indicated minimum value, to represent a scenario for this cost component.

Another major cost is for the land required for the CEA operation, for non-production space (e.g., restrooms, administrative offices and parking). Following Eaves and Eaves (2018), we assume that non-production land area is equal to 0.56 times the area of the production facility. Overall, the PF require about half the land area required for a GH for the production levels assumed. The lower land requirement (and therefore cost) and higher energy use (for both lighting and cooling) for the PF are the key cost components for comparison with GH operations. We calculated the cost of purchasing the land required for operating the two types of CEA operations based on per-acre values of commercial land parcels offered for sale in the two focal metropolitan regions.

To calculate the annualised cost of investment in CEA operations, we summed the total value of investment in structures, equipment and land and then assumed that this entire amount would be financed with a ten-year loan at an annual interest rate of 6.2%, which is the weighted average cost of capital (WACC) for US farming or agriculture operations reported by Damodaran (2018) for January 2018. Although this assumes no equity investment in the operation, this is roughly equivalent to charging a 6.2% opportunity cost per year for equity invested in the business. (Additional details are in Table 11.)

3.2 Cumulative Energy Demand (CED)

CED is expressed in MJ of total energy per kg of functional unit (e.g., 1 kg saleable lettuce delivered to the wholesale market). This comprises the energy used for the production, transport, and the use of production inputs including structures and equipment. For field production, this includes the energy in diesel and gasoline used in farm equipment and the electricity used in pumping water used in production. Diesel and gasoline use are reported directly in Tourte et al. (2017) but because gasoline use is quite small (less than 8 litres) we converted only the diesel fuel use to its energy equivalent using the standard factors of Btu per litre and MJ per Btu. Electricity for water pumping is calculated based on the diesel equivalent required to pump an acre-inch of water and the equivalent number of kWh per unit of diesel fuel. Energy use for GH and PF operations was calculated based on the energy modeling approach discussed above and includes energy in natural gas used for heating and electricity for lighting and cooling. Energy used in transportation is calculated based on the estimated amount of diesel fuel required to transport lettuce from the production location to the wholesale markets in New York and Chicago. (Additional details are in Tables 13 and 14.)

3.3 Global Warming Potential (GWP)

GWP is expressed in terms of kg CO_2 equivalent per kg of functional unit (e.g., 1 kg lettuce delivered to the wholesale market). This comprises the CO_2 generated for the production, transport, and the use of production inputs including structures and equipment. Similar to a number of previous studies, we ignore the potential impacts of changes in soil carbon for the field production operation. Natural gas (NG) use is converted to CO_2 equivalent using a fixed conversion factor of 0.0503 MT CO_2 per GJ energy in NG. Electricity CO_2 is based on total kWh used in production multiplied by state-specific emissions factors for California (Field), New York and Indiana obtained from the US Energy Information Agency. Emissions from transportation are calculated based on the diesel fuel required to transport lettuce from the production site to the wholesale markets, using a standard diesel conversion factor of 2.7 kg CO_2/litre (10.21 kg CO_2/gallon). (Additional details are in Table 15.)

3.4 Water Use (WU)

Water use is expressed in terms of litres water per functional unit (e.g., 1 kg lettuce delivered to the wholesale market). This comprises only the water used for the production process, not water for evaporative cooling or in transportation. As noted above, water use is estimated as 20.9 litres/kg for both the GH and PF operations (Harbick & Albright, 2016) and a 135.4 litres/kg for field-based operations in California (Tourte et al., 2017).

3.5 Scenario Analysis

Many assumptions are required to assess the comparative economic and environmental assessment of the three systems under study. Key assumptions used in our study are based on published data or industry sources that are specific to the supply chains analyzed, such as product yields, input use (especially energy), labor costs and land costs. Although comprehensive sensitivity analysis is often recommended for LCA studies (Bjorklund, 2002; Beccali et al., 2010), we undertake a less broad *scenario analysis* to assess whether our findings for the landed costs are robust. We compare our estimate of average landed costs for the CEA supply chains as described above to a "best case" scenario that assumes the best currently feasible productivity and lowest costs based on published literature and industry contacts. Specifically, for both the CEA operations in both cities, we assume 20% increase in yields per growing area, and a 40% lower cost for structures and equipment ($\$322.9/m^2$ rather than $\$538.2/m^2$), based on information from industry contacts. For New York, we assess a lower land value in New York City, using the value for the Chicago-area operation, which is consistent with locating further from the New York wholesale market (which also implies higher transportation costs for the "best case"). We also assume lower per-kWh electricity costs in New York by assuming the lower value reported for

Indiana. As noted above, we assumed energy-optimizing for the GH lighting in the average performance case and this assumption is applied in the "best case" scenario also. Although this approach does not allow us to assess the distribution of costs or for any possible (e.g., optimal) configuration of a CEA operation, it provides substantive evidence about the likely comparative performance with field-based production for many configurations of CEA operations.

4 Findings

4.1 Landed Costs Findings

Our analysis indicates that the total landed costs for CEA supply chains to provide lettuce to the Chicago and New York City metro areas are markedly larger than those with field-based production in the Salinas Valley of California (Table 2). Lettuce produced and delivered from the GH has a landed cost 158% to 163% higher than that of field lettuce from California, despite much higher transportation costs for the field-produced lettuce. Lettuce produced in a PF has a landed cost 153% to 157% higher than field produced lettuce. The differences between CEA supply chains and field production are smaller in the Chicago market (despite lower transportation costs from California) due to lower land values and lower rates per kWh for electricity. Similar to Eaves and Eaves (2018), we find that GH and PF can have similar landed costs in both locations; higher energy costs for PF are offset by lower land requirements.

In addition to the overall cost differences, the structure of costs for these supply chains are quite different. Field production costs are quite low and packaging (including harvesting) and shipment costs account for 67 to 70% of landed costs, whereas they comprise less than 12% of landed cost for GH and PF operations. For the CEA GH, labour and management, energy and structures account for more than 80% of landed costs, and transportation costs are minimal. Labour costs are notably higher for CEA supply chains, in part due to additional labor required for

Table 2 Total landed cost for delivery of 1 kg lettuce to wholesale produce markets in New York City and Chicago from field-based production, a CEA greenhouse (GH) and a CEA plant factory (PF)

	New York City wholesale market, Hunt's point			Chicago international produce market		
Cost category	Field	GH	PF	Field	GH	PF
	$/kg delivered saleable lettuce					
Production supplies	0.17	0.29	0.27	0.17	0.29	0.27
Labour and management	*0.09*	*4.31*	*4.31*	*0.09*	*4.31*	*4.31*
Packaging	0.98	0.76	0.76	0.98	0.76	0.76
Utilities other than water and energy	0.00	0.03	0.03	0.00	0.03	0.03
Miscellaneous	0.50	0.10	0.10	0.50	0.10	0.10
Water (direct production only, not evaporative cooling)	0.02	0.08	0.08	0.00	0.08	0.08
Energy for operations	*0.04*	*0.46*	*1.36*	*0.04*	*0.41*	*0.89*
Structure, equipment, land, growing and delivery equipment	*0.08*	*2.05*	*0.90*	*0.08*	*1.00*	*0.40*
Transportation from production to market (fuel)	*1.17*	*0.00*	*0.00*	*0.87*	*0.04*	*0.04*
Total landed cost	3.04	8.09	7.82	2.72	7.03	6.89

Source: Authors' own calculations

Note: *Field* indicates field-based production in Salinas Valley, California, *GH* indicates a CEA greenhouse located in the same metropolitan area as the wholesale market, and *PF* indicates a CEA Plant Factory located in the same metropolitan area as the wholesale market

production, but also due to the administrative staff required for management and marketing that are typically lower and spread over much larger volumes for field-based operations. These results suggest that greater productivity of CEA GH labour and utilities—as well as locations that optimise trade-offs between land and transportation costs—will be necessary for costs to be more comparable between field and CEA lettuce supply chains.[2]

4.2 Environmental Impacts Findings

The environmental impacts analysis indicates that CEA supply chains have larger energy use and greenhouse gas emissions than those based on field production (Table 3). GH supply chains have markedly lower energy demand and GWP than PF supply chains in both studied locations, primarily due to the energy required for lighting and cooling. GH supply chains delivering to New York have estimated GWP only 3% larger than field-based supply chains, but the difference is much larger in Chicago due to higher energy use in production and longer transportation distances. More generally, CED and GWP per kg lettuce for the GH and

Table 3 Environmental impacts for the delivery of 1 kg lettuce to wholesale produce markets in New York City and Chicago from field-based production, a CEA greenhouse and a CEA plant factory

	New York City wholesale market, Hunt's point			Chicago international produce market		
	Field	GH	PF	Field	GH	PF
CED (MJ/kg lettuce)	18.52	23.83	42.52	14.24	29.19	44.74
GWP (kg CO_2.eq/ kg lettuce)	1.29	1.33	2.72	0.99	2.07	4.62
WU (liters/kg lettuce)	201.43	20.86	20.86	201.43	20.86	20.86

Source: Authors' own calculations

Note: *Field* indicates field-based production in Salinas Valley, California, *GH* indicates a CEA greenhouse in the same metropolitan area as the wholesale market, and *PF* indicates a CEA Plant Factory in the same metropolitan area as the wholesale market

the PF for New York are lower than for the same supply chains serving Chicago due to the assumed shorter transportation distance and lower energy use for heating, cooling and lighting. The model results show that the average CED and GWP values are lower for field-based production than for CEA. However, during the months of June, July, and August, the CED and GWP values for CEA are only 28 to 33% of the average values, bringing them well below the field-based values. CED and GWP values for CEA in non-summer months are high, but this is also during a period of the year when field-based product is much less available. Water use for production (but not cooling) is significantly larger per kg lettuce for the field-based production system. Overall, these results suggest that no one production system and location will always be preferred for all environmental outcomes.

4.3 Comparison to Previous Results

A number of previous studies have assessed the environmental impacts of lettuce supply chains, often using LCA methods. Our results can usefully be compared to the results of the previous studies, although the basic method, system boundaries and data sources often differ (Table 4). In general, our assessed values for the three production systems are consistent with those reported in previous studies. Our field production system reports higher CED than previous studies, in part because of the long distances the product is transported. Our GWP values are consistent with previous study values, despite that fact that most previous studies accounted for the embodied effects of inputs, structures and equipment. Similarly, our CED values for GH production appear to be lower than those of previous studies because those studies considered embodied energy. We report values of GWP for the PF consistent with previous studies.

Table 4 Comparison of environmental outcomes from current and previous studies of lettuce production and supply chains

Production setting	CED (MJ/kg)	GWP (kg CO_2-eq/kg)	WU (liters/kg)	References
Field	*14.24–18.52*	*0.99–1.29*	*135.4*	*Author calculations*
Literature field, including post-farm	5.67–7.00[a]	0.25–3.75	42–97	Bartzas, Zaharaki, and Komnitsas (2015), Emery and Brown (2016), Gunady, Biswas, Solah, and James (2012), Rothwell et al. (2016), Stoessel et al. (2012)
Literature field, production only	2.98	0.14–2.30	83–160	Bartzas et al. (2015), Emery and Brown (2016), Gunady et al. (2012), Foteinis and Chatzisymeon (2016), Hospido et al. (2009), Romero-Gámez, Audsley, and Suárez-Rey (2014), Rothwell et al. (2016)
Greenhouse (GH)	*23.83–29.19*	*1.33–2.07*	*20.9*	*Author calculations*
Literature GH, including post-farm	38.67	0.52–2.62	20–36	Hospido et al. (2009), Rothwell et al. (2016)
Literature GH, production only	3.15–3.47	0.21–2.46	–	Bartzas et al. (2015), Hospido et al. (2009), Rothwell et al. (2016)
Plant Factory (PF)	*42.52–44.74*	*2.72–4.62*	*20.9*	*Author calculations*
Literature PF, production only	–	2.30–6.20	–	Shiina et al. (2011)

Source: Authors' own calculations and cited references
[a]Non-renewable energy only for maximum value

4.4 Scenario Analysis Findings

As is the case with most studies comparing the costs and environmental outcomes of alternative supply chain configurations, the nature of these configurations (size, location, and production technology) can vary considerably. Our "best case" scenario represents likely lowest cost values for both GH and PF in the two locations. Assuming the productivity and costs for the "best case" scenario considerably lowers landed costs for both GH (20 to 31%) and PF (17 to 27%) but this does not change the basic result that field production in California is far less costly (Table 5). This suggests that our findings with regard to the average assumed costs and productivity are likely to be robust. However, lower cost for land purchase do affect the relative costs of GH and PF (Table 5). The best case scenario would reverse the cost rankings, with GH operations indicating lower costs than PF. This shift occurs because the larger land footprint for the GH makes its landed cost more sensitive to assumptions regarding land values.

Table 5 Comparisons of the baseline and best case total landed cost of 1 kg lettuce delivered to wholesale markets in New York and Chicago metropolitan areas

Location, CEA operation	Baseline scenario	Best case scenario	Best case less baseline scenario	Field production to indicated location	Best case less field production
	$/kg delivered lettuce				
New York				3.04	
GH	8.09	5.59	−2.50		2.55
PF	7.82	5.69	−2.13		2.65
Chicago				2.72	
GH	7.03	5.63	−1.39		2.91
PF	6.89	5.69	−1.20		2.97

Source: Authors' own calculations

Note: All values in $/kg lettuce delivered to wholesale market in each metropolitan area. The Baseline Scenario represents average CEA performance in the two metropolitan areas, as reported in Table 2. The Best Case Scenario modifies assumptions to represent current best possible performance. For New York, the Best Case Scenario assumes the land value, electricity rates and transportation costs for Chicago. For both locations, the unit costs for structure and equipment is assumed to be 40% lower than in the Baseline and productivity per unit production area is assumed to be 20% higher

5 Discussion

Our analyses are broadly consistent with evidence available from the limited previous work on similar topics. Field-based lettuce supply chains have lower landed costs because of lower per kg land, equipment, structure, labour, and energy inputs costs, despite much higher transportation costs due to the distance from wholesale market customers. Thus, the underlying cost structures for field-based and CEA supply chains are quite different, and within relevant potential ranges for improvement, it appears that at present there are limited management options for CEA operations to achieve costs approaching those of field-based operations. Labour is a substantial cost in CEA production and opportunities to lower costs with as automation technologies are further developed should be included in future analyses. However, we acknowledge that comparable or lower landed costs alone are not required for a successful CEA business, especially if consumers are willing to pay price premiums either because of the "local" nature of the food or its perceived environmental friendliness. Current CEA businesses producing lettuce exist in many US metropolitan areas, and although there has been no formal study of their financial performance, there is continued interest in CEA investment, which suggests the potential for profitability despite much higher costs.

Our analyses also shed light on the relative costs of the two types of CEA operations. Similar to Eaves and Eaves (2018), we found that under our baseline conditions, PF can have comparable or lower costs compared to GH. The underlying rationale is that higher energy costs are offset by lower land costs due to the smaller footprint required in a PF to achieve the same level of production. However, this means the decision to invest in GH versus PF is sensitive to the costs of both energy and land, as well as the cost per unit area for structures and equipment.

Combining the analysis of economic and environmental outcomes is not common in previous LCA-based studies comparing the performance of different production systems. However, our analysis suggests that informed decision making on the part of supply chain actors and consumers can benefit from information about both of these dimensions because the two sets of indicators will not result in the same rank for the alternatives. On a cost basis—i.e., assuming that buyers consider the product of different systems as essentially identical—field-based supply chains are preferred. From an environmental perspective, CEA have considerably higher CED and somewhat higher GWP, but these are much smaller for GH operations than for PF despite roughly similar landed costs. Locating the CEA operation further from urban customers tends to reduce costs (due particularly to lower land values) but increases transportation costs and negative environmental impacts. However, both CEA systems use far less water per unit than field-based production. Thus, buyers and ultimate consumers may face relevant trade-offs between cost and environmental outcomes, and the environmental outcomes themselves can be sensitive to factors such as location within a metropolitan area for CEA operations. The mix of fuels used to generate electricity can also affect the nature of GWP, although in our study such differences were of limited importance.

Although our findings are a substantive contribution to the knowledge of the potential role and impacts of CEA supply chains serving metropolitan areas, our work could be extended in five principal ways. First, consideration of the embodied energy and environmental effects through LCA would allow more direct comparisons with similar studies using that approach and provide a more comprehensive estimate of environmental impacts. Second, additional scenario analysis would help to identify more specifically the importance of individual assumptions about costs and environmental impacts. Third, further work could usefully identify the scale, configuration and location of operations that minimise total cost within a given metropolitan area. Costs appear particularly sensitive to land values, so analysis of the location within a metropolitan area and of rooftop greenhouse operations (e.g., Nadal et al., 2017) would be of particular relevance. Fourth, additional cities could be analysed, because the climate conditions were relatively similar in New York and Chicago (New York is ASHRAE climate zone 4A, and Chicago is 5A), and the equipment configurations and energy use would be quite different in drier and hotter climate zones. Finally, additional assessment of profitability (rather than just cost) through consideration of product selling prices, consumer perceptions of, and preferences for, field and CEA lettuce, and revenue streams would provide an improved context for assessing the potential of CEA operations to provide lettuce (and other leafy greens) to metropolitan areas of the US.

6 Conclusions

Our analysis of three supply chains to provide lettuce to two US metropolitan areas indicates that at present the lowest landed-cost option is a supply chain based on field production rather than non-automated GH or PF in urban locations. Because the landed cost differences are larger (nearly double even in the "best case" scenario) this suggests that modifications to reduce the costs of non-automated, urban CEA systems to the level of field production will present major challenges. In addition, the studied configurations and locations of CEA supply chains operating within metropolitan urban areas may have higher energy use and GWP, although all the CEA operations analysed used less water per kg of lettuce than field production. Thus, the rankings based on costs and environmental outcomes do not always align. Although the configuration of a CEA supply chain affects environmental impacts, it is inappropriate to claim that "local" CEA supply chains for lettuce are broadly more environmentally friendly than field-based production, even when field lettuce is shipped long distances. Additional analyses of alternative scales, locations and more automated CEA configurations as well as seasonal field-based production closer to metropolitan areas could provide further insights to supply chain actors.

The future development of CEA supply chains is likely to continue despite higher costs, due to the ability of CEA systems to control selected quality aspects for leafy greens (e.g., production of micro-greens for the high-end restaurant segment) and their flexibility as suppliers to certain market segments. Although it is beyond the scope of our analysis, differentiation of CEA-produced leafy greens from those produced by more conventional field-based methods to receive substantially higher prices would seem necessary for business success given much higher costs. As the number of CEA operations increases, additional evidence will become available about their role and status in supplying metropolitan food needs. Further analysis of the scaling-up effects (potentially both positive through agglomeration economies and negative through competition for scarce resources) will be appropriate as growth proceeds.

Appendix

Table 6 Characteristics and assumptions for energy modeling of GH and PF production systems

Energy system characteristic or assumption	Value	Units	Comment or value
GH production area	4460	m^2	
GH growing area	90%		% of total GH production area
PF production area	802	m^2	
PF layers	10		
PF growing area	50%		% of total PF production area
New York TMY3 station			LaGuardia airport
Chicago TMY3 station			Midway airport
Day temp set point	24	°C	
Night temp set point	19	°C	
RH low set point	50%		
RH high set point	70%		
GH transmittance (frame/glazing)	70%		
Shade cloth reduction	60%		
Supplemental light efficacy[a]	2.1	μmol/J	
Infiltration rate	0.5	ACH	
DLI target	17	$Mol/m^2/day$	
Heating system			Natural gas boiler
GH cooling system			Evaporative pads
PF cooling system			Chiller
Heating efficiency	80%		
Evaporative pad effectiveness	80%		
Chiller COP	5.5		
Average crop spacing	48	$Head/m^2$	

Source: Authors' own calculations and assumptions

[a]The same efficacy value is assumed for both the GH and PF. The specified value is the best efficacy value for High Pressure Sodium (HPS) lighting often used in GH and among the higher efficacy values for broad-spectrum LED lighting used in PF

Table 7 Detailed calculations for field-based lettuce production operating costs per acre

Category	Quantity used	Units of quantity	Price/unit	Units of price	Cost/acre, $
Production supplies					*1234*
Seed (package)	157.50	Thousand	1.4	/thousand	221
Herbicide	1.00	Acre	92	/acre	92
Insecticide	1.00	Acre	282	/acre	282
Fertiliser					
Compost	2.00	Ton	55	/ton	110
Potassium sulphate	150.00	Lb	0.86	/lb	129
7-7-0-7	30.00	Gal	2.03	/lb	61
28-0-0-5	20.00	Gal	2.28	/lb	46
20-0-0-5	37.00	Gal	1.73	/lb	64
Fungicide	1	Acre	230	/acre	230
Packaging costs					*6975*
Harvest-field pack	900	Carton	6	/carton	5400
Cool/palletise	900	Carton	1	/carton	900
Market/sales fee	900	Carton	0.75	/carton	675
Miscellaneous costs					*1173*
Soil sample	1	Acre	8	/acre	8
Laser level	0.5	Acre	165	/acre	82.5
Haul/spread compost	1	Acre	20	/acre	20
List bed 3-row 80″	1	Acre	23	/acre	23
Ground application	1	Acre	15	/acre	15
Plant thinning—automated	1	Acre	115	/acre	115
Air application	3	Acre	20	/acre	60
Pest control advisor/ certified crop advisor	1	Acre	30	/acre	30
Machinery repair	1	Acre	165	/acre	165.0
Liability insurance	1	Acre	20	/acre	20.0
Food safety program	1	Acre	40	/acre	40.0
Regulatory program	1	Acre	40	/acre	40.0
Office expense	1	Acre	350	/acre	350.0
Field sanitation	1	Acre	12	/acre	12.0
Property taxes	1	Acre	28	/acre	28.0
Property insurance	1	Acre	2	/acre	2.0
Investment repairs	1	Acre	96	/acre	96.0
Interest on operating costs @ 4.5%	1.00	Acre	66	/acre	66
Production labour					*622*
Equipment operator labour	10.51	Hours	21.85	/hour	229.6
Irrigation labour	13	Hours	17.8	/hour	231.4
Non-machine labour	9.52	Hours	16.9	/hour	160.9
Utilities					*532*

(*continued*)

Table 7 (continued)

Category	Quantity used	Units of quantity	Price/unit	Units of price	Cost/acre, $
Water—pumped	14.00	Acre-inch	18	/acre-inch	252.00
Fuel—gas	2	Gal	3.25	/gal	6.5
Fuel—diesel	87.861	Gal	2.7	/gal	237.2
Lube	1	Acre	36	/acre	36.0
Miscellaneous costs					*2430*
Land rent	1.80	Acre	1350	/acre	2430

Source: Authors' own calculations and assumptions and Tourte et al. (2017)
Note: Yield per acre is assumed to be 900 cartons each weighing 11.4 kg (25 lbs), for a total weight of 10,631 kg less 30% shrink for 7144 saleable yield. One acre equals 0.4046 hectare

Table 8 Detailed calculations for field-based lettuce production structure and equipment costs per acre

Category	Investment cost, $	Annualised costs $/acre[a]
Structures	72,000	3
Shop building 2400 ft^2	72,000	3
Production equipment	559	3,059,790
Fuel tanks—overhead	1	**3,059,790**
Shop tools	1	10,975
Drip system	89	20,000
Sprinkler system	48	341,884
Sprinkler pipe	131	370,495
205HP crawler	74	1,139,000
Disc—offset 25′	19	350,000
Subsoiler—16′	15	48,769
Triplane—16′	9	42,454
Chisel—heavy 26′	18	38,000
Ring roller—heavy 18′	6	51,218
Lilliston rolling 3-row	4	15,552
Bed shaper 3-row	8	18,000
150HP 4WD tractor	48	44,412
Row crop planter	13	225,000
Cultivator 3-row	2	54,887
Fertiliser Bar 20′	2	9500
Drip tape laying machine 3-row	4	13,000
Pickup 3/4 ton	15	16,117
Saddle tanks 300 gallons	1	50,000
Spray boom 20′	1	1660
Ring-roller 25′	11	2900
Drip tape extraction sled	10	29,000
120HP 2WD tractor	29	30,000

Source: Authors' own calculations and assumptions and Tourte et al. (2017)
[a]Calculated using an assumption of 20% capital recovery per year, divided by 250 acres for two crops per year

Table 9 Detailed calculations for field-based lettuce transportation costs

Wholesale market area and cost category	Quantity	Units of quantity	Price/ unit	Units of price	Cost/ shipment
New York City area					*8329*
Fuel (diesel)	800	Gallons	2.96	$/gallon	2368
Driver labour	44	Hours	21.90	$/hour	964
Overhead and other costs			150%	% of direct	4997
Chicago area					*6184*
Fuel (diesel)	800	Gallons	591	Gals fuel	1751
Driver labour	44	Hours	33	Hours	723
Overhead and other costs			150%	% of direct	3710

Source: Authors' own calculations and assumptions

Note: One-way distance to New York is 3000 miles and to Chicago 2218 miles. Assumes a backhaul proportion of 75%, and fuel use of 5 miles per gallon of diesel fuel

Table 10 Annual operations costs for greenhouse and plant factory operations

Cost category	Units of quantity	Quantity used		Cost/year, $	
		GH	PF	GH	PF
Production supplies				*130,707*	*124,257*
Seed (package)	Packages	637	637	70,070	70,070
Horticubes	Cases	550	550	35,750	35,750
Beneficial insects	Packages	83	83	3320	3320
Fertiliser					
Blended mix	Pounds	7813	7813	7813	7813
$CaNO_3$	Pounds	7813	7813	3594	3594
Additions	Pounds	105	105	105	105
Fungicide/ pesticide	Gallons	105	0	6300	0
Sanitiser	Gallons	103	103	3605	3605
Sticky traps	Packages	5	0	150	0
Packaging costs				*346,105*	*346,105*
Box	Box	126,302	126,302	315,755	315,755
Labels	Roll	607	607	30,350	30,350
Miscellaneous costs				*46,900*	*46,900*
Advertising, mailings, flyers	Campaigns	1	1	200	200

(continued)

Table 10 (continued)

Cost category	Units of quantity	Quantity used		Cost/year, $	
		GH	PF	GH	PF
Continuing education	Meetings	40	40	10,000	10,000
Internet service	Months	12	12	2400	2400
Laboratory fees	Tests	1000	1000	20,000	20,000
Office supplies	Months	12	12	2400	2400
Postage	Months	12	12	2400	2400
Marketing materials & promotions	Promos	2	2	2000	2000
Record keeping	Months	12	12	3000	3000
Software	Programs	5	5	2500	2500
Subscriptions	Subscriptions	10	10	1000	1000
Marketing & trade shows	Trade show	2	2	1000	1000
Utilities other the energy and water				*12,960*	*12,960*
Mobile phones	Months	240	240	12,000	12,000
Telephone	Months	24	24	960	960
Labour and management				*1,961,041*	*1,961,041*
Seed/transplant/harvest/package	Hours	78,812	78,812	1,024,558	1,024,558
Delivering to market	Hours	9094	9094	109,124	109,124
Production management	Hours	6062	6062	78,812	78,812
Sales manager	Positions	1	1	75,000	75,000
Admin assistant	Positions	1	1	45,000	45,000
Executive level	Positions	1	1	100,000	100,000
Outside services	$	1	1	75,999	75,999
Fringe benefits	%	30%	30%	452,548	452,548
Water cost				*32,578*	*32,578*
Water for production (not cooling)	Gallons	6,031,727	6,031,727	32,578	32,578

Source: Authors' own calculations and assumptions

Note: Assumes that unit costs are not location specific (i.e., are the same for New York and Chicago). Yields for both GH and PF operations are 454,685 kg per year

Table 11 Total investment costs for structures, land and equipment for greenhouse and plant factory operations

	New York		Chicago	
Cost category	GH	PF	GH	PF
Structures	2400,000	632,160	2400,000	632,160
Production area, ft^2	43,200	4320	43,200	4320
Production levels	1	10	1	10
Non-production grow area, ft^2	4800	4320	4800	4320
Total production-related area, ft^2	48,000	8640	48,000	8640
Cost of structures & equipment, \$/ft^2	50	50	50	50
Ratio PF to GH Costs[a]	1.00	1.46	1.00	1.46
Land	4,077,410	1,931,405	523,355	247,905
Production area, ft^2	48,000	8640	48,000	8640
Factor for packing, parking and bathrooms	0.558	0.558	0.558	0.558
Parking, packing, bathrooms, ft^2	26,784	26,784	26,784	26,784
Total land area required, ft^2	74,784	35,424	74,784	35,424
Ft^2/acre	43,560	43,560	43,560	43,560
Acres required	1.72	0.81	1.72	0.81
Value of land, \$/acres	2,375,000	2,375,000	304,843	304,843
Growing and delivery equipment costs	466,800	466,800	466,800	466,800
Back pack sprayer	1600	1600	1600	1600
Carbon dioxide generator	7680	7680	7680	7680
Cooler	20,000	20,000	20,000	20,000
Delivery truck with AC	110,000	110,000	110,000	110,000
Fertiliser mixing pump	480	480	480	480
Meters and sensors				
EC	2560	2560	2560	2560
pH	800	800	800	800
Thermometer	400	400	400	400
Monitors				
Humidity	480	480	480	480
CO_2	2000	2000	2000	2000
Growing system	320,000	320,000	320,000	320,000
Scale	800	800	800	800
Total structures, land and equipment	6,477,410	2,563,565	2,923,355	880,065
Annual cost, \$/year	933,531	407,381	455,749	181,063

Source: Authors' own calculations and assumptions

Note: Annual cost assumes a 6.2% interest rate for 10 years based on Weighted Average Cost of Capital in farming and agriculture from Damodaran (2018)

[a]Assumes the ratio between GH and PF reported by Eaves and Eaves (2018)

Table 12 Detailed calculations for CEA lettuce transportation costs

Wholesale market area and cost category	Quantity	Units of quantity	Price/ unit	Units of price	Cost/ shipment
New York City area					*777*
Fuel (diesel)	312	Gallons	2.49	$/gallon	777
Chicago area					*20,449*
Fuel (diesel)	6909	Gallons	2.96	$/gallon	20,449

Source: Authors' own calculations and assumptions

Note: Labour costs are included in other operations costs, Table 10. One-way distance to New York is 2.1 miles (3.5 km) and to Chicago 46.5 miles (74.8 km). Deliveries are assumed to be made 10 times per week with a refrigerated reefer truck with diesel fuel use of 7 miles per gallon

Table 13 Energy use and cost calculations for New York metropolitan area greenhouse and plant factory operations

	Greenhouse				Plant factory			
Month, cost category	Heating (GJ)	Lighting (GJ)	Cooling (GJ)	Total energy (GJ)	Heating (GJ)	Lighting (GJ)	Cooling (GJ)	Total energy (GJ)
Jan	1006	648	14	1668	113	1119	457	1689
Feb	1258	425	13	1696	99	1011	411	1521
Mar	1017	310	14	1340	80	1119	459	1658
Apr	626	193	14	833	47	1083	445	1575
May	445	111	14	570	27	1119	461	1608
Jun	157	162	14	332	19	1083	453	1555
Jul	85	130	14	230	16	1119	480	1615
Aug	85	103	14	202	16	1119	474	1609
Sep	239	229	14	482	30	1083	452	1565
Oct	506	282	14	802	44	1119	462	1625
Nov	495	601	14	1109	66	1083	446	1595
Dec	830	681	14	1525	99	1119	458	1676
Total	6748	3876	165	10,789	657	13,177	5456	19,290
Natural gas (NG) cost								
GJ per MMBtu	0.948				0.948			
MMBtu	6396				622			
Mcf per MMBtu	0.964				0.964			
Mcf NG	6168				600			
NG cost, $/Mcf	6.79				6.79			
NG cost, $/year	41,878				4075			
Electricity cost								

(*continued*)

Table 13 (continued)

Month, cost category	Greenhouse Heating (GJ)	Greenhouse Lighting (GJ)	Greenhouse Cooling (GJ)	Greenhouse Total energy (GJ)	Plant factory Heating (GJ)	Plant factory Lighting (GJ)	Plant factory Cooling (GJ)	Plant factory Total energy (GJ)
kWh per GJ		277.8	277.8	277.8		277.8	277.8	277.8
kWh used		1,076,534	45,841	1,122,375		3,660,208	1,515,586	5,175,794
Electricity cost, $/kWh		0.1060	0.1060	0.1060		0.1060	0.1060	0.1060
Electricity cost, $/year		114,113	4859	118,972		387,982	160,652	548,634

Source: Authors' own calculations and assumptions

Note: Costs for electricity will also include a "demand charge" for the GH calculated as 379 kW times 12 times $10.77 per kW, equal to $48,982 per year. The "demand charge" for the PF is calculated as 516 kW times 12 times $10.77 per kW, equal to $66,688 per year

Table 14 Energy use and cost calculations for Chicago metropolitan area greenhouse and plant factory operations

	Greenhouse				Plant factory			
Month, cost category	Heating (GJ)	Lighting (GJ)	Cooling (GJ)	Total energy (GJ)	Heating (GJ)	Lighting (GJ)	Cooling (GJ)	Total energy (GJ)
Jan	1451	685	16	2152	137	1119	459	1715
Feb	1264	492	15	1770	104	1011	417	1532
Mar	1039	391	16	1445	74	1119	464	1657
Apr	629	354	16	999	38	1083	451	1573
May	558	156	16	730	27	1119	467	1614
Jun	228	148	16	391	14	1083	459	1556
Jul	93	120	16	230	8	1119	478	1605
Aug	107	156	16	279	11	1119	478	1608
Sep	214	165	16	395	16	1083	459	1558
Oct	467	437	16	920	41	1119	468	1628
Nov	657	501	16	1174	60	1083	450	1594
Dec	1063	695	16	1774	110	1119	462	1691
Total	7770	4298	189	12,257	643	13,177	5512	19,332
Natural gas (NG) cost								
GJ per MMBtu	0.948				0.948			
MMBtu	7365				609			
Mcf per MMBtu	0.964				0.964			
Mcf NG	7102				588			
NG cost, $/Mcf	7.80				7.80			
NG cost, $/year	55,394				4584			

(*continued*)

Table 14 (continued)

	Greenhouse				Plant factory			
Month, cost category	Heating (GJ)	Lighting (GJ)	Cooling (GJ)	Total energy (GJ)	Heating (GJ)	Lighting (GJ)	Cooling (GJ)	Total energy (GJ)
Electricity cost								
kWh per GJ		277.8	277.8	277.8		277.8	277.8	277.8
kWh used		1,193,857	52,589	1,246,445		3,660,208	1,531,089	5,191,297
Electricity cost, $/kWh		0.0637	0.0637	0.0637		0.0637	0.0637	0.0637
Electricity cost, $/year		76,049	3350	79,399		233,155	97,530	330,686

Source: Authors' own calculations and assumptions

Note: Costs for electricity will also include a "demand charge" for the GH calculated as 379 kW times 12 times $11.19 per kW, equal to $50,892 per year. The "demand charge" for the PF is calculated as 520 kW times 12 times $11.19 per kW, equal to $69,826 per year

Table 15 Detailed calculations of CO_2 equivalent emissions, field, greenhouse and plant factory operations

Emissions category and calculation information	New York			Chicago		
	Field	GH	PF	Field	GH	PF
Emissions from natural gas						
GJ from NG	0	6748	657	0	7770	643
CO_2 per GJ from NG, MT/GJ	0.0503	0.0503	0.0503	0.0503	0.0503	0.0503
CO_2 from NG, MT/year	0.0	339.4	33.0	0.0	390.8	32.3
Emissions from electricity						
GJ from electricity	1.87	4041	18,633	1.87	4487	18,689
CO_2 per GJ from electricity, MT/GJ	0.0662	0.0645	0.0645	0.0662	0.1069	0.1069
CO_2 from electricity, MT/year	0.1238	260.7	1202.4	0.1238	479.6	1997.4
Emissions from diesel						
Diesel used, gallons/year	888	312	312	679	6909	6909
CO_2 from diesel, kg/gallon	10.21	10.21	10.21	10.21	10.21	10.21
CO_2 from diesel, kg/year	9065	3186	3186	6936	70,537	70,537
Miles traveled per year	3750	2184	2184	2773	48,360	48,360
CH_4 emissions, g/mile	0.0051	0.0051	0.0051	0.0051	0.0051	0.0051
CH_4 emissions, g/year	19.1	11.1	11.1	14.1	246.6	246.6
CH_4 to CO_2 conversion	25	25	25	25	25	25
CO_2 equivalents emissions as CH_4, kg/year	0.5	0.3	0.3	0.4	6.2	6.2
N_20 emissions factor, g/mile	0.0048	0.0048	0.0048	0.0048	0.0048	0.0048
N_20 emissions, g/year	18.0	10.5	10.5	13.3	232.1	232.1
N_20 to CO_2 conversion	298	298	298	298	298	298
CO_2 equivalent emissions as N_20, kg/year	5.4	3.1	3.1	4.0	69.2	69.2
CO_2 equivalent from diesel, MT/year	9.0709	3.1889	3.1889	6.9403	70.6119	70.6119
Total CO_2 emissions, MT/year	*9.1*	*603.3*	*1238.6*	*7.1*	*941.0*	*2100.3*
Production, kg lettuce per year	7144	454,685	454,685	7144	454,685	454,685
CO_2 emissions, kg CO_2/kg lettuce	1.29	1.33	2.72	0.99	2.07	4.62

Source: Authors' own calculations and assumptions

Note: Diesel emissions are for field production and transportation for all three operations. Calculations ignore a small amount of gasoline used in field production (2 gallons, 7.5 litres)

Notes

1. At present, we estimate that at least 90% of the light in VF would need to come from supplemental light, so in practice most light in a vertical farm would probably need to come from supplemental sources.
2. Note that we have not accounted for potential differences in prices that wholesalers (or consumers) are willing to pay for lettuce produced in the same metropolitan area. This may make the profitability differences smaller than the landed cost differences that are our focus here.

References

Albright, L. D., Both, A. J., & Chiu, A. J. (2010). Controlling greenhouse light to a consistent daily integral. *Transactions of ASAE, 43*(2), 421–431. https://doi.org/10.13031/2013.2721

ASHRAE. (2004). *Energy standard for buildings except low-rise residential buildings. ANSI/ASHRAE/IESNA Standard 90.1-2004*. Atlanta, GA: American National Standards Institute/American Society of Heating, Refrigerating and Air-Conditioning Engineers/Illuminating Engineering Society of North America.

ASHRAE. (2017). *2017 ASHRAE handbook: Fundamentals*. Atlanta, GA: American Society of Heating, Refrigeration and Air-Conditioning Engineers.

Athena Sustainable Materials Institute. (2017). LCA, LCI, LCIA, LCC: What's the difference? Retrieved from http://www.athenasmi.org/resources/about-lca/whats-the-difference/

Bartzas, G., Zaharaki, D., & Komnitsas, K. (2015). Life cycle assessment of open field and greenhouse cultivation of lettuce and barley. *Information Processing in Agriculture, 2*, 191–207.

Beccali, M., Cellura, M., Iudicello, M., & Mistretta, M. (2010). Life cycle assessment of Italian citrus-based products. Sensitivity analysis and improvement scenarios. *Journal of Environmental Management, 91*(7), 1415–1428. https://doi.org/https://doi.org/10.1016/j.jenvman.2010.02.028.

Björklund, A. E. (2002). Survey of approaches to improve reliability in LCA. *The International Journal of Life Cycle Assessment, 7*(2), 64. https://doi.org/10.1007/BF02978849

Crawley, D. B., Lawrie, L. K., Winkelmann, F. C., Buhl, W. F., Huang, Y. J., Pedersen, C. O., et al. (2001). EnergyPlus: Creating a new-generation building energy simulation program. *Energy and Buildings, 33*(4), 319–331. https://doi.org/10.1016/S0378-7788(00)00114-6

Damodaran, A. (2018). Cost of capital by sector, United States. Retrieved December 15, 2018, from http://people.stern.nyu.edu/adamodar/New_Home_Page/datafile/wacc.htm

Eaves, J., & Eaves, S. (2018). Comparing the profitability of a greenhouse to a vertical farm in Quebec. *Canadian Journal of Agricultural Economics/Revue Canadienne d'agroeconomie, 66*(1), 43–54. https://doi.org/10.1111/cjag.12161

Emery, I., & Brown, S. (2016). Lettuce to reduce greenhouse gases: A comparative life cycle assessment of conventional and community agriculture. In S. Brown, K. McIvor, & E. Hodges Snyder (Eds.), *Sowing seeds in the city: Ecosystem and municipal services* (pp. 161–169). Dordrecht: Springer Netherlands. https://doi.org/10.1007/978-94-017-7453-6_12

Foteinis, S., & Chatzisymeon, E. (2016). Life cycle assessment of organic versus conventional agriculture. A case study of lettuce cultivation in Greece. *Journal of Cleaner Production, 112,* 2462. https://doi.org/10.1016/j.jclepro.2015.09.075. Oxford: Elsevier Sci Ltd.

Gómez, M. I., Mattson, N., & Nishi, I. (2017, November 2). Interactive business tool for CEA vegetables: Lettuce & Tomato Presentation at the Cornell CEA Entrepreneur Conference. Retrieved from http://cea.cals.cornell.edu/research/marketing/Cost%20Study.pdf

Graamans, L., Baeza, E., van den Dobbelsteen, A., Tsafaras, I., & Stanghellini, C. (2018). Plant factories versus greenhouses: Comparison of resource use efficiency. Agricultural Systems, 160, 31–43. https://doi.org/https://doi.org/10.1016/j.agsy.2017.11.003

Gunady, M. G. A., Biswas, W., Solah, V. A., & James, A. P. (2012). Evaluating the global warming potential of the fresh produce supply chain for strawberries, romaine/cos lettuces (Lactuca sativa), and button mushrooms (Agaricus bisporus) in Western Australia using life cycle assessment (LCA). *Journal of Cleaner Production, 28*, 81–87. https://doi.org/10.1016/j.jclepro.2011.12.031

Harbick, K., & Albright, L. (2016). Comparison of energy consumption: Greenhouses and plant factories. *Acta Horticulturae,* (1134), 285–292. https://doi.org/10.17660/ActaHortic.2016.1134.38

Hospido, A., Milà, I., Canals, L., Mclaren, S., Truninger, M., Edwards-jones, G., et al. (2009). The role of seasonality in lettuce consumption: A case study of environmental and social aspects. *The International Journal of Life Cycle Assessment, 14*(5), 381–391. https://doi.org/10.1007/s11367-009-0091-7

Kozai, T., Niu, G., & Takagaki, M. (Eds.). (2015). *Plant factory: An indoor vertical farming system for efficient quality food production*. Academic Press. 422 pp. ISBN 9780128017753.

Mougeot, L. J. A. (2000). Urban agriculture: Definition, presence, potentials and risks. In N. Bakker, M. Dubbeling, S. Gündel, U. Sabel-Koschella, & H. De Zeeuw (Eds.), *Growing cities, growing food: Urban agriculture on the policy agenda. A reader on urban agriculture* (pp. 99–117). Feldafing, Germany: DSE/ETC.

Nadal, A., Llorach-Massana, P., Cuerva, E., Lïpez-Capel, E., Montero, J. I., Josa, A., et al. (2017). Building-integrated rooftop greenhouses: An energy and environmental assessment in the Mediterranean context. *Applied Energy*, 187. https://doi.org/10.1016/j.apenergy.2016.11.051

Newbean Capital. (2015, March). Indoor crop production feeding the future. Retrieved March 23, 2018, from https://indoor.ag/whitepaper/

Notarnicola, B., Sala, S., Anton, A., McLaren, S. J., Saouter, E., & Sonesson, U. (2017). The role of life cycle assessment in supporting sustainable agri-food systems: A review of the challenges. *Journal of Cleaner Production, 140*, 399–409. https://doi.org/10.1016/j.jclepro.2016.06.071

Romero-Gámez, M., Audsley, E., & Suárez-Rey, E. M. (2014). Life cycle assessment of cultivating lettuce and escarole in Spain. *Journal of Cleaner Production, 73*, 193–203.

Rothwell, A., Ridoutt, B., Page, G., & Bellotti, W. (2016). Environmental performance of local food: Trade-offs and implications for climate resilience in a developed city. *Journal of Cleaner Production, 114*, 420–430.

Shiina, T., Hosokawa, D., Roy, P., Orikasa, T., Nakamura, N., & Thammawong, M. (2011). Life cycle inventory analysis of leafy vegetables grown in two types of plant factories. *Acta Horticulturae*, 115. https://doi.org/10.17660/ActaHortic.2011.919.14

Stoessel, F., Juraske, R., Pfister, S., & Hellweg, S. (2012). Life cycle inventory and carbon and water food print of fruits and vegetables: Application to a Swiss retailer. *Environmental Science & Technology, 46*(6), 3253–3262. https://doi.org/10.1021/es2030577

Tourte, L., Smith, R. F., Murdock, J., & Sumner, D. A. (2017). *Sample costs to produce and harvest iceberg lettuce—2017: Central coast—Monterey, Santa Cruz, and San Benito counties.* University of California Agriculture and Natural Resources, Cooperative Extension and Agricultural Issues Center, UC Davis Department of Agricultural and Resource Economics.

USDA. (2017). *Vegetables 2016 summary.* Washington, DC: National Agricultural Statistics Service.

Wilcox, S., & Marion, W. (2008). *Users manual for TMY3 data sets.* Golden, CO: National Renewable Energy Laboratory. Technical Report NREL/TP-581-43156.

Evaluating the Benefits of Collaborative Distribution with Supply Chain Simulation

M. Rabe, M. Poeting, and A. Klueter

1 Introduction

According to the OECD, the earth's population is expected to increase from 7 billion to over 9 billion people by 2050 and it is likely that cities will absorb the major part of the total world population growth. The increasing trend towards urbanisation can be illustrated by the following figures: In 1970, 1.3 billion people, or 36% of the world population, lived in urban areas. By 2015 that share had reached 54% and this rising trend is expected to continue in the coming decades (BBSR, 2017; United Nations, 2012). By 2050, nearly 80% of the world population is

---

M. Rabe • A. Klueter
Department IT in Production and Logistics, TU Dortmund University, Dortmund, Germany
e-mail: markus.rabe@tu-dortmund.de; astrid.klueter@tu-dortmund.de

M. Poeting (✉)
Institute of Transport Logistics, TU Dortmund University, Dortmund, Germany
e-mail: moritz.poeting@tu-dortmund.de

E. Aktas, M. Bourlakis (eds.), *Food Supply Chains in Cities*,
https://doi.org/10.1007/978-3-030-34065-0_3

projected to live in cities. As a consequence, challenges such as air pollution and transport congestion will accumulate within urban areas (Sigman, Hilderink, Delrue, Braathen, & Leflaive, 2012).

Urbanisation has both positive and negative environmental consequences. On one hand, the concentration of activities in urban areas allows for extensive interaction between parties involved in the urban environment and economies of scale. This means that urbanisation can lead to higher economic growth. On the other hand, large concentrations of economic activities also result in higher levels of air pollution. Therefore, urbanisation requires adapted transportation policies to avoid major complications in the transport system, negative environmental implications that result from traffic congestion, and a rise of greenhouse gas (GHG) emissions (BBSR, 2017; Sigman et al., 2012). Since 2013, GHG-emissions from the EU-28 transport sector have been increasing, especially in the road transport sector. In comparison with 2014, EU-28 emissions in 2015 had increased by almost 2%, mainly due to higher emissions from road transport, followed by aviation. In 2015, transport (including aviation and shipping) contributed 25.8% of total GHG-emissions in the EU-28. Road transport caused around 73% of total GHG emissions from transport. Of these emissions, 44.5% were contributed by passenger cars, while 18.8% were caused by heavy-duty vehicles (Fig. 1).

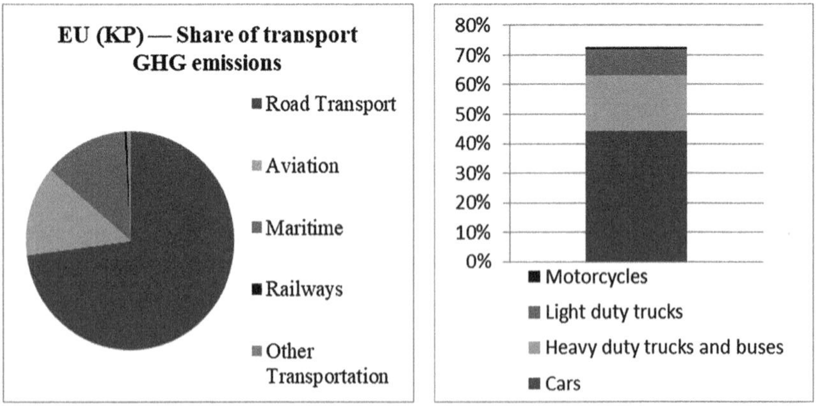

Fig. 1 The 2015 share of transport in the European Union. (Source: EEA, 2017)

To meet the target of reducing global GHG emissions by 80%, which is assumed to be necessary to keep global warming below 2 °C, the European Commission published a White Paper on Transport in 2011. The White Paper put forward several non-binding long-term targets for the transport sector, with an overall goal to reduce transport GHG emissions by at least 60% by 2050, compared with the level of 1990 (European Commission, 2011). However, emissions from transport (including aviation but excluding international shipping) increased more than 23% compared with 1990, and, therefore, they need to fall by 68% until 2050 to meet the 60% GHG emission reduction target (EEA, 2017).

Besides environmental effects, the rising population in metropolitan areas also represents a new challenge for logistics in cities in terms of supplying the citizens with goods for their daily needs. The distribution of goods, such as food and other fast-moving consumer goods (FMCG), places a heavy burden on the urban infrastructure. In theory, there are various approaches that tackle the challenges of distributing goods in urban areas. In practice, the impact of these measures is difficult to quantify and the implementation of concepts on real pilots is costly and time-consuming. However, only measures that have proven their advantages have a real chance to be put into practice. Simulation can be used effectively for this purpose.

Traditionally, simulation has been applied to various sectors, such as manufacturing, healthcare, services, and defence (Jahangirian, Eldabi, Naseer, Stergioulas, & Young, 2010) and is recognised as the second-most used technique in the field of Operations Management after mathematical modelling (Pannirselvam, Ferguson, Ash, & Siferd, 1999). Simulation can be seen as a Decision Support Tool in logistics, supply chain management, and manufacturing, both at a strategic and at a tactical/operational level (Tako & Robinson, 2012). Typical examples of problems that can be modelled and evaluated through the use of simulation are inventory planning, production scheduling, and delivery planning.

Hence, supply chain simulation technology enables us to evaluate the impact of specific measures on real application examples. This contributes to identifying promising measures and convincing decision makers to implement these in real applications. The overall goal of research in the context of urban transport is to model and quantify measures for cost-efficient and environmentally friendly distribution inside cities through the use of discrete event supply chain simulation.

In this article we give an overview of city logistics measures and how simulation contributes to the assessment of city logistics measures on the example of collaborative transport and UCCs. Therefore, the following section gives a literature review on measures for sustainable transport within the urban area and supply chain simulation in the context of city logistics. Next, we present a simulation approach to generate models for urban transport based on real data, using the supply chain simulation tool SimChain. Then, we present experimentation and results on a case study inside the metropolitan area of Athens. The results from this case study are discussed and a conclusion is given to summarise the findings.

2 Literature Review

Urban logistics has been investigated by researchers for several years. However, the subject is still evolving because of the continuously changing influences, i.e., through e-commerce, increasing sensitivity to environmental issues, and technological changes that enable new delivery scenarios, such as electric vehicles, drones, and driverless vehicles. However, despite its importance and increasing relevance, the literature on urban logistics is highly fragmented (Lagorio, Pinto, & Golini, 2016). To give an up-to-date overview, this chapter focusses on urban logistics measures that can be widely found in current literature. Furthermore, the state of simulation for supply chains in the urban context is presented.

2.1 Urban Logistics

Taniguchi (2001) defines city logistics as "the process for totally optimising the logistics and transport activities by private companies with support of advanced information systems in urban areas considering the traffic environment, the traffic congestion, the traffic safety and the energy savings within the framework of a market economy." In this context, the major stakeholders are shippers, freight carriers, administrators, and residents who are involved in city logistics (Taniguchi, Thompson, & Yamada, 2014).

Shippers and freight carriers are widely driven by economic aspects but are acting within the boundaries of policies and regulations established by the municipality. Therefore, administrators are able to influence the urban transport system by establishing regulations and enacting laws which are also influenced by the cities' residents. In addition, technological factors play an essential role in creating the way and means for the processing of transport in cities.

Policies and Regulations

Public authorities usually introduce mitigation measures against the externalities of freight transport with the aim to make the urban environment more attractive and sustainable for their inhabitants and on the other side to regulate freight transport. Some of the main measures are night deliveries, time windows for daytime deliveries, multiple-use lanes, congestion charging, truck toll systems, lorry-dedicated routes, low emission zones (LEZ), and environmental zones (Geroliminis & Daganzo, 2005). Those policies were implemented in different urban areas across Europe for reducing traffic congestion and air pollution.

The main difficulties appearing in urban areas are related to the dynamic characteristics of urban traffic congestion, the difficulty of finding parking spaces, access restrictions, and other regulations (Muñuzuri & Cortés, 2012). Therefore, the need for more efficient and effective logistics management solutions can be addressed by not only optimising costs but also fully tackling environmental issues, i.e., noise and air pollution.

E-Commerce Practices

Growing e-commerce is a major factor for increasing traffic in urban areas, especially on the last-mile of delivery. Due to the development of new business models, it is projected that this trend will continue for many years. With a large number of enterprises carrying out e-commerce, logistics has been largely influenced by expanding markets and the development of new technologies (Yu, Wang, Zhong, & Huang, 2016). Current practices focus on self-pick-up facilities located near central points (e.g., inside of supermarkets or near metro or train stations), where the final customers can pick up their orders on a 24/7 basis. Smart parcel lockers that facilitate express deliveries are already located in many countries and a growing number of retailers adopt them, including Amazon. These solutions are aimed primarily for non-food deliveries since the temperature control requirements for food are more difficult to manage. Other practices in same-day e-commerce deliveries include the use of taxis for last-mile deliveries in the urban context. This practice does not significantly affect the traffic congestion and other environmental key performance indicators (KPIs) in the urban context while, at the same time, it increases customer satisfaction.

Pickup Points

To keep customer satisfaction in e-commerce at a high level, a full-coverage supply of goods in cities must be guaranteed. At the same time, effectiveness in processes and maximization of productivity are part of many company's aspirations. Pick-up and drop-off points, for example, are used to reduce the risk of missed deliveries and improve shipment consolidation. Another similar perspective is the use of mobile depots as an opportunity to strengthen existing policies in restricting daytime access to the city centre. A mobile depot is a trailer fitted with a loading dock, warehousing facilities and an office. The trailer is used as a mobile inner city base from where last-mile deliveries and first-mile pick-ups are done. Mobile depots are expected to contribute to reduced congestion and emission concentrations. They are considered useful alternatives when adequate individual stock receipts are not an option. Additionally, mobile depots allow for reallocating land from parking spaces to spaces for placement of depots and might provide local authorities with an incentive to allow freight transport in public transit lanes and pedestrian streets outside periods with high traffic volumes (Bjerkan, Sund, & Nordtømme, 2014).

Urban Consolidation Centres

An urban consolidation centre (UCC) is a logistics facility that is situated in relatively close proximity to the urban area, which serves a city centre, an entire town, or a specific site such as shopping centres, airports, hospitals, or major construction sites (Browne, Sweet, Woodburn, & Allen, 2005). As shown in Fig. 2, deliveries to these locations are dropped off at the UCC and subsequently sorted, consolidated, and delivered to their final destination - often using environmentally friendly vehicles (e.g., small vans, electric vehicles, or cargo-bikes).

The combination of a sharp reduction in average shipment sizes with the growth in just-in-time manufacturing and distribution systems, dynamic management of inventories, and e-commerce initiatives suggest that today's shared urban freight facilities will be de-consolidation centres where large deliveries are transferred to smaller vehicles for the final leg of their trips (Regan & Golob, 2005). Nevertheless, the investments needed for the development of such consolidation centres and the obstacles that the city environment brings forth have not yet allowed for generating such facilities in a great extent. Implementing a UCC alone does not guarantee a positive environmental impact. Instead, the change to cleaner

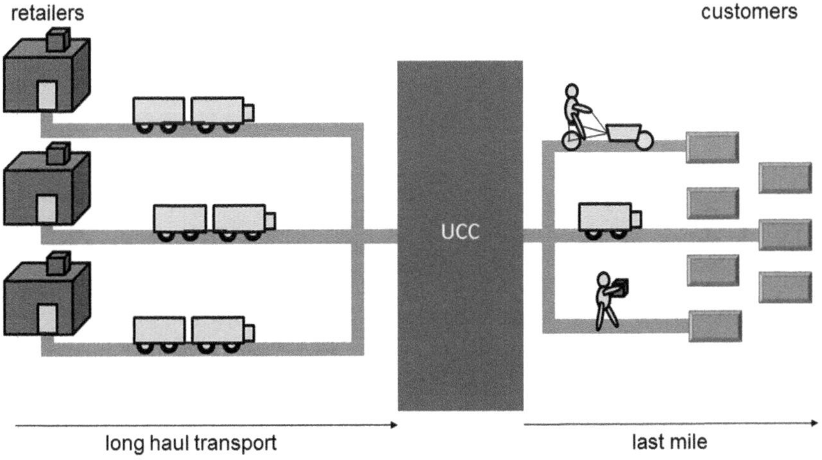

Fig. 2 Schematic operating principle of a UCC. (Source: Authors)

vehicles was determined to be a key enabler for emission reduction. Moreover, a high participation of suppliers or recipients is needed to capture enough volume (critical mass) to achieve high capacity utilisation of the delivery vehicles and fewer journeys in the urban area. This requires actively promoting the UCC concept (Clausen, Geiger, & Pöting, 2016).

Low-Emission Vehicles

One measure that has a direct impact on the reduction of emissions in transport is the use of low-emission vehicles, i.e., electric, hybrid-CNG[1]-, and LNG[2]-powered vehicles for inner-city courier shipments instead of conventional diesel and gasoline engine vehicles. However, the costs of electric vehicles remain significantly higher than conventional vehicles (Quak, Nesterova, & Van Rooijen, 2016). Furthermore, a significant number of countries are introducing cycle logistics as an alternative in distributing light goods in city centres (Schliwa, Armitage, Aziz, Evans, & Rhoades, 2015). A vehicle's maximum payload is rarely needed for small-scale deliveries such as daily-use FMCG goods, media products, documents, spare parts, or laboratory samples. In these cases a cargo bike can be applicable, especially for parcel companies (Gruber, Kihm, & Lenz, 2014). In addition, low-emission vehicles can be combined with pooling strategies. For instance, a UCC located in a city centre in collaboration with electric vehicles or cargo bikes could perform the first-mile, last-mile, and express services in an economically competitive way.

Information Systems

Within the context of city logistics, information systems (IS) have a strong potential for improvement by acting as a tool for managing various KPIs across logistics processes and, thus, pinpointing areas for optimisation and savings. Moreover, they could be the basis for supporting the development of more sustainable logistics processes and products. Previous research studies have discussed the critical role of IS in the greening of supply chains and logistics (Dao, Langella, & Carbo, 2011;

Melville, 2010). IS can enable firms to standardise, monitor, capture, and utilise data and metadata that help them to measure and evaluate internal and external performance in both financial and environmental terms (Björklund, Martinsen, & Abrahamsson, 2012; Dao et al., 2011; Hervani, Helms, & Sarkis, 2005; Melville, 2010). Big Data Analytics and Cloud Computing would be possibly adopted to enhance the e-commerce logistics in terms of system level, operational level, and decision-making level that may be real-time and intelligent in the next decade (Yu et al., 2016). In the realms of supply chains, they should also facilitate both collaboration and information exchange by improving information flows among supply chain partners (Banker, Bardhan, & Asdemir, 2006). Despite the increasing interest on these platforms, business and academic studies that discuss specifically their design are limited over the last few years. These studies adopt mainly the Design Science Approach, to shed light on the design process, and then develop and evaluate the respective artefact (Bensch, Andris, Gahm, & Tuma, 2014; Gräuler, Teuteberg, Mahmoud, & Marx, 2013; Hilpert, Thoroe, & Schumann, 2011). Previous contributions suggest the need for insights into how this kind of systems operate in practice, how they are implemented, and recommendations for their effective implementation and use (Melville, 2010; Watson, Boudreau, & Chen, 2010).

Collaborative Logistics

Collaborative logistics is coming into increasing focus for corporate logistics managers as the potential benefits lying in such practices are becoming apparent. The increasing complexity of processes encourages companies to work in teams or to form partnerships, and collaboration is a promising approach to tackle these challenges (Rabe, Klueter, Clausen, & Poeting, 2016).

Today, logistics networks coexist with several different sectors (e.g. automotive, health) and with different goods (e.g. parts, medicines). For example, a cold-chain truck for pharmaceuticals would not usually be shared to transport fresh groceries, despite the fact that both require cold-chain services. However, due to increasing time and cost pressures, there is a growing shift towards collaboration between and within sectors, resulting in new possibilities for network sharing. Key opportunities in this context are the optimisation of load capacity for standard network trucks with enhanced capabilities (e.g., temperature control), cost savings with better resource efficiency and capacity utilisation, increased handling capabilities of different logistics volumes at peak times through multi-purpose infrastructure, and added flexibility with new logistics channels (DHL, 2017).

Transport pooling and supply chain flow management do not only enhance a company's operational and economic performance, but also add extra value to the final product (Angerhofer & Angelides, 2006; Ramanathan, 2014; Vanovermeire, Sörensen, Van Breedam, Vannieuwenhuyse, & Verstrepen, 2014). Furthermore, supply chain pooling given by horizontal cooperation among several independent supply chains creates a new common supply chain network that could reduce the costs and the CO_2 emissions from transport (Chen, Hao, Li, & Yiu, 2012). However, most forms of horizontal cooperation require a neutral coordinator whose tasks and duties are similar to the current service offered by a logistic service provider (Leitner, Meizer, Prochazka, & Sihn, 2011).

Nowadays, the importance of reducing overall costs instead of just tackling individual node costs is also underlined both by private companies and academic researchers. Many examples exist on collaborative or shared logistics in the literature; some of them are about grocery distribution (Caputo & Mininno, 1996), distribution in rural areas (Hageback & Segerstedt, 2004), freight carriers (Krajewska, Kopfer, Laporte, Ropke, & Zaccour, 2008), or railway transportation (Sherali & Lunday, 2011).

Recently, many firms are using collaborative logistics solutions, although our literature review indicates that none of them has adopted a generic approach aligning all three levels of corporate decision making (operational, tactical, and strategic). Also, existing solutions focus on specific networks and, thus, they do not only give a narrow scope of reality, but they also limit the interpretation of the outcomes. In addition, the intense data exchange that needs to take place for that collaboration to happen as well as the hard barriers the firms put in such data sharing, do not leave a margin for any advanced practices. To remain competitive, shippers must reduce their costs and in particular transportation costs. Collaboration and integration of their demands may help to reach this goal. This can be addressed by solving pick-up and delivery problems with multiple time windows with paired demands, where vehicles have to transport loads from pickup locations to delivery locations while respecting capacity and time constraints (Manier, Manier, & Al Chami, 2016).

In addition to the industrial interest in collaborative logistics, academic research has also focused on this field. The research on collaborative logistics covers different topics and includes analysis of the drivers, impediments, and facilitators (Cruijssen, Dullaert, & Fleuren, 2007), impact analysis and performance measurement (Frisk, Göthe-Lundgren, Jörnsten, & Rönnqvist, 2010; Liao & Kuo, 2014), and computational intelligence algorithms for dealing with collaborative distribution (Shang & Cuff, 1996; Sprenger & Mönch, 2012). Most of these efforts use surveys and analytical approaches as a research method and highlight the need for further research based on real-world cases. Besides, most of the previous research focuses on the examination of a specific collaborative logistics practice, and the comparison of various alternatives is underrepresented. Moreover, extant literature still lacks a general conceptual classification to guide practitioners in setting up horizontal cooperation, while not all forms of horizontal cooperation are applicable to any given sector or company (Cruijssen, Cools, & Dullaert, 2007).

According to Lehoux, Audy, D'amours, and Rönnqvist (2009), the appropriate incentives should be provided to the companies to participate in coalition and to maintain collaboration. An important issue is how the potential savings that arise from collaboration should be allocated among the participants. It is common in the literature of collaborative logistics to model the situation with cooperative games (Guajardo & Rönnqvist, 2015; Lozano, Moreno, Adenso-Díaz, & Algaba, 2013). Indeed, Vanovermeire et al. (2014) studied how savings could be allocated under the concept of game theory. Moreover, they examined the viability of implementing collaborative strategies in practice, providing participants with the opportunity to adopt a flexible attitude, i.e., allow for changing the terms of their deliveries.

A crucial factor for an efficient collaboration is the sharing of information among partners (Lindawati, van Schagen, Goh, & de Souza, 2014; Wang & Kopfer, 2014). Wang and Kopfer (2014) proposed a decentralised request exchange mechanism for common planning among the members of a coalition, while only vehicle routes are considered for exchange. According to Wang, Kopfer, and Gendreau (2014), the partners have to share their private information, if they want to have potential savings. All the participants should be aware of the other players' actions to achieve an efficient collaboration by outsourcing or subcontracting to other companies, by exchanging requests for transportation needs, and by sharing capacity with other players. Nowadays, there is sufficient information (customer orders, final demand, customer location, pallet-vehicle movement, GPS data of the trucks, inventory level, prices, etc.) to enable an effective collaboration among logistics systems. Therefore, big data plays an important role in supporting the players' decisions (Ilie-Zudor et al., 2015).

2.2 Supply Chain Simulation

In recent years, the use of simulation in the supply chain is getting more and more frequent. Simulation of the supply chain's operation and transportation system is very helpful in identifying problems with low cost and in a short time period. The new collaborative techniques and Collaborative Transport Management (CTM) systems that have been arising during the last years can be evaluated, validated, and optimised through such techniques (Chan & Zhang, 2011). Keeping in mind the changing market demand and the needs arising in e-commerce logistics, fresh food deliveries, and the FMCG sector, the efficiency and flexibility of global distribution holds the key to success at an international level.

A simulation model offers the ability to consider all relevant characteristics and challenges of the real system in detail. A variety of simulation methods and environments in logistics exist that focus on specific (practical) applications. Physical simulations are used to analyse the physical characteristics and movements of individual vehicles or groups of a few vehicles, e.g., driving and manoeuvring operations of vehicles. To model traffic flows of several vehicles microscopically, specific simulation environments are used to consider all relevant processes and effects of infrastructure use, e.g., different types of users (motorised private transport, goods transport, public transport, pedestrians, and cyclists), driving characteristics, traffic rules, and signalling patterns. Some applications also integrate emission models to further evaluate the impact of traffic on the environment (Fellendorf & Vortisch, 2010). Both types of simulation methods provide the ability to map the behaviour of vehicles at a high level of detail; they are not focusing on modelling a logistic system. However, specific simulation environments exist for transport applications. These software solutions enable the analysis of logistics hubs as well as transport and supply networks. For logistics hubs, several applications have been developed for different modes of transport and types of shipment: container terminals, terminals for combined transport, air freight terminals as well as cross docks, forwarding agencies, parcel hubs, and warehouses (De Koster, Le-Anh, & Van Der Meer, 2004; Manivannan, 1998; Nsakanda, Turcotte, & Diaby, 2004). Depending on the objective and the context of analysis, different types of simulation methods are applied: from easy-to-use flow models of handling activities to more complex simulation environments considering operational details such as specific control strategies for handling equipment, buffers, or operational workforce (Liu & Takakuwa, 2010). Different solutions also exist to support the complex planning tasks in supply chains and transport networks (Kleijnen, 2005; Lee, Cho, Kim, & Kim, 2002; Terzi & Cavalieri, 2004; Van Donk & Van Der Vaart, 2005). An extended literature review on supply chain modelling and simulation is given by Oliveira, Lima, and Montevechi (2016). Research in the field typically focuses on model design under specific challenges, e.g., a high number of levels in the network or a high complexity of the related products (Long, 2014).

New solutions for the challenges of urban logistics are constantly being sought. In the Netherlands, approaches for underground logistics concepts were developed. Van Duin (1998) identified weaknesses in these approaches by simulation. Another approach uses simulation to design automated underground transport systems around Schiphol Airport with automated guided vehicles (Van der Heijden et al., 2002). A further study by van Duin, Kortmann, and van den Boogaard (2014) examines whether the distribution of goods to the city of Amsterdam by water could be a possible solution for the future, taking into account the lively traffic of recreational crafts and touring boats.

One of the first publications on research that used simulation to explore the potential of horizontal collaboration in urban logistics was published by McDermott (1975). The core of the work is the solution to a vehicle routing problem in New York City whereby the savings potential of forwarders through horizontal cooperation is determined. The City Business District is divided into ten zones, each with a consolidation centre. For a simulated day, the required number of tours, the number of vehicles, the distance covered, and the working time per zone are determined, from which the assumed costs can be calculated. Thus, the savings potential through consolidation can be shown. The majority of recent publications discuss the solution of Multi Depot Vehicle Routing problems to determine the potential savings on the route and also good locations for the depots.

Van Duin, van Kolck, Anand, Tavasszy, and Taniguchi (2012) deal with the problem of declining distribution centres and how to make their use more attractive. The model is based on a vehicle routing problem with time windows, which is solved by a genetic algorithm. The aim of the model is to investigate the viability of distribution centres from an environmental and economic point of view in a fictitious scenario using a multi-agent model. Various input data is used for the simulation, such as costs for using the distribution centre, tolls, or government grants. It is shown that the use of distribution centres makes sense if the deliveries are not time-critical (just-in-time delivery). Although the results show that distribution centres are not financially worthwhile, the authors point out that this is a sensible concept for avoiding congestion and emissions.

3 Simulation Approach

As described above, the problem basis in the field of city logistics is broadly diversified. On the one hand, there is a need for action, but also a multitude of measures that the actors can take to improve the conditions in the urban environment. From the perspective of logistics service providers, political regulations form the conditions for their business activities. From their economic point of view, however, it is most important to determine the potential for using new practices and technologies and their impact on the situation in the inner cities. On this basis, politicians can also play a supportive role, e.g., by setting incentives such as subsidising UCC or restricting access for trucks to the city centres. Although UCCs worldwide have improved urban freight distribution and reduced externalities, other UCC initiatives have not materialised due to problems such as business model limitations. Björklund, Abrahamsson, and Johansson (2017) identified seven critical factors of viable city logistics business models: the ability to scale up and down the UCC solution; an ability to continuously develop and adapt to a dynamic environment; the important entrepreneurial role of the initiator as well; the acknowledgment of society; the ability to innovate new services; logistics and supply chain management competence; and the ability to take full advantage of advanced IT. All seven factors describe continuously redeveloped business models seeking to seize new and unexpected opportunities, yet also indicate that city logistics systems require local authorities and municipalities to act as initiators, enablers, and customers.

The aim of this contribution is to examine suitable measures from the operators' point of view and to discuss conclusions on their practical feasibility of implementation. Thus, our research uses comprehensive datasets from real distribution systems to generate simulation models for the distribution of goods in urban areas. The models serve to determine a wide range of key figures to assess the impact of specific city logistics measures, i.e., UCCs and collaboration. The research is designed to compare the results of the current operations (as-is) and the distribution process conducted with the utilization of a UCC (to-be). Certain KPIs such as travelled distances, loading factors, and emissions are used to evaluate the execution of the delivery trips. Then, an in-depth assessment is conducted for both to identify the benefits of collaboration for each case in terms of costs as well as environmental and social impact. For experimentation, the simulation tool SimChain is used, which is a discrete event simulation (DES) for supply chain simulation based on Siemens PlantSimulation. Technically, this tool mainly consists of three major parts: a graphical user interface for model configuration, a database in which all configuration data and simulation results are stored, and a DES framework for the supply chain. All model elements are generated automatically in PlantSimulation using building blocks from a predefined template library. The structure and the principle modelling approach of SimChain are described in detail by Gutenschwager and Alicke (2004).

The experimentation part includes as-is scenarios for validation and to-be scenarios that extend the current situation by specific measures of the respective to-be scenario. For the to-be case, infrastructure-sharing practices (i.e., performing the deliveries via a UCC) are examined. In this case, the 3PL companies and the retailers transfer their goods to the UCCs with vehicles of high capacity whereby UCCs are usually located close to the city centre. Then, smaller freight vehicles of the UCC perform the last-mile delivery to the final customers. A series of KPIs are calculated by the simulation model for both 3PL companies and retailers, to compare their operations without and with the utilisation of the UCC. Based on these KPIs, the scenarios are compared and assessed. The aim of the assessment is to identify whether the implementation of a UCC can offer additional benefits to the operation of the 3PL companies and retailers, and which environmental and social impact it might bring for the city centre.

Major research limitations arise from the fact that in most cases essential data are only partially known to the operators themselves as they are not documented in their IT systems. Partially due to the given dataset, extended efforts on data pre-processing were inevitable. As a result, the comparison of scenarios using the processed data has proven itself as reasonable, while an evaluation of absolute values seems to be questionable. In addition, there was no exact information about the customer locations, which was a major issue of the primary data. The simulation tool assumes geo-coordinates for all locations, but only ZIP codes were available in the records. For this reason, specific geo-coordinates had to be generated. This was done within the data pre-processing phase with the help of a python tool. ZIP codes were processed through the Google

Maps API to identify the centre of ZIP code areas. The geo-coordinates of those centres were returned by the python tool. After pre-processing, which included parsing the ZIP code data, correcting syntax errors (e.g., due to Greek letters), and adding missing values from secondary data and assumptions from expert interviews, the data were used to generate the simulation model.

4 Experimentation and Results

This section assesses freight operation schemes for urban areas, i.e., UCCs and collaboration. The examined case study includes the current freight delivery schemes adopted by 3PL operators and retailers using a case based in Athens compared with the use of a UCC in the outskirts of the city. The metropolitan area of Athens, Greece, has been selected for our case study, because it is the most populous and largest urban area in Greece and because there is immense road traffic caused by transport in the centre of Athens with narrow roads.

During the cross-docking process inside a UCC, the deliveries are grouped by the ZIP Codes of the recipients' delivery points. Thus, it is expected that the vehicles that perform the last-mile delivery are visiting an increased number of delivery points in nearby areas. Two different scenarios have been developed and examined, which are described below. For each scenario, an as-is and a to-be case are examined. In the as-is case, the 3PL companies execute individually without adopting any collaborative schemes, while in the to-be case, the distribution flows are consolidated via one or two UCCs.

4.1 First UCC Scenario

In the first scenario, five hubs are located in the suburban area serving a total amount of 8537 end customers in the urban area. The operating 3PL companies delivering inside the city are mainly located and operate hubs in the western area of Athens. The UCC's location in the to-be case was determined as a two-echelon location routing problem in a previous publication (Gruler & Juan, 2017). However, in a real application, the stakeholders must decide which location is suitable for a UCC considering market-oriented factors (e.g., the development of the area), business factors (e.g., state subsidies, costs for new construction, or rental development), and infrastructure factors (e.g., transport connections) (Rabe, Klueter, & Wuttke, 2018). Figure 3 shows the locations of the hubs and the UCC in the to-be case.

For the 3PL business case, simulation experiments are performed to evaluate the as-is situation using the four 3PL datasets. The four as-is scenarios include the operations of four 3PLs serving 186 up to 1650

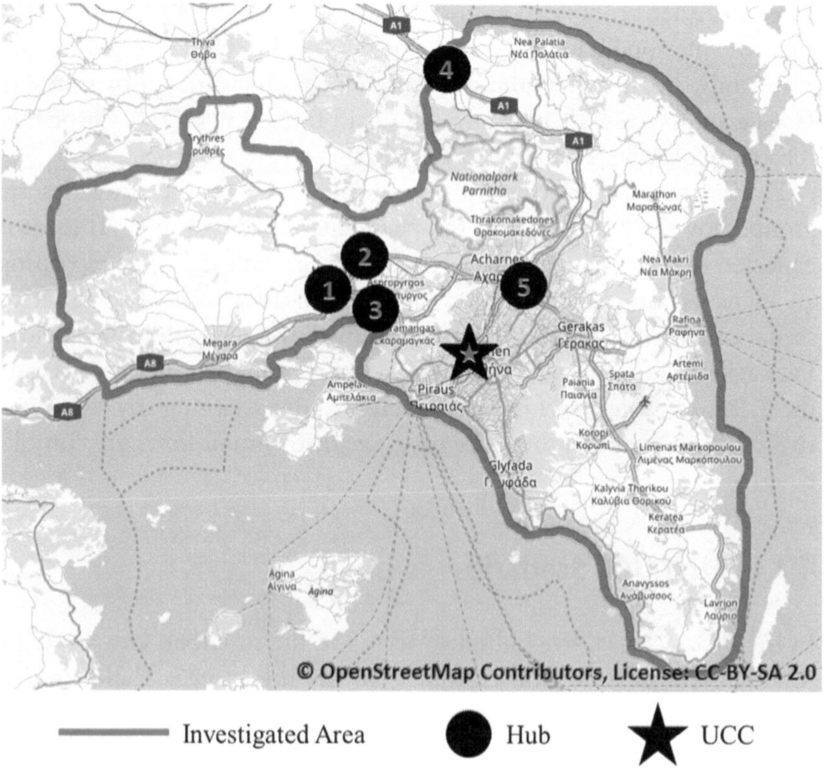

Fig. 3 First UCC scenario to be evaluated. (Source: Rabe et al., 2018)

customers with a total of 15,935 orders. The observed period is 01 July 2014 to 07 January 2015. The to-be case includes collaboration of all four 3PL using a common UCC.

4.2 Second UCC Scenario

For the second scenario, the western part of Athens is excluded, as the north-eastern part has the greatest potential for optimization. The large geographic reach of this area combined with the limited customer network results in a loading factor of 75% or less. For this reason, the initiation of a UCC could increase the loading factor of the trucks and reduce the total distance travelled, since the bundling of goods flows requires fewer vehicles. Each UCC supplies a certain area, defined by ZIP codes. In this way, the customer orders could be clearly divided into the individual regions and assigned to the UCCs in the order list. This also ensures that the tours from the hubs lead to the right UCC. Figure 4 illustrates the locations of the two UCCs and the associated areas.

In the second UCC scenario, the four as-is scenarios include the operations of four 3PLs serving 42 up to 1732 customers with a total of 8503 orders. The observed period is also 01 July 2014 to 07 January 2015. The to-be scenario also includes collaboration of all four 3PL using a common UCC.

4.3 Parameters and Assumptions

In addition to existing primary data, various assumptions have been made to compensate missing information and to generate the model (Rabe et al., 2018): For the stock keeping units (SKUs), the unit for SKU to calculate truck capacity and utilization is considered to be one box. A regular Euro-pallet is assumed to accommodate 50 boxes. The dimensions of one box are defined as 45 cm × 30 cm × 21.5 cm (17.7″ × 11.8″ × 8.5″). This assumption was made in order to be able to calculate the volume per transport order in detail and thus the total loading factor of each truck, since usually a transport order is less than a full pallet. We

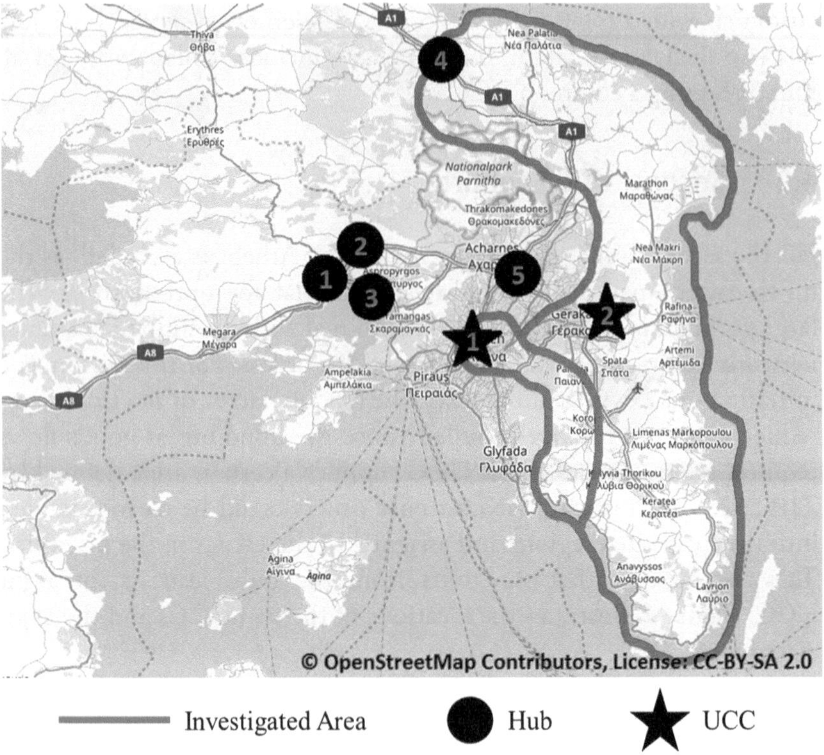

Fig. 4 Second UCC scenario (selected areas) to be evaluated. (Source: Rabe et al., 2018)

assume that customers do not share boxes; each box is unambiguously assigned to a customer. For the 3PL as-is case, the delivery is operated using small trucks with a capacity of 750 boxes. For the to-be scenario, we also consider articulated trucks with a capacity of 1650 boxes. If a transport order exceeds the maximum capacity of a vehicle, another vehicle is sent to fulfil the order on the same day. The functionality to generate additional vehicles was implemented to meet the requirements of last-mile delivery for the 3PL companies. The distance for each trip is calculated using the Euclidean distance multiplied with a factor of 1.3 as an estimation for actual driven distance (Rabe et al., 2018):

$$d_{x,y} = 1.3 \;.\; \|x - y\|_2 \tag{1}$$

The calculations are validated with actual geographic data for the case of the Athens Metropolitan Area. For a van, the service time per stop (unloading time) is set to 12 minutes fixed plus 1.8 seconds per box in accordance with 3PLs and transport experts. The fixed time includes time for parking and documentation. The average driving speed for a van is set to 48.3 km/h (Jung, Kim, Kim, Hong, & Park, 2017).

4.4 Key Performance Indicators

In order to evaluate the different collaboration scenarios, certain KPIs addressing both economic and environmental aspects are used, i.e., the number of tours, number of drops, drops per trip, total distance travelled, and loading factor. To determine the social impact, the distance travelled in high density areas and the share of distance in high density areas is calculated. A lower share of transports in high density areas goes along with a reduction of noise, traffic congestion, and emissions. The KPIs that are evaluated are described in detail by Rabe et al. (2018).

5 Results

In this section, the main findings of the simulation study are presented. A detailed evaluation of the Greece case study is given by Rabe et al. (2018).

For the first UCC scenario, the simulation results in Table 1 show that the total number of deliveries made increased by 2709 trips (+42%). The to-be case comprises two-phase delivery runs from the depots of the 3PL companies to the UCC and deliveries from the UCC to the customer (last mile delivery). The number of vehicles in the to-be case was higher, because two different vehicle types are used to carry out the delivery run. The vehicles used to enter the city centre and deliver the last mile fell to 325 vehicles (−5%). The average distance travelled per tour was also reduced from 54 km to 31 km.

Table 1 Simulation results for the second UCC scenario (selected areas)

Scenario	Description	# Tours	# Drops	Distance (km)	# Boxes
No collaboration	3PL A	572	13,065	61,036.3	331,486
	3PL B	205	2950	7110.6	91,101
	3PL C	93	239	3139.1	2729
	3PL D	1234	11,400	64,139.6	857,961
	Overall	2104	27,654	135,425.6	1,283,277
With 2 UCCs	3PL A to UCC1	157		5151.2	
	3PL B to UCC1	127		1841.2	
	3PL C to UCC1	46		1102.7	
	3PL D to UCC1	279		7310.7	
	3PL A to UCC2	206		8892.3	
	3PL B to UCC2	78		975.0	
	3PL C to UCC2	17		688.5	
	3PL D to UCC2	365		15,738.2	
	UCC1 to customers	808	12,122	7990.3	543,031
	UCC2 to customers	1090	15,532	59,960.0	740,246
	UCC overall	1898	27,654	67,950.2	1,283,277
	Overall	3173		109,651.1	

Source: Rabe et al. (2018)

The results indicate an increase of traffic in the city which rises from 29.8% to 42.8% due to the use of the UCC. This can be explained by the prescribed location of the UCC, which is not as usual outside the urban area, but in the inner peripheral area. Therefore, supplying the UCC creates additional transports through the city area instead of avoiding them.

In the second scenario, the numbers of tours increased because of the above-mentioned two-echelon network structure. However, as these tours are much shorter now, the overall distance decreases. The drops per tour, boxes per tour, and loading factor in the 3PL_C scenario are comparatively low (Table 2). This is due to sparsely available data in the observed postal code area, which only provides information about one to four customers per day.

Compared to the first scenario, the traffic volume in high density areas has not increased in the to-be case. On the contrary, it even fell slightly from 26.7% to 25.3%. This can be explained by the fact that the second UCC is outside the defined area, and the transports to this UCC and

Table 2 Post-calculated KPIs for the second UCC scenario (selected areas)

Scenario	Description	Drops per tour	Distance per tour (km)	Boxes per tour	Loading factor (percent)
No collaboration	3PL A	22.8	106.7	579.5	77.3
	3PL B	14.4	34.7	444.4	59.3
	3PL C	2.6	33.8	29.3	3.9
	3PL D	9.2	52.0	695.3	92.7
	Overall	13.1	64.4	609.9	81.3
With 2 UCCs	3PL A to UCC1		32.8		
	3PL B to UCC1		14.5		
	3PL C to UCC1		24.0		
	3PL D to UCC1		26.2		
	3PL A to UCC2		43.2		
	3PL B to UCC2		12.5		
	3PL C to UCC2		40.5		
	3PL D to UCC2		43.1		
	UCC1 to customers	15.0	9.9	672.1	89.6
	UCC2 to customers	14.3	55.0	679.2	90.6
	UCC overall	14.6	35.8	676.1	90.2

Source: Rabe et al. (2018)

from this UCC to the customers only lead through the city with a minor proportion (Rabe et al., 2018). Thus, the results show how important it is to choose the right location for the UCC.

6 Conclusions

This chapter demonstrates the state of the art in supply chain simulation for the urban environment. City logistics measures that can improve the operation of logistics service providers, i.e., UCCs and collaboration, are implemented in a simulation model to evaluate possible business opportunities. The experimentation results from the example of Athens show that supply chain simulation can be effectively used to study the impact of city logistics measures on practical business cases. The results are analysed and compared to identify benefits of the implemented measures in terms of cost, environment, and society.

Two different business scenarios including real data of 3PL companies in Greece have been developed in which the principle of UCCs was implemented. A series of KPIs were calculated to compare the results of the as-is and to-be cases in order to identify whether this collaborative practice can be beneficial in terms of both economic and environmental aspects. As both simulation results have shown, utilizing UCCs will lower the number of vehicle trips needed to meet the customer's demand and also decreases the required distance travelled. Moreover, the saved distance mainly results from savings in urban areas. Thus, the CO_2 emissions are significantly reduced in sensitive areas. Another advantage of the UCC scenarios is the shorter average length of a trip due to the increased physical proximity of the UCCs and the majority of customers. The physical proximity can be increased even more by adding more UCCs to the scenario as seen in the second scenario and will yield even more savings in terms of driven distance and average distance per trip.

The simulation has mainly been limited by the available raw data from the companies, e.g., there was no information about the time window in which customers were supplied. Thus, customers' orders were fulfilled on a daily basis, which is a quite imprecise assumption for the FMCG sector. As it was not possible to receive reliable information, the investment costs and the costs of operating a UCC in Athens were not taken into account. An attempt to estimate the cost of deliveries could be made by considering the distance travelled, the type of vehicle, and the personnel costs. However, there was no information in the evaluated case about the type of vehicle and the costs of personnel available.

Overall it can be concluded that availability, quality, and the level of detail of data are major challenges for the investigation and data cleansing is an essential part of the effort for a supply chain simulation study. Considering a UCC, the position of the plant in relation to the city is a critical strategic decision that can well be supported by a simulation study. Future studies will focus on similar scenarios utilizing other means of transport, e.g., electric vehicles or cargo-bikes.

Acknowledgements This work is partially supported by European Union's Horizon 2020 research and innovation programme within the "Rethinking Urban Transportation through advanced tools and supply chain collaboration"

(U-TURN) project under grant agreement No 635773. All datasets used for the experiments were provided by ELTRUN E-Business Research Centre of Athens University of Economics and Business (AUEB) and OPTILOG. The authors are thankful with AUEB and OPTILOG for conducting several interviews, esp. with the EEL Working Group participants, and for sharing the data sets gathered for further research activities.

Notes

1. Compressed Natural Gas.
2. Liquified Natural Gas.

References

Angerhofer, B. J., & Angelides, M. C. (2006). A model and a performance measurement system for collaborative supply chains. *Decision Support Systems, 42*, 283. https://doi.org/10.1016/j.dss.2004.12.005

Banker, R. D., Bardhan, I., & Asdemir, O. (2006). Understanding the impact of collaboration software on product design and development. *Information Systems Research, 17*, 352. https://doi.org/10.1287/isre.1060.0104

BBSR—Bundesinstitut für Bau-, Stadt- und Raumforschung. (2017). *CO_2-neutral in cities and neighbourhoods—The European and international perspective*. Retrieved June 30, 2018, from http://www.bbsr.bund.de/BBSR/EN/Publications/OnlinePublications/2017/bbsr-online-10-2017-dl.pdf?__blob=publicationFile&v=3

Bensch, S., Andris, R. J., Gahm, C., & Tuma, A. (2014). IT outsourcing: An IS perspective. In *Proceedings of the Annual Hawaii International Conference on System Sciences*. https://doi.org/10.1109/HICSS.2014.520

Bjerkan, K. Y., Sund, A. B., & Nordtømme, M. E. (2014). Stakeholder responses to measures green and efficient urban freight. *Research in Transportation Business and Management, 11*, 32. https://doi.org/10.1016/j.rtbm.2014.05.001

Björklund, M., Abrahamsson, M., & Johansson, H. (2017). Critical factors for viable business models for urban consolidation centres. *Research in Transportation Economics, 64*, 36. https://doi.org/10.1016/j.retrec.2017.09.009

Björklund, M., Martinsen, U., & Abrahamsson, M. (2012). Performance measurements in the greening of supply chains. *Supply Chain Management, 17*, 29. https://doi.org/10.1108/13598541211212186

Browne, M., Sweet, M., Woodburn, A., & Allen, J. (2005). *Urban freight consolidation centres*. Final report. London. Retrieved from http://ukerc.rl.ac.uk/pdf/RR3_Urban_Freight_Consolidation_Centre_Report.pdf

Caputo, M., & Mininno, V. (1996). Internal, vertical and horizontal logistics integration in Italian grocery distribution. *International Journal of Physical Distribution & Logistics Management, 26*, 64. https://doi.org/10.1108/09600039610149101

Chan, F. T. S., & Zhang, T. (2011). The impact of collaborative transportation management on supply chain performance: A simulation approach. *Expert Systems with Applications, 38*, 2319. https://doi.org/10.1016/j.eswa.2010.08.020

Chen, X., Hao, G., Li, X., & Yiu, K. F. C. (2012). The impact of demand variability and transshipment on vendor's distribution policies under vendor managed inventory strategy. *International Journal of Production Economics, 139*, 42. https://doi.org/10.1016/j.ijpe.2011.05.005

Clausen, U., Geiger, C., & Pöting, M. (2016). Hands-on testing of last mile concepts. *Transportation Research Procedia, 14*, 1533. https://doi.org/10.1016/j.trpro.2016.05.118

Cruijssen, F., Cools, M., & Dullaert, W. (2007). Horizontal cooperation in logistics: Opportunities and impediments. *Transportation Research Part E: Logistics and Transportation Review, 43*, 129. https://doi.org/10.1016/j.tre.2005.09.007

Cruijssen, F., Dullaert, W., & Fleuren, H. (2007). Horizontal cooperation in transport and logistics: A literature review. *Transportation Journal*. https://doi.org/10.2307/20713677

Dao, V., Langella, I., & Carbo, J. (2011). From green to sustainability: Information technology and an integrated sustainability framework. *Journal of Strategic Information Systems, 20*, 63. https://doi.org/10.1016/j.jsis.2011.01.002

De Koster, R. B. M., Le-Anh, T., & Van Der Meer, J. R. (2004). Testing and classifying vehicle dispatching rules in three real-world settings. *Journal of Operations Management, 22*, 369. https://doi.org/10.1016/j.jom.2004.05.006

DHL. (2017). *Sharing economy logistics: Rethinking logistics with access over ownership*. DHL Trend Research.

EEA. (2017). *Greenhouse gas emissions from transport*. European Environment Agency. https://doi.org/IND-111-en

European Commission. (2011). White paper on transport. *Transport*. https://doi.org/10.2832/30955

Fellendorf, M., & Vortisch, P. (2010). Microscopic traffic flow simulator VISSIM. In J. Barceló (Ed.), *International series in operations research and management science* (Vol. 145, pp. 63–93). New York, NY: Springer. https://doi.org/10.1007/978-1-4419-6142-6_2

Frisk, M., Göthe-Lundgren, M., Jörnsten, K., & Rönnqvist, M. (2010). Cost allocation in collaborative forest transportation. *European Journal of Operational Research, 205*, 448. https://doi.org/10.1016/j.ejor.2010.01.015

Geroliminis, N., & Daganzo, C. (2005). *A review of green logistics schemes used in cities around the world*. UC Berkeley Center for Future Urban Transport: A Volvo Center of Excellence.

Gräuler, M., Teuteberg, F., Mahmoud, T., & Marx, G. (2013). Requirements prioritization and design considerations for the next generation of corporate environmental management information systems: A foundation for innovation. *International Journal of Information Technologies and Systems Approach, 6*, 98. https://doi.org/10.4018/jitsa.2013010106

Gruber, J., Kihm, A., & Lenz, B. (2014). A new vehicle for urban freight? An ex-ante evaluation of electric cargo bikes in courier services. *Research in Transportation Business and Management, 11*, 53. https://doi.org/10.1016/j.rtbm.2014.03.004

Gruler, A., & Juan, A. A. (2017). *A simulation-optimization approach for the two-echelon location routing problem arising in the creation of urban consolidation centres*. In S. Wenzel & T. Peter (Eds.), Simulation in Produktion und Logistik 2017 (pp. 129–138). Kassel: kassel university press.

Guajardo, M., & Rönnqvist, M. (2015). Operations research models for coalition structure in collaborative logistics. *European Journal of Operational Research, 240*, 147. https://doi.org/10.1016/j.ejor.2014.06.015

Gutenschwager, K., & Alicke, K. (2004). Logistik management. In T. Spengler, S. Voß, & H. Kopfer (Eds.), *Logistik management* (pp. 161–178). Heidelberg, Germany: Physica. https://doi.org/10.1007/978-3-7908-2362-2

Hageback, C., & Segerstedt, A. (2004). The need for co-distribution in rural areas – A study of Pajala in Sweden. *International Journal of Production Economics, 89*, 153. https://doi.org/10.1016/j.ijpe.2003.10.006

Hervani, A. A., Helms, M. M., & Sarkis, J. (2005). Performance measurement for green supply chain management. *Benchmarking: An International Journal, 12*, 330. https://doi.org/10.1108/14635770510609015

Hilpert, H., Thoroe, L., & Schumann, M. (2011). Real-time data collection for product carbon footprints in transportation processes based on OBD2 and smartphones. In *Proceedings of the Annual Hawaii International Conference on System Sciences*. https://doi.org/10.1109/HICSS.2011.356

Ilie-Zudor, E., Ekárt, A., Kemeny, Z., Buckingham, C., Welch, P., & Monostori, L. (2015). Advanced predictive-analysis-based decision support for collaborative logistics networks. *Supply Chain Management, 20*, 369. https://doi.org/10.1108/SCM-10-2014-0323

Jahangirian, M., Eldabi, T., Naseer, A., Stergioulas, L. K., & Young, T. (2010). Simulation in manufacturing and business: A review. *European Journal of Operational Research, 203*, 1. https://doi.org/10.1016/j.ejor.2009.06.004

Jung, S., Kim, J., Kim, J., Hong, D., & Park, D. (2017). An estimation of vehicle kilometer traveled and on-road emissions using the traffic volume and travel speed on road links in Incheon City. *Journal of Environmental Sciences (China), 54*, 90–100. https://doi.org/10.1016/j.jes.2015.12.040

Kleijnen, J. P. C. (2005). Supply chain simulation tools and techniques: A survey. *International Journal of Simulation and Process Modelling, 1*(1), 82–89.

Krajewska, M. A., Kopfer, H., Laporte, G., Ropke, S., & Zaccour, G. (2008). Horizontal cooperation among freight carriers: Request allocation and profit sharing. *Journal of the Operational Research Society, 59*, 1483. https://doi.org/10.1057/palgrave.jors.2602489

Lagorio, A., Pinto, R., & Golini, R. (2016). Research in urban logistics: A systematic literature review. *International Journal of Physical Distribution & Logistics Management, 46*, 908. https://doi.org/10.1108/IJPDLM-01-2016-0008

Lee, Y. H., Cho, M. K., Kim, S. J., & Kim, Y. B. (2002). Supply chain simulation with discrete-continuous combined modeling. *Computers and Industrial Engineering, 43*, 375. https://doi.org/10.1016/S0360-8352(02)00080-3

Lehoux, N., Audy, J. F., D'amours, S., & Rönnqvist, M. (2009). Issues and experiences in logistics collaboration. *IFIP Advances in Information and Communication Technology*. https://doi.org/10.1007/978-3-642-04568-4_8

Leitner, R., Meizer, F., Prochazka, M., & Sihn, W. (2011). Structural concepts for horizontal cooperation to increase efficiency in logistics. *CIRP Journal of Manufacturing Science and Technology, 4*, 332. https://doi.org/10.1016/j.cirpj.2011.01.009

Liao, S. H., & Kuo, F. I. (2014). The study of relationships between the collaboration for supply chain, supply chain capabilities and firm performance: A case of the Taiwans TFT-LCD industry. *International Journal of Production Economics, 156*, 295. https://doi.org/10.1016/j.ijpe.2014.06.020

Lindawati, L., van Schagen, J., Goh, M., & de Souza, R. (2014). Collaboration in urban logistics: Motivations and barriers. *International Journal of Urban Sciences*. https://doi.org/10.1080/12265934.2014.917983

Liu, Y., & Takakuwa, S. (2010). Enhancing simulation as a decision-making support tool for a cross-docking center in a dynamic retail-distribution environment. In A. Tolk, S. Y. Diallo, I. O. Ryzhoy, L. Yilmaz, S. Buckley & J. A. Miller (Eds.), *Proceedings of the 2010 Winter Simulation Conference*. Piscataway: IEEE. https://doi.org/10.1109/WSC.2010.5678863

Long, Q. (2014). An agent-based distributed computational experiment framework for virtual supply chain network development. *Expert Systems with Applications, 41*, 4094. https://doi.org/10.1016/j.eswa.2014.01.001

Lozano, S., Moreno, P., Adenso-Díaz, B., & Algaba, E. (2013). Cooperative game theory approach to allocating benefits of horizontal cooperation. *European Journal of Operational Research, 229*, 444. https://doi.org/10.1016/j.ejor.2013.02.034

Manier, H., Manier, M. A., & Al Chami, Z. (2016). *Shippers' collaboration in city logistics*. IFAC-Papers Online. https://doi.org/10.1016/j.ifacol.2016.07.904, 49, 1880.

Manivannan, M. (1998). Simulation of logistics and transportation systems. In J. Banks (Ed.), *Handbook of simulation* (pp. 571–604). New York: Wiley.

McDermott, D. R. (1975). An alternative framework for urban goods distribution: Consolidation. *Transportation Journal, 15*(Fall 75), 29–39.

Melville, N. P. (2010). Information systems innovation for environmental sustainability. *MIS Quarterly, 34*(1), 1–21.

Muñuzuri, J., & Cortés, P. (2012). Recent advances and future trends in city logistics. *Journal of Computational Science, 3*, 191. https://doi.org/10.1016/j.jocs.2012.05.003

Nsakanda, A. L., Turcotte, M., & Diaby, M. (2004). Air cargo operations evaluation and analysis through simulation. In R. G. Ingalls, M. D. Rossetti, J. S. Smith & B. A. Peters (Eds.), *Proceedings of the 2004 Winter Simulation Conference. Piscataway: IEEE.* https://doi.org/10.1109/WSC.2004.1371531

Oliveira, J. B., Lima, R. S., & Montevechi, J. A. B. (2016). Perspectives and relationships in supply chain simulation: A systematic literature review. *Simulation Modelling Practice and Theory, 62*, 166. https://doi.org/10.1016/j.simpat.2016.02.001

Pannirselvam, G. P., Ferguson, L. A., Ash, R. C., & Siferd, S. P. (1999). Operations management research: An update for the 1990s. *Journal of Operations Management, 18*, 95. https://doi.org/10.1016/S0272-6963(99)00009-1

Quak, H., Nesterova, N., & Van Rooijen, T. (2016). Possibilities and barriers for using electric-powered vehicles in city logistics practice. *Transportation Research Procedia, 12*, 157. https://doi.org/10.1016/j.trpro.2016.02.055

Rabe, M., Klueter, A., Clausen, U., & Poeting, M. (2016). An approach for modeling collaborative route planning in supply chain simulation. In *Proceedings of the 2016 Winter Simulation Conference*. https://doi.org/10.1109/WSC.2016.7822264

Rabe, M., Klueter, A., & Wuttke, A. (2018). Evaluating the consolidation of distribution flows using a discrete event supply chain simulation tool: Application to a case study in Greece. In M. Rabe, A. A. Juan, A. Skoogh, S. Jain, & B. Johansson (Eds.), *Proceedings of the 2018 Winter Simulation Conference*. Piscataway: IEEE.

Ramanathan, U. (2014). Performance of supply chain collaboration—A simulation study. *Expert Systems with Applications, 41*, 210. https://doi.org/10.1016/j.eswa.2013.07.022

Regan, A. C., & Golob, T. F. (2005). Trucking industry demand for urban shared use freight terminals. *Transportation, 32*, 23. https://doi.org/10.1007/s11116-004-2218-9

Schliwa, G., Armitage, R., Aziz, S., Evans, J., & Rhoades, J. (2015). Sustainable city logistics—Making cargo cycles viable for urban freight transport. *Research in Transportation Business and Management, 15*, 50. https://doi.org/10.1016/j.rtbm.2015.02.001

Shang, J. S., & Cuff, C. K. (1996). Multicriteria pickup and delivery problem with transfer opportunity. *Computers and Industrial Engineering, 30*, 631. https://doi.org/10.1016/0360-8352(95)00181-6

Sherali, H. D., & Lunday, B. J. (2011). Equitable apportionment of railcars within a pooling agreement for shipping automobiles. *Transportation Research Part E: Logistics and Transportation Review, 47*, 263. https://doi.org/10.1016/j.tre.2010.09.005

Sigman, R., Hilderink, H., Delrue, N., Braathen, N. A., & Leflaive, X. (2012). *OECD environmental outlook to 2050*. OECD Publishing. https://doi.org/10.1787/9789264122246-en

Sprenger, R., & Mönch, L. (2012). A methodology to solve large-scale cooperative transportation planning problems. *European Journal of Operational Research., 223*, 626. https://doi.org/10.1016/j.ejor.2012.07.021

Tako, A. A., & Robinson, S. (2012). The application of discrete event simulation and system dynamics in the logistics and supply chain context. *Decision Support Systems, 52*, 802. https://doi.org/10.1016/j.dss.2011.11.015

Taniguchi, E. (2001). City logistics. *Infrastructure Planning Review, 18*, 1. https://doi.org/10.2208/journalip.18.1

Taniguchi, E., Thompson, R. G., & Yamada, T. (2014). Recent trends and innovations in modelling city logistics. *Procedia—Social and Behavioral Sciences, 125*, 4. https://doi.org/10.1016/j.sbspro.2014.01.1451
Terzi, S., & Cavalieri, S. (2004). Simulation in the supply chain context: A survey. *Computers in Industry, 53*, 3. https://doi.org/10.1016/S0166-3615(03)00104-0
United Nations. (2012). *World urbanization prospects: The 2011 revision.* New York: United Nations Department of Economic and Social Affairs/Population Division. https://doi.org/10.2307/2808041, 24, 883.
Van der Heijden, M. C., Van Harten, A., Ebben, M. J. R., Saanen, Y. A., Valentin, E. C., & Verbraeck, A. (2002). Using simulation to design an automated underground system for transporting freight around Schiphol airport. *Interfaces, 32*, 1. https://doi.org/10.1287/inte.32.4.1.49
Van Donk, D. P., & Van Der Vaart, T. (2005). A case of shared resources, uncertainty and supply chain integration in the process industry. *International Journal of Production Economics, 96*, 97. https://doi.org/10.1016/j.ijpe.2004.03.002
van Duin, J. H. R. (1998). Simulation of underground freight transport systems. In C. Borrego & L. Sucharov (Eds.), *Urban transport and the environment for the 21st century IV* (pp. 149–158). Southampton, UK: WIT Press.
van Duin, J. H. R., Kortmann, R., & van den Boogaard, S. L. (2014). City logistics through the canals? A simulation study on freight waterborne transport in the inner-city of Amsterdam. *International Journal of Urban Sciences, 18*, 186. https://doi.org/10.1080/12265934.2014.929021
van Duin, J. H. R., van Kolck, A., Anand, N., Tavasszy, L. A., & Taniguchi, E. (2012). Towards an agent-based modelling approach for the evaluation of dynamic usage of urban distribution centres. *Procedia—Social and Behavioral Sciences.* https://doi.org/10.1016/j.sbspro.2012.03.112
Vanovermeire, C., Sörensen, K., Van Breedam, A., Vannieuwenhuyse, B., & Verstrepen, S. (2014). Horizontal logistics collaboration: Decreasing costs through flexibility and an adequate cost allocation strategy. *International Journal of Logistics Research and Applications, 17*, 339. https://doi.org/10.1080/13675567.2013.865719
Wang, X., & Kopfer, H. (2014). Collaborative transportation planning of less-than-truckload freight: A route-based request exchange mechanism. *OR Spectrum, 36*(2), 1–21. https://doi.org/10.1007/s00291-013-0331-x

Wang, X., Kopfer, H., & Gendreau, M. (2014). Operational transportation planning of freight forwarding companies in horizontal coalitions. *European Journal of Operational Research, 237*(3), 1133–1141. https://doi.org/10.1016/j.ejor.2014.02.056

Watson, R. T., Boudreau, M.-C., & Chen, A. J. (2010). Information systems and environmentally sustainable development: Energy informatics and new directions for the IS community. *MIS Quarterly, 34*(1), 23–38. https://doi.org/10.2307/20721413

Yu, Y., Wang, X., Zhong, R. Y., & Huang, G. Q. (2016). E-commerce logistics in supply chain management: Practice perspective. *Procedia CIRP, 52*, 179. https://doi.org/10.1016/j.procir.2016.08.002

An Approximate Dynamic Programming Approach for a Routing Problem with Simultaneous Pick-Ups and Deliveries in Urban Areas

Mustafa Çimen, Çağrı Sel, and Mehmet Soysal

1 Introduction

In today's competitive environment, time and quality-based competition contributes to negative environmental externalities. The word "green" refers to the quality of not being harmful to the environment or depleting natural resources.[1] Nowadays, organisations cooperate with customers and suppliers to reduce the impacts of their operations to the environment and

M. Çimen (✉)
Management Science, Hacettepe University, Ankara, Turkey
e-mail: mcimen@hacettepe.com.tr

Ç. Sel
Industrial Engineering, Karabük University, Karabuk, Turkey
e-mail: cagrisel@karabuk.edu.tr

M. Soysal
Operations Management, Hacettepe University, Ankara, Turkey
e-mail: mehmetsoysal@hacettepe.com.tr

E. Aktas, M. Bourlakis (eds.), *Food Supply Chains in Cities*,
https://doi.org/10.1007/978-3-030-34065-0_4

become green in their provision of goods and services. All operations in a value chain including procurement of required inputs, producing goods and services, distribution to the market and after sales services are managed with the sense of environmental awareness.

'Sustainability' is the ability to sustain long-term ecological balance. However, environmental awareness is not enough for a supply chain to be sustainable on its own. Sustainability is also rooted in economic growth and social progress. Governments and private enterprises which respect sustainable development take precautions for achieving green growth.

During the last decade, the food and beverage industry has come to the top of the research agenda due to the growing concerns on several environmental and social issues, e.g. increasing global food losses and food waste, impacts of food perishability issues in terms of ecology, food transportation issues on reducing carbon footprint and the relative climate impacts. Food production and distribution had significant gains from the development of the new fast-growing "sustainable supply chain management" concept. The gains can be exemplified from numerous qualitative and quantitative studies existing in the literature as follows:

1. Production plans can be prepared for meeting the demand of distribution centres to reduce the number of vehicles required daily, which saves time and resources (e.g. Bard & Nananukul, 2009; Bilgen, 2010; Sel & Bilgen, 2014).
2. Different packaging formats can be treated as separate flows in distribution networks, which improves traceability (e.g. Ioannou, 2005).
3. Return flows for several items such as soft-drink bottles, recyclable containers or returnable transport items can be managed in closed distribution networks, which eliminates waste (e.g., Accorsi, Cascini, Cholette, Manzini, & Mora, 2014; Soysal, 2016; Zeimpekis, Bloemhof-Ruwaard, & Bourlakis, 2014).
4. Daily routes can be planned by grouping customers into zones, which reduces total travelled distance (e.g., Villarreal et al., 2009; Vlontzos & Pardalos, 2017).
5. Several logistics system scope issues such as the fuel consumption, carbon footprint and demand uncertainty are incorporated into the inventory routing problems, which reduces uncertainty and improves

sustainability indicators (e.g. Bektaş & Laporte, 2011; Brandenburg & Rebs, 2015; Soysal, Bloemhof-Ruwaard, & Van der Vorst, 2014).

A particular decision problem that is confronted in food and beverage industry is distribution planning in urban areas. Since a high ratio of the population reside in urban residential areas in especially developed countries, an accordingly high ratio of food and beverage distribution operations take place in urban areas. Planning these operations requires to consider a particular challenge associated with urban settlements: traffic. Traffic congestion is frequently observed in urban areas, especially in specific arcs/roads and specific hours of the day (i.e., rush hours). Varying traffic conditions in different hours of the day results in so-called "time-dependent travel times/speeds" (Soysal & Çimen, 2017). Decision makers need to take this additional set of parameters into account in order to make a more accurate distribution plan, which significantly increase problem complexity.

The focus of our research is on urban distribution in soft-drink supply chain due to its following two characteristics which differ from the many other food supply chains: (1) Unlike many other food products, the quality decay of soft-drink products over time is at a negligible level, which allows to ignore food perishability-related challenges. (2) In the soft-drink industry, returnable packages (bottles) are often used due to economic and environmental benefits, which forces companies not only to manage direct deliveries but also plan backhauls. Accordingly, in terms of distribution management, there is a necessity to address the vehicle routing problems with simultaneous pickups and deliveries, and time-dependent vehicle speeds, particularly confronted in the soft-drink distribution in urban areas.

This study provides a state-of-art assessment of the literature on the soft-drink supply chain to reveal the recent gains, particularly from a sustainability point of view. Furthermore, a real-life case is addressed on a common reverse logistics problem of collecting/reusing the reusable empty soda bottles encountered in the soft-drink industry. The problem has been formulated and solved by means of an adaptation of a recent Approximate Dynamic Programming based optimisation algorithm.

To conclude, the contribution of this study is twofold: (1) presenting a structured review of the current state of the literature on supply chain management in soft-drink industry; and (2) presenting an adaptation of an Approximate Dynamic Programming based algorithm developed for vehicle routing problems by Çimen and Soysal (2017) for use in basic reverse logistics problems frequently faced in urban soft-drink distribution operations.

The rest of the study is organised as follows. The next section presents a brief taxonomy of the literature. The subsequent sections present a state of art assessment on the soft-drink supply chain to reveal the research directions towards greening the supply chain and sustainability. This section is followed by the case study which is a real-life vehicle routing problem with deliveries and pickups of a carbonated soft-drink company. The last section discusses the overall results and presents conclusions.

2 A Brief Taxonomy of the Literature on Supply Chain Practices in the Soft-Drink Industry

The reviewed papers are selected based on a Web of Science search with "soft-drink" keyword and each of the following keywords: (1) logistics and supply chain management, (2) planning and scheduling problems, (3) production-distribution problems, (4) lot-sizing problems, (5) vehicle routing problems, and (6) inventory management. As a result, 38 research papers written in English have been accessed and taxonomic analyses of these studies are briefly discussed from the viewpoint of sustainability. Note that, a state-of-art assessment of the literature on supply chain operations in soft-drink industry specifically in urban areas could not be made, since none of the 38 studies incorporate any urban-specific assumptions into their mathematical models. In fact, none of these studies account for any urban-specific feature in either theoretical or numerical part of their studies, except for two. Temme et al. (2014) investigate whether environmental externalities are significantly associated with the urbanization grade in their numerical analyses. Euchi and Frifita (2017) mention that one of the scenarios in their numerical study fits to a setting with urban area assumption, however, the feature that fits to that of urban areas is that the majority of the customers are located close to each other in their data. The analyses and discussions on the literature, therefore, is

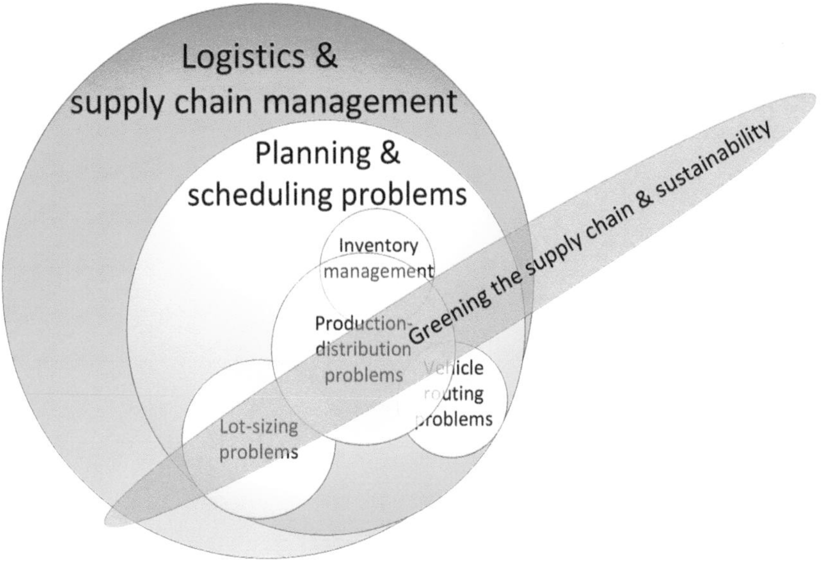

Fig. 1 The interactions between the research fields reviewed. (Source: Author)

made with a broader perspective. Figure 1 presents the interactions between the keywords searched and, therefore, the research fields reviewed where the size of the circle indicates the breadth of the topic.

Figure 2 presents the number of research papers published per year. Table 1 presents the outlets that show interest to the topic. Journal of Cleaner Production with five research papers comes first in the list based on the number of published research papers. Other journals such as "European Journal of Operational Research", "International Journal of Production Economics", and "Computers & Operations Research" are also pioneering journals in the field. These journals are commonly categorised in the following titles; "Industrial and Manufacturing Engineering", "Computers/Management Science and Operations Research", "Modelling and Simulation", "Business, Management and Accounting" and "Renewable Energy, Sustainability and the Environment". The researchers in the fields of "Engineering", "Operations research management science", "Business economics" and "Computer science" have shown relatively more interest on the topic (see Table 2).

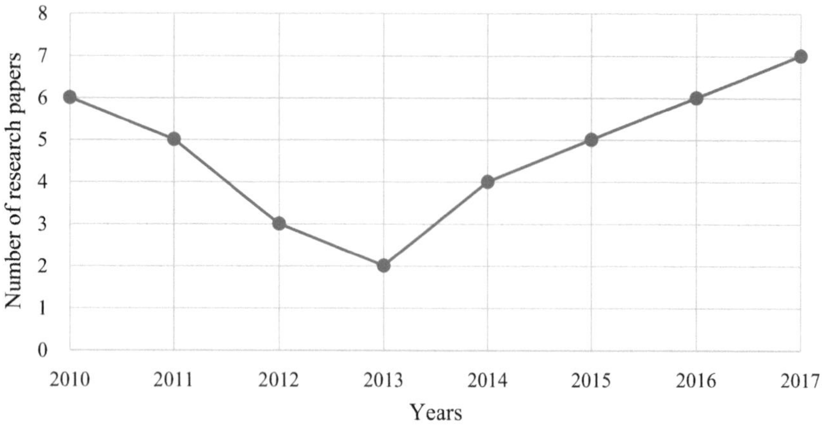

Fig. 2 Number of research papers per year. (Source: Author)

Table 1 Journals publishing the research papers

Journals	Records	Research papers
Journal of Cleaner Production	5	Almeida, Rodrigues, Agostinho, and Giannetti (2017), Luz, de Francisco, and Piekarski (2015), Nardi, da Silva, Ribeiro, and de Oliveira (2017), Silva, Medeiros, and Vieira (2017), Simon, Ben Amor, and Foldenyi (2016)
European Journal of Operational Research	4	Briskorn, Zeise, and Packowski (2016), Dong, Xu, and Evers (2012), Meyr and Mann (2013), Pureza, Morabito, and Reimann (2012)
International Journal of Production Economics	3	Ferreira, Clark, Almada-Lobo, and Morabito (2012), Ramanathan and Muyldermans (2010), Wan and Sanders (2017)
Computers & Operations Research	3	Ferreira, Morabito, and Rangel (2010), Santos, Massago, and Almada-Lobo (2010), Toledo, de Oliveira, Pereira, Franca, and Morabito (2014)

(continued)

Table 1 (continued)

Journals	Records	Research papers
Expert Systems with Applications	2	Mohamadghasemi and Hadi-Vencheh (2011), Ramanathan and Muyldermans (2011)
Strategic Management Journal	2	Zhou and Wan (2017a, 2017b)
Advances in Engineering Software	1	Cha and Roh (2010)
Computers & Chemical Engineering	1	Sel, Bilgen, Bloemhof-Ruwaard, and van der Vorst (2015)
Decision Sciences	1	Wan and Dresner (2015)
Food Chemistry	1	Iammarino et al. (2016)
International Journal of Production Research	1	Coelho, Munhoz, et al. (2016)
International Journal of Advanced Manufacturing Technology	1	Modak, Panda, and Sana (2016)
International Journal of Operations & Production Management	1	Ramanathan (2012)
International Journal of Production Research	1	Almeder and Almada-Lobo (2011)
Journal of Applied Research and Technology	1	Maldonado, Rangel, and Ferreira (2014)
Journal of Manufacturing Systems	1	Sel and Bilgen (2014)
Life Science Journal Acta Zhengzhou University overseas edition	1	Shahbazkhan, Shahriari, and Najafi (2012)
Management Decision	1	Euchi and Frifita (2017)
Mathematical Problems in Engineering	1	Toledo, Kimms, Franca, and Morabito (2015)
OR Spectrum	1	Bilgen and Günther (2010)
Packaging Technology and Science	1	Silvenius et al. (2013)
Public Health Nutrition	1	Temme et al. (2014)
Robotics and Computer-Integrated Manufacturing	1	Cha, Roh, and Lee (2010)
Transportation Research Part D-Transport and Environment	1	Soysal (2016)
Water Resources Management	1	Ercin, Aldaya, and Hoekstra (2011)

Source: Author

Table 2 Research areas in the field

Research areas	Records
Engineering	24
Operations research/management science	18
Business economics	11
Computer science	10
Environmental sciences ecology	6
Science technology other topics	5
Food science technology	4
Nutrition dietetics	3
Mathematics	2
Water resources	1
Transportation	1
Robotics	1
Public environmental occupational health	1
Life sciences biomedicine other topics	1
Chemistry	1
Automation control systems	1
Agriculture	1

Source: Author

Table 3 presents the impacts of the research papers to the relevant literature in terms of total citations received as of May 2018. The research papers are ranked in a descending order of the number of citations received, i.e. Ercin et al. (2011), Bilgen and Günther (2010), Ramanathan and Muyldermans (2010) and so forth.

Table 4 lists the keywords used in the research papers. Such an analysis allows us to enlighten the prominent keywords on the topic. The study of Ercin et al. (2011), which is the most cited research in this field, addresses corporate water footprint assessment for a carbonated beverage company. This research focuses on evaluating the environmental impacts of PET-Bottle Sugar-Containing Carbonated Beverage.

Several types of other practices exist in the literature for improving the sustainability of supply chain operations in the soft-drink industry, such as taking food perishability or greenhouse gas emissions into account (e.g., Sel et al., 2015; Soysal, 2016), selection of packaging materials between aluminium, PET, glass options, and deciding on collection systems such as kerbside collection, deposit-refund system for recycling & reuse (e.g. Almeida et al., 2017; Molina-Besch, 2016; Simon et al., 2016).

Table 3 The most cited research papers

No.	Authors	Total citations
1	Ercin et al. (2011)	53
2	Bilgen and Günther (2010)	31
3	Ramanathan and Muyldermans (2010)	23
4	Pureza et al. (2012)	22
5	Ramanathan (2012)	21
6	Cha and Roh (2010)	21
7	Ferreira et al. (2012)	18
8	Ferreira et al. (2010)	17
9	Silvenius et al. (2013)	16
10	Ramanathan and Muyldermans (2011)	15
11	Temme et al. (2014)	11
12	Cha et al. (2010)	11
13	Toledo et al. (2014)	10
14	Meyr and Mann (2013)	10
15	Simon et al. (2016)	9
16	Sel et al. (2015)	7
17	Almeder and Almada-Lobo (2011)	7
18	Sel and Bilgen (2014)	6
19	Dong et al. (2012)	6
20	Santos et al. (2010)	5
21	Wan and Dresner (2015)	4
22	Soysal (2016)	3
23	Luz et al. (2015)	3
24	Mohamadghasemi and Hadi-Vencheh (2011)	3
25	Almeida et al. (2017)	2
26	Wan and Sanders (2017)	1
27	Toledo et al. (2015)	1
28	Maldonado et al. (2014)	1

Source: Author

The academic interest on the topic particularly in the last years demonstrate the importance of the greening and sustainability issues in the soft-drink industry.

It is also worth to mention that some of the aspects of supply chain systems that may lead to greener results, such as respecting backward flows do not receive sufficient interest from researchers. From our keyword analyses, it can be observed that around only one tenth of the papers in the field address reverse logistics, which demonstrates a possible gap in the literature.

Table 4 The keywords used in the research papers

Research paper	Keywords					
Bilgen and Günther (2010)	Production and distribution planning	Consumer goods industry	Block planning	Mixed-integer linear programming		
Cha and Roh (2010)	Modeling and simulation	Simulation framework	Discrete event simulation	Discrete time simulation	Process planning	Shipbuilding
Cha et al. (2010)	Modeling	Simulation	Simulation framework	Process planning	Shipbuilding	Block erection
Ferreira et al. (2010)	Relax and fix heuristics	Production lot-scheduling models	Soft-drink industry	Mixed integer programming		
Pureza et al. (2012)	Vehicle routing with multiple deliverymen	Vehicle routing with time windows	Mixed integer programming	Tabu search optimisation	Ant colony optimisation	Beverage distribution
Ramanathan and Muyldermans (2010)	Promotions	Collaboration	Demand structure	Structural equation modelling		
Santos et al. (2010)	Capacitated lot-sizing and scheduling	Feasible solutions	Nested domains	Genetic algorithms		
Almeder and Almada-Lobo (2011)	Lot-sizing	Production planning	Scheduling	Mixed integer linear programming		

Ercin et al. (2011)	**Water footprint**	Corporate water strategy	Impact assessment	Sugar	**Soft-drink**
Mohamadghasemi and Hadi-Vencheh (2011)	Multiple criteria ABC inventory classification	Ordering policies	Fuzzy rules		
Ramanathan and Muyldermans (2011)	Demand-factor model	Structural equation modelling	Information exchange	Supply chain collaboration	
Dong et al. (2012)	Transshipments	Asymmetric information	Incentive contracts	Supply chain management	
Ferreira et al. (2012)	Production lot-scheduling models	Asymmetric travelling salesman problem	**Soft-drink industry**		
Ramanathan (2012)	Sales demand	Reference demand model	Decision making	Information exchange	Collaboration
Shahbazkhan et al. (2012)	Supply chain management	Lean	Agile	Lean-agile supply chain	
Meyr and Mann (2013)	Scheduling	Heuristics	Simultaneous lot-sizing and scheduling	Production	
Silvenius et al. (2013)	**Food waste**	Food packaging	**Food losses**	Life-cycle assessment	

(*continued*)

Table 4 (continued)

Research paper	Keywords					
Maldonado et al. (2014)	Production planning	Integrated lot-sizing and scheduling models	Asymmetric travelling salesman problem	Multi-commodity flow		
Sel and Bilgen (2014)	Supply chain management	Production and distribution planning	Mixed integer linear programming	Hybrid approach	Simulation	Rolling horizon heuristics
Temme et al. (2014)	Food consumption	**Greenhouse gas emissions**	**Sustainability**	Nutrients		
Toledo et al. (2014)	Genetic algorithms	Mathematical programming	Math-heuristics	**Soft-drink industry**	Production planning	Lot-sizing and scheduling
Luz et al. (2015)	Life cycle inventory	Model	Evaluation	Industry innovation		
Sel et al. (2015)	Integrated planning and scheduling	Mixed integer linear programming	Constraint programming	Hybrid approach	**Perishability**	Dairy supply chain
Wan and Dresner (2015)	Cost	Demand	Modularity	Product variety		
Briskorn et al. (2016)	Production	Cyclic production scheme	Stable cycle length	Stochastic demand	Service level requirement	
Coelho, Munhoz, et al. (2016)	Vehicle routing problems	**Reverse logistics**	**Green supply chain management**	GPU computing	Variable neighbourhood search	

Iammarino et al. (2016)	Radiostrontium	Milk, dairy products, animal feed	Validation	Liquid scintillation	Radioactivity	Beta emitters
Modak et al. (2016)	**Recycling**	Duopolies retailers	**Closed-loop supply chain**			
Simon et al. (2016)	**Beverage packaging**	**Post-consumer bottle**	Life cycle assessment	Policy making	Kerbside collection	Deposit-refund system
Soysal (2016)	**Closed-loop inventory routing**	**Greenhouse gas emissions**	**Energy consumption**	**Closed-supply chains**		
Almeida et al. (2017)	**Materials selection**	Aluminum cans	PET bottles	Glass bottles	**Soft-drinks packages**	
Euchi and Frifita (2017)	Metaheuristic	One-to-many-to-one	Variable neighbourhood descent	Vehicle routing		
Nardi et al. (2017)	**Sustainability**	**Reverse logistics**	**Returnable bottles**	Factor analysis		
Silva et al. (2017)	**Cleaner production**	PDCA cycle	Cans loss index	**Waste**	Cost	
Wan and Sanders (2017)	Product variety	Inventory levels	Forecast bias	Vertical integration		
Zhou and Wan (2017a, 2017b)	Vertical integration	Coordination	Product variety	Information asymmetry	Stockout	

Source: Author

Only the research papers which have given keywords are listed and, the keywords which are related to sustainability issues are highlighted in the table

3 A State-of-the-Art Assessment of the Literature

Today's concerns force organisations to manage their supply chains in a view of environmental (e.g., pollution, global warming and resulting climate change, and depletion of natural resources) and social (e.g. overpopulation, urban sprawl, and public health issues) facts. M. Christopher (2016) provides following definitions in his Logistics & Supply Chain Management book: (1) "*Logistics is the process of strategically managing the procurement, movement and storage of materials, parts and finished inventory (and the related information flows) through the organisation and marketing channels in such a way that current and future profitability are maximised through* ***the cost-effective fulfilment of orders***"; (2) as a wider concept than logistics, the supply chain is "*the management of upstream and downstream relationships with suppliers and customers in order to deliver superior customer value* ***at less cost to the supply chain as a whole***". For these early definitions, we may criticise the highlighted phrases, i.e. "the cost-effective fulfilment of orders" and "at less cost to the supply chain as a whole" in terms of pointing out only the cost-driven objectives without positioning the environmental and social concerns within the definitions of logistics and sustainable supply chain management. However, it is almost impossible to ignore the environmental and social aspects of the logistics operations as they have close links with the cost performance of the activities.

There is a vast amount of research papers in the field which do not incorporate the greening and sustainability issues into their studies. Chang and Lin (2010) make an editorial discussion on forecasting (e.g. factor selection, data mining technology, the new tools and techniques, the linkage between inventory planning and the performance indices for supply chain management). Ramanathan and Muyldermans (2010, 2011) and Ramanathan (2012) study on promotional sales and their effects on a leading soft-drink company in the UK. They identify the demand factors for forecasting. Conceptual models and structural equation models are introduced to make promotional planning. Shahbazkhan et al. (2012) study lean-agile supply chain in soft-drink industry and identify the key factors for developing managerial competence. Wan and Dresner (2015) study the dynamic interplay between product demand, operational costs, and product variety decisions. Similarly, Zhou and Wan (2017a, 2017b) study product variety, sourcing complexity, and the bottleneck of coordination in the soft-drink bottler companies. As a more recent study, Iammarino et al. (2016) consider the contamination issues in food products (sugary drinks, juice, milk etc.). They provide a risk assessment study on radioactive materials which might be used as environmental pollution indicators.

There is a growing need for research on identifying environmental and social indicators and defining how to understand and measure sustainability in soft-drink industry. This will be a promising research direction in the first place which could yield a series of research potential for the green supply chain management and sustainability in soft-drink industry (see, Akkerman, Farahani, & Grunow, 2010; Sel & Bilgen, 2015; Soysal, Bloemhof-Ruwaard, Meuwissen, & van der Vorst, 2012).

In what follows, we approach the related studies of the literature with the following rough classification: (1) production and inventory problems, (2) production-distribution problems, (3) supply chain problems with sustainability concerns, and (4) other related topics. Table 5 presents the related studies in each category.

A significant gap that is worth to mention in literature on sustainable supply chain management in soft-drink industry is the incorporation of time-dependent vehicle speeds into distribution/routing problems, which is a frequent condition in urban distribution. Management of logistics operations in urban areas requires decision support tools that are able to take variable vehicle speed into account due to traffic congestion (Malandraki & Daskin, 1992). Vehicle speed variations influence many logistics key performance indicators including total travel time, total fuel consumption and therefore total emitted emissions (Franceschetti, Honhon, Van Woensel, Bektaş, & Laporte, 2013; Gendreau, Ghiani, & Guerriero, 2015). Accordingly, researchers address routing problems with time-dependent vehicle speeds (Chen, Hsueh, & Chang, 2006; Kok, Hans, & Schutten, 2012; Kuo, 2010). These problems do not assume that vehicles retain their vehicle speeds during whole trip in all

Table 5 The research papers classified in each category

Category	Research paper	Main contribution	Methodology	Application
Production and inventory problems	Almeder and Almada-Lobo (2011)	Analysing the synchronisation problem of a secondary scarce resource in lot-sizing and scheduling of unrelated parallel machines	Two models based on the general lot-sizing and scheduling formulation and the capacitated lot-sizing formulation	An illustrative lot-sizing and scheduling example that may be confronted in bottling of soft-drinks
	Briskorn et al. (2016)	A decision support tool to develop a cyclic production schema	Three-level nested solution approach	Case studies from beverage and pharmaceutical companies, randomly generated datasets
	Cha et al. (2010) and Cha and Roh (2010)	Introducing a combined discrete event and discrete time simulation framework on the bottle filling operations of a soft-drink company	A combined discrete event and discrete time simulation framework	A simulation example from a soft-drink company is analysed for the combined discrete event and discrete time simulation model developed for process planning in shipbuilding
	Dong et al. (2012)	Addressing the complex transhipment collaboration issues in a multi-level supply chain with the presence of asymmetric information and decentralised decision making	A transhipment model and transhipment contracting mechanisms	A representative supply chain derived from the soft-drink industry
	Ferreira et al. (2010)	Lot-sizing and scheduling the one-stage/one-machine production process of small-scale soft-drink plants	Relax and fix heuristics	A Brazilian regional small-scale soft-drink plant
	Ferreira et al. (2012)	Introducing alternative formulations for synchronised two-stage lot sizing and scheduling in soft-drink production	Four single-stage models based on the general lot-sizing and scheduling formulation, the asymmetric travelling salesman formulation	A real-world bottling plant
	Maldonado et al. (2014)	Developing a multi commodity flow-based lot sizing and scheduling model for soft-drink production planning	Three models based on the sub-tour elimination constraints used in the asymmetric traveling salesman formulations	Instances generated with data from the literature
	Meyr and Mann (2013)	Introducing new solution heuristics for the general lot-sizing and scheduling problem of parallel production lines	Decomposition approaches	Soft-drink production process is provided as an example of the problem type
	Santos et al. (2010)	Handling the infeasibility drawbacks of genetic algorithms for the capacitated lot sizing and scheduling problem	Genetic algorithms	An illustrative example from soft-drink bottling and food manufacturing industries
	Toledo et al. (2014)	Proposing a solution approach to a two-level production planning and scheduling problem with interdependence and synchronisation issues	A combined genetic algorithm/ mathematical programming approach	A real-world problem that arises in soft-drink production

	Toledo et al. (2015)	Presenting a mathematical formulation and benchmark problems for the synchronised and integrated two-level lot sizing and scheduling problem	The generalized mathematical model	A set of instances based on data provided by a soft-drink company
	Mohamadghasemi and Hadi-Vencheh (2011)	Presenting the suitable policies for the items of class B in ABC inventory classification	Fuzzy logic	The soft-drink production line's raw material warehouse in a food factory
	Wan and Sanders (2017)	Investigating how firms can increase product variety while maintaining inventory levels	A moderated mediation analysis	Data collected from the distribution network of a soft-drink company operating in a make-to-stock inventory management system
Production-distribution problems	Bilgen and Günther (2010)	Addressing planning issues arising in the fast moving consumer goods industry	Mixed-integer linear programming	Case study from a company producing fruit juices and soft-drinks
	Coelho, Munhoz, et al. (2016)	Developing a heuristic approach to the single vehicle routing problem with deliveries and selective pickup	An integrated CPU–GPU heuristic inspired on variable neighbourhood search	A soft-drink distribution is exemplified to illustrate the need of a selective pickup component
	Euchi and Frifita (2017)	Presenting a specific variant of vehicle routing problem with simultaneous pickup and delivery	A hybrid genetic algorithm with variable neighbourhood search	Real case of soft-drink distribution in Tunisia
	Pureza et al. (2012)	Presenting a variant of the vehicle routing problem with time windows for which a number of extra deliverymen can be assigned to each route to reduce service times, for example, in the distribution of beverage and soft-drink	A mathematical programming formulation, a tabu search and an ant colony optimisation heuristic	A set of classic examples from the literature
	Sel and Bilgen (2014)	Implementation of the mixed-integer linear programming based rolling horizon heuristics and a hybrid simulation approach to the production allocation and distribution planning problem	Mixed-integer linear programming, fix-and-relax/fix-and-optimise heuristics and a hybrid simulation approach	An industrial case from the soft-drink industry

(*continued*)

Table 5 (continued)

Category	Research paper	Main contribution	Methodology	Application
Supply chain problems with sustainability concerns	Almeida et al. (2017)	Introducing environmental accounting based on emergy as a tool to assist materials selection	Emergy analysis	The case of soft-drinks packaging in Brazil
	Ercin et al. (2011)	Corporate water footprint accounting and impact assessment	The water footprint estimation	A hypothetical case of the water footprint of a sugar-containing carbonated beverage
	Luz et al. (2015)	A model to assess the relationship between life cycle inventory and innovation	Life cycle assessment	Life cycle inventory of aluminium packaging for soft-drinks is used as an indicator to construct the model
	Modak et al. (2016)	Cooperative and non-cooperative models of two-echelon closed-loop supply chain consisting of a manufacturer and duopolies of retailers with recycling facilities	Mathematical model	The beverage containers, beer and soft-drink bottles are highlighted aspects linked to recycling
	Nardi et al. (2017)	A methodology to view and monitor the economic, social, and environmental impacts of operational and strategic decisions in reverse logistics industries	Factor analysis	A non-public company engaging in reverse logistics activities that is authorized to produce and sell soft-drinks for one of the largest brands in the world
	Sel et al. (2015)	A hybrid mixed-integer linear/constraint programming decomposition strategy	Mixed-integer linear programming, constraint programming	An illustrative case study on production planning and scheduling of yoghurt, produced by a similar make-and-pack production process with soft-drink
	Silva et al. (2017)	Implementation of a cleaner production program to reduce an organisation's cans loss index	The plan-do-check-action	The data on the loss rate of cans and the cost of this loss was obtained through institutional documents of the soft-drink company surveyed
	Silvenius et al. (2013)	Investigating the environmental impacts of food waste and the influence of packaging alternatives	Life cycle assessment	Three case studies for ham, dark bread and a fermented soy-based drink
	Simon et al. (2016)	Providing a holistic view of environmental and human health implications of beverage packaging materials	Life cycle assessment	The collection of post-consumer bottles
	Soysal (2016)	Enhancing the traditional models for the closed-loop inventory routing to make them more useful with pickup and deliveries for the decision makers in closed-loop supply chains	A probabilistic mixed-integer linear programming model	A case study on the distribution operations of a soft-drink company
	Temme et al. (2014)	Evaluating the greenhouse gas emission of diets including soft-drinks	Descriptive analyses	Diets in Netherlands

Other related topics	Iammarino et al. (2016)	Risk assessment about an important contaminant relating to food supply chains, present in the main food products, e.g., wine, beer and sugary drinks	Statistical analysis	The samples collected from 13 Italian regions and from Japan (Fukushima), during the period October 2011–October 2015
	Ramanathan (2012)	Improving the forecast accuracy by collaborative forecasting	Reference demand model	A case study approach to study various demand factors of soft-drink products of the UK based company
	Ramanathan and Muyldermans (2010)	Identifying demand factors for promotional planning and forecasting	Structural equation modelling	A case of a soft-drink company in the UK
	Ramanathan and Muyldermans (2011)	Identifying the underlying structure of demand during promotions	Structural equation modelling	Electronic point of sales data of a soft-drink manufacturer
	Shahbazkhan et al. (2012)	Identifying and evaluating effective factors on lean-agile supply chain	Fuzzy logic	A company's soft-drink production
	Wan and Dresner (2015)	Closing the product variety decision loop by jointly capturing the impact of product variety on demand and operational cost	A model with three equations; order quantity, cost and product variety	The network of a U.S. based soft-drink bottler (the bottling and distribution affiliate of a large soft-drink firm)
	Zhou and Wan (2017a, 2017b)	Evaluate vertical relationships in value chains where one stage competes on product variety under great uncertainty and the other stage competes on scale	A quantitative case study approach	Operations data at about 300 distribution centres within a major soft-drink bottler

Source: Author

arcs, but consider several speed profiles based on traffic congestion occurred in different time zones (Figliozzi, 2012).

The results of the studies on varying distribution/routing problems confirm that significant cost, energy and emission savings could be obtained in delivery operations by means of quantitative models that respect time-dependent vehicle speeds (Çimen & Soysal, 2017; Figliozzi, 2011; Jabali, Van Woensel, & De Kok, 2012; Soysal, Bloemhof-Ruwaard, & Bektaş, 2015). However, according to our review, none of the 38 papers on soft-drink supply chain management incorporate time-dependent vehicle speeds into their mathematical models. There is a clear need for additional researches on soft-drink supply chain management, which provide decision support models/tools that respect time-dependent vehicle speeds, in order to aid decision makers in the industry.

Our brief literature review shows that the research in the field often ignores the hierarchical relation between tactical and operational level decisions and the corresponding interaction between production planning and scheduling problems. However, operational schedules have to be aligned with tactical plans. Therefore, it might be beneficial to handle operational and tactical decisions simultaneously (see Bilgen & Çelebi, 2013; Maravelias & Sung, 2009; Sel et al., 2015). The production process should be considered in a broad view for getting beyond the myopic focus on the bottling process. The machine scheduling and line balancing might be error-prone problems since multiple products with different labels/sizes and many-to-many *production* environments in parallel mixing tanks and bottling machines may complicate the production processes. Similar to the case between tactical and operational level decisions, the synchronisation of production and distribution operations are also crucial in supply chain control of the soft-drink industry.

Perishability, which is receiving a growing attention in food logistics management refers the continuous decay and/or complete loss of a product's value over time. Although juices and soft-drinks are mostly long-lasting products, they still have a limited shelf-life which restricts their storage duration and delivery conditions. Additionally, the fresh juices and soft-drink syrups, which are among the main ingredients of soft-drinks (i.e. raw materials and/or intermediate products), are subject to continuous loss of value during the production process. The perishable juices and soft-drinks might have been a risk for human health and adhering to thermal regulations might be a must for production and distribution activities. The ingredients must be produced and stored in special conditions restricting the intermediate storage duration, i.e. the time between the two mixing and bottling stages. Therefore, quality decay aspects of raw materials and semi-finished products have to be respected in decision making processes (see, Akkerman & van Donk, 2009; Akkerman, Van Donk, & Gaalman, 2007; Sel, Bilgen, & Bloemhof-Ruwaard, 2017). The decision support models for food logistics should also respect shelf-life considerations, for example, as a decreasing function of the economic value of fruit juices and soft-drinks over time (see, Lütke Entrup, 2005; Sel et al., 2015; Soysal, Bloemhof-Ruwaard, Haijema, & van der Vorst, 2015).

Another recent concept is process flexibility. Process flexibility refers the ability to produce a variety of products in a production resource. Many aspects of the modern competitive industries force companies to be flexible, particularly in terms of responding varying demands of different product types. Process flexibility allows companies to adjust their production schedules to meet various demand schemes with better using the production capacities. However, process flexibility forms an interrelated set of production decisions, which results in significantly complex decision problems. There is an important gap in the literature in terms of inventory optimisation for stochastic process flexibility systems. Although a few studies addressed the research area (e.g., Çimen & Kirkbride, 2017; Jordan & Graves, 1995; Simchi-Levi & Wei, 2015), there is still a vast set of problems for which decision aid tools should be developed for particularly food logistics.

Studies that have a motivation to improve sustainability performance may incorporate less frequently addressed sustainability indicators that are affected especially from urban supply chain operations, such as labour rights, city congestion, traffic noise, etc., which will provide more holistic views of sustainability thinking.

An emerging concept, "Extended Producer Responsibility", refers a policy approach under which producers are responsible for the treatment or disposal of post-consumer products. This concept requires sellers to plan forward and backward hauls simultaneously. Such a logistics management has potential to contribute to the three pillars of sustainable development: economic, environmental and social. For instance, collecting empty bottles of soft-drinks enables to decrease bottling costs through recycling, to reduce environmental impacts of waste, and to prevent misspending of scarce production factors used for producing new bottles. However, most of the addressed urban supply chain networks in the field do not involve backward flows of goods. Applications on soft-drink industry could benefit from additional research attempts on reverse logistics in urban areas, which has been widely addressed in many other industries.

4 Solving Real Problems with Quantitative Techniques: A Case Study

Practical applications of the theoretical knowledge are at least as important as the theoretical developments themselves, since all theoretical developments are supposed to have a (direct or indirect) positive effect on daily lives of people at some point in the future. We, therefore, here present an example of how a real-life problem often confronted by decision makers in companies can be solved by using the approaches presented in the academic literature.

For this purpose, we have selected a green vehicle routing problem often encountered in the soft-drinks industry operating in urban areas. Soft-drink companies usually reuse the empty bottles returned from customers/dealers. Reusing these bottles not only help companies to reduce their costs, but also contributes to a higher sustainability performance through reusing materials and reducing waste. However, collecting the empty bottles from the dealers while delivering new products turns the problem into a pick-up and delivery problem, where each dealer has both a supply of empty bottles to pick-up and a demand for new products. Additionally, many of the dealers are usually located in urban areas, which necessitates incorporating time-dependent vehicle speeds into the problem and, therefore, increases the problem complexity.

Many companies manage this pick-up and delivery process by means of crossdocking points to serve dealers in urban areas. Incorporating crossdocking points also contributes to improving economic and environmental performance of the supply chain, since it benefits from the use of environmentally friendly vehicles (light, small, electric, etc.) for serving nodes subject to traffic congestion, while larger vehicles are employed for non-urban travels. On the non-urban vehicle routes, the new products are delivered to crossdocking points from the central depot and the empty bottles are picked-up from the crossdocking points back to the central depot simultaneously. In terms of urban distribution, crossdocking points act as the central depot for the dealers assigned to the corresponding crossdocking point.

In the following subsections, first, we describe a common vehicle routing problem considered with simultaneous deliveries of products, and pickups of returnable/refillable soda bottles. Second, we adopt an Approximate Dynamic Programming (ADP) heuristic presented by Çimen and Soysal (2017) to solve such problems. Finally, we present the data and the real operational settings of a carbonated soft-drink company and show the resultant delivery plans.

4.1 Problem Description

The problem is a reverse logistics problem of collecting/reusing the empty soda bottles to reduce packaging waste. Here, bi-directional flow of goods, i.e. forward and backward, are considered for routing the transportation vehicles in a closed-loop supply chain. The vehicle routing problem with simultaneous delivery and pickup is one of the special cases of the classical vehicle routing problem, where retailers/customers may send and receive goods simultaneously (see Parragh, Doerner, & Hartl, 2008).

For the addressed problem, we will solve two separate routing problems. The first problem is the pick-up and delivery problem of serving the crossdocking points, and the second one is for serving dealers in urban areas by a given crossdocking point. Each of these problems can be defined on a complete graph G = (N, A), where N = {0, 1,2, …, n} is the set of nodes in which node 0 represents the central depot (or the cross-docking point in urban distribution problem) and the remaining nodes represent the crossdocking points (or dealers in urban distribution problem). $A = \{(i,j) : i,j \in N, i \neq j\}$ is the set of arcs and each arc $\{i,j\} \in A$ has a nonnegative distance, d_{ij}. Each served point *i* has a non-negative demand β_i and a non-negative return R_i. At the resource node (central depot or crossdocking point), there are a limited number of homogeneous vehicles each with a certain capacity. Each vehicle route starts and ends at the resource node. At any point, the total load of the vehicle cannot exceed its capacity.

For this problem setting, our purpose is to derive a set of routes that minimises the total expected routing cost, where each node is visited exactly once by any of the vehicles, the demands of each node are satisfied, and the returns of each node is collected. We include fuel and wage costs in the routing cost function. Wage cost depends on the total travel time of the vehicles, i.e., the total time that the vehicle's drivers work. Incorporating fuel cost estimation enables to account for emissions from transportation. Fuel consumption/emissions are dependent on vehicle type, vehicle speed, vehicle load, and travel distance (Demir, Bektaş, & Laporte, 2011; Ligterink, Tavasszy, & de Lange, 2012). Although many alternatives are also available, here we use the model suggested by Hickman, Hassel, Joumard, Samaras, and Sorenson (1999) for its simplicity. However, we would like to note that the proposed solution approach is flexible that any other respected emission estimation approach could be used. According to Hickman et al. (1999), the total amount of generated emission F (CO_2) per travelled distance (in km) will be:

$$F = K + av + bv^2 + cv^3 + \frac{d}{v} + \frac{e}{v^2} + \frac{f}{v^3}$$

where v is the average speed (km/h) of the vehicle, K is a constant, a, b, c, d, e and f are parameters that change according to the vehicle type.

Following emission estimation, we estimate the related fuel consumption by using a fuel conversion factor for transport activities.

4.2 Modelling Approach

The vehicle routing problem with simultaneous pick-up and deliveries is a variant of the classical vehicle routing problem which is well-known to be NP-hard (Çatay, 2010). As the problem size increases, the memory and time requirements of the classical optimisation techniques also increase rapidly. This renders the use of classical optimisation approaches infeasible in most of the real-life problems. Approximations and/or heuristics are often used to overcome this obstacle. Here, we will use an ADP approach for solving the addressed problem.

The solution algorithm is adopted from the one presented by Çimen and Soysal (2017). The idea behind the ADP approach lies between the processes of Dynamic Programming and simulation methods. The approach gathers sample information on the potential returns of decisions through a simulation. At the end of each iteration of this simulation, the collected sample information is used for improving the preferences on the decision alternatives. Theoretically, a simulation with an infinite number of iterations will converge the value functions to the optimal. However, a limited number of iterations will also provide promising solutions, since the simulation will visit the most-frequently confronted states for a given policy at the beginning of its iterations and gather information on them more quickly.

For the addressed problem, the sample information is collected via testing different routes in simulation. For each arc, the return/cost yielded by the routes, in which the arc is travelled, is used for updating the value function of the arc. Note that the yielded return/cost and the value functions respect time-dependent travel speeds (i.e., travel durations and emissions are calculated by taking time-dependent travel speed realizations into account). Then, these value functions are used for determining new promising routes in the following iterations of the simulation. Finally, the route with the lowest observed cost is selected for the decision maker. See Algorithm 1 for a detailed pseudocode of the used algorithm.

Algorithm 1 Pseudocode of the Adopted ADP Heuristic

Step 1: Initialisation
Initialise the value function estimations ($Q(i,j) = 0$) for all arcs $(i,j) \in A$
Initialise the visiting frequencies ($\phi(i,j) = 0$) for all arcs $(i,j) \in A$
Set the values of the heuristic parameters: probability of random actions (ε), eligibility trace decay value (λ), Simulation Length and Policy Simulation Length.
For (sim_counter = 1 **to** Simulation Length)
Reset simulation parameters:
Initialise eligibility traces ($e_{i,j} = 0$) for all arcs $(i,j) \in A$
Set time to initial time: $t = 0$
Set load to vehicle capacity: $l = L$
Set number of unused vehicles to total number of vehicles: $r = V$
Set current node to depot: $i = 0$
Generate a random variable, ε_1, uniformly between [0,1]
While all nodes are not visited, **do**
Step 2: Selecting Action
Select the next node to travel, j, using an ε-greedy policy:
Generate a random variable, ε_2, uniformly between [0,1].
If $\varepsilon_1 > \varepsilon$ and $\varepsilon_2 > \varepsilon$ then $j \leftarrow \underset{j}{\operatorname{argmin}} \{Q(i,j)\}$
Else $j \leftarrow$ a random node.
End If
Step 3: Observing Random Output of Simulation
Calculate the travel duration, T, where the speed is assumed to be $\hat{\varsigma}_{i,j}$:

$$T \leftarrow d'_{i,j} / \hat{\varsigma}_{i,j}$$

$$t \leftarrow t + d'_{i,j} / \hat{\varsigma}_{i,j}$$

Add service time of j (ST_j) to duration: $T \leftarrow T + ST_j$
Calculate the emissions and the travel cost observation $\hat{\theta}_{i,j,z}$ for this arc
Step 4: Updating value function estimations
Set eligibility trace for current arc: $e_{i,j} \leftarrow 1$
Calculate temporal difference: $\delta \leftarrow \hat{\theta}_{i,j} - Q(i,j)$
Calculate the step-size value, α:
If $\delta < 0$, **then** $\alpha = 1$
Else $\alpha = 1 / (\max(1, \phi(i,j))$
End If
Update the **temporary** value function estimations, $Q'(i',j')$ and eligibility traces, $e_{i',j'}$ for all arcs $(i',j') \in A$ using the travel cost observation:
$Q'(i',j') \leftarrow \alpha\, \delta\, e_{i',j'} + Q'(i',j')$
e_{i',j'} $\leftarrow \lambda\, e_{i',j'}$

Step 5: Moving to the Next Stage
$\phi(i,j) \leftarrow \phi(i,j)) + 1$
$i \leftarrow j$
End While
Update value function estimations, $Q(i',j')$ for all arcs $(i',j') \in A$ using the temporary value function estimations:
$Q(i',j') \leftarrow Q'(i',j')$
Check whether the latest policy generates the lowest-average-cost-so-far. If so, record the policy. If not, do nothing.
End For

Note that the performance of the algorithm is subject to tuning a number of parameters. The setting of ε, the probability of random actions, sets the balance between exploration and exploitation. Higher ε means more exploration of the action space, while lower ε results in learning more about the greedy actions. Another parameter, eligibility trace decay value (λ), is the weight of the effects of earlier decisions and the last one on the latest output. Low λ settings refer to higher effect of the latest decision on the latest output. Simulation Length and Policy Simulation Length are also important parameters. Higher values would result in better estimations, however, would also require higher computation times.

4.3 Data and Results

Here, we present an implementation of the presented ADP algorithm on the real-life settings (i.e. distances) of a soft-drink company's distribution problem. We have selected a carbonated soft-drink company with production facilities in Bursa, which deals with filling/packaging mineral water (soda) in the reusable bottles and delivery to retailers operating in Turkey.

The company owns a soda filling facility (the central depot) which serves 42 dealers in 10 cities. We have arbitrarily selected a dealer in each city as the crossdocking point to serve the dealers in that city. In addition, as an exemplar case for the urban food distribution, we have addressed the delivery plans between the crossdocking point in İstanbul and the dealers in the city, where 19 (including the crossdocking point) of the 42 dealers of the company are located. The addressed transportation network is represented in Fig. 3. The distances are calculated through Google Maps[2] (see Tables 8 and 9 in Appendix A). The demand and return amounts are changing day-by-day throughout the year, therefore, realistic arbitrary amounts are generated in this example. Since the company does not currently have dedicated crossdocking points, it is not possible to observe service times. So that, service times are also arbitrarily generated. Tables 6 and 7 present the generated demand and the return amounts in number of bottles and service times in hours for each crossdocking point and the dealers in İstanbul.

Two homogeneous large vehicles are used for non-urban delivery and three smaller ones are used for urban delivery in İstanbul. Capacities are 15,000 and 5000 bottles for large and small vehicles, respectively.

The fuel consumptions of vehicles are estimated based on the travelled distance and the average speed, using the aforementioned emission estimation model. The required model parameters are taken from Hickman et al. (1999), and are as follows: (1) large vehicles are assumed to be heavy goods vehicles with gross vehicle weights from 32 to 40 ton ($K = 1576$; $a = -17.6$; $b = 0$; $c = 0.00117$; $d = 0$; $e = 36{,}067$ and $f = 0$), and (2) small vehicles are assumed to be heavy goods vehicles with gross vehicle weights

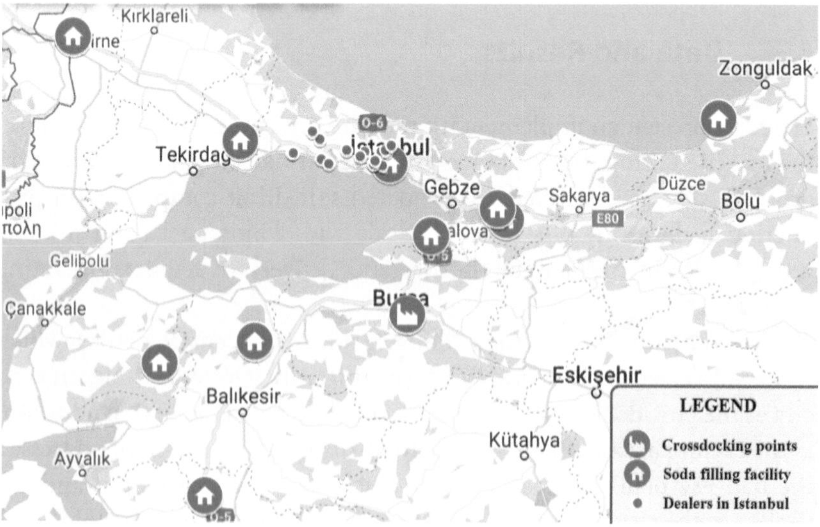

Fig. 3 The soda distribution network of the carbonated soft-drink company. (Source: Google Maps website)

Table 6 The demand and the return amounts per replenishment cycle and service times for each crossdocking point

Crossdocking point	Demand (in bottles)	Return (in bottles)	Service time (in hours)
1	2298	2167	5.16
2	1857	1719	1.51
3	1285	878	1.86
4 (İstanbul)	10,511	10,009	17.32
5	619	489	0.37
6	869	1277	1.55
7	500	534	0.52
8	1602	2023	3.45
9	723	528	1.31
10	959	1204	1.40

Source: Author

Table 7 The demand and the return amounts per replenishment cycle and service times for each dealer in İstanbul

Dealers	Demand (in bottles)	Return (in bottles)	Service time (in hours)
1	776	727	0.50
2	484	309	1.28
3	671	891	1.74
4	345	201	1.02
5	537	541	0.81
6	1000	796	1.77
7	591	627	0.41
8	418	342	0.62
9	711	548	1.31
10	335	440	0.72
11	322	452	0.73
12	817	1096	1.67
13	480	364	0.75
14	257	294	0.68
15	661	648	0.44
16	968	819	1.64
17	149	126	0.09
18	600	586	0.61

Source: Author

from 7.5 to 16 ton, ($K = 871$; $a = -16$; $b = 0.143$; $c = 0$; $d = 0$; $e = 32{,}031$ and $f = 0$). The CO_2 emissions are estimated by assuming that each litre of fuel consumption generates 2.63 kg CO_2 (Defra, 2007). Fuel price and wage are taken as 1.6 €/l and 0.24 €/min, respectively.

Average speed of vehicles is assumed to be 90 km/h for non-urban travel. For urban travels, time-dependent travel speeds are employed due to traffic congestion in the city. It is assumed that the delivery travels begin at 8 am, where a one-hour traffic congestion exists. This congested period is followed by a *free flow* period lasting for eight hours. Then, between 5 pm and 6 pm, another heavy congestion occurs. The rest of the day after 6 pm is again assumed to be free flow period. Average speed of vehicles are 50 km/h and 20 km/h for free flow and congested periods respectively.

The described problems are solved using the aforementioned ADP algorithm. The algorithm is run using the following arbitrary settings: $\varepsilon = 0.1$; $\alpha = 0.1$; $\lambda = 0.5$ and Simulation Length = 1,000,000 iterations for each problem, in line with the settings of Çimen and Soysal (2017). The resultant routes are obtained in approximately 4 hours on a Pentium (R) i7 computer with 2.20 GHz CPU and 8 GB RAM.

Figure 4 presents the resulting feasible vehicle routes for serving the crossdocking points in terms of wage and fuel (which represents CO_2 emissions) costs with pick-up and deliveries obtained from the ADP based decision support algorithm. Per replenishment cycle, the resultant cost is found as 2040.86 €, emissions are 1938.77 kg and total distance is 2282.56 km. The resultant vehicle routes are as follows (numbers represent crossdocking point identities):

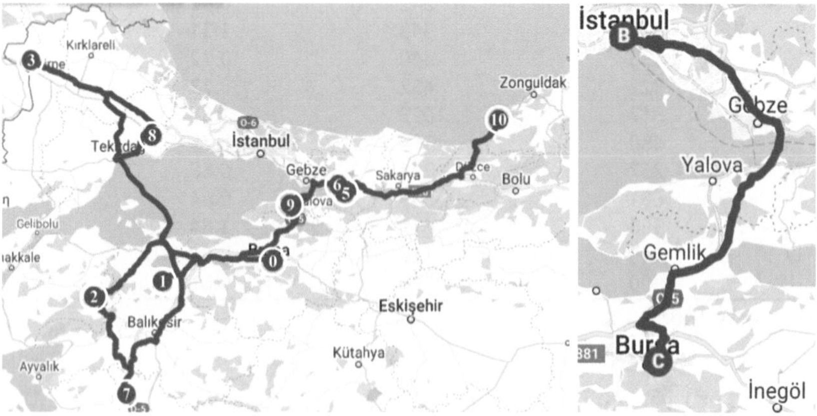

Fig. 4 The resulting two vehicle routes for non-urban travels with pickup and deliveries. (Source: Google Maps website)

Vehicle 1: 0-9-6-5-10-1-8-3-7-2-0.
Vehicle 2: 0-4-0.

Figure 5 presents the resulting feasible vehicle routes for serving the dealers in İstanbul in terms of wage and fuel (which is a proxy for CO_2 emissions) costs with pick-up and deliveries obtained from the ADP based decision support algorithm. The resultant cost is found as 427.50 €, emissions are 142.20 kg and total distance is 308.49 km. The resulting vehicle routes are as follows (numbers represent crossdocking point number):

Vehicle 1: 0-10-4-11-14-6-7-5-16-17-0.
Vehicle 2: 0-9-13-15-3-2-8-1-18-0.
Vehicle 3: 0-12-0.

Note that although this problem involves 42 nodes, the proposed algorithm can be used for tackling with larger-sized routing problems. In addition, the algorithm is quite flexible that it could be adopted for solving other variants of decision making problems in vehicle routing field, including but not limited to problems with stochastic vehicle speeds or problems constrained with time windows.

5 Conclusion

In this study, we provide a literature review on supply chain management problems encountered in the soft-drinks industry. In addition, we study a frequently confronted operational vehicle routing problem of the soft-drinks industry operating in urban areas. We aim to find feasible routes with an objective comprises wage costs and fuel costs (which represents CO_2 emissions) for simultaneous soft-drink deliveries and pickups for collecting/reusing empty soda bottles.

The study attracts attention to a great potential of green supply chain management and sustainability practices in the field. The study could be useful not only for researchers studying on the topic, but also for practitioners and decision makers working in the soft-drinks industry. The

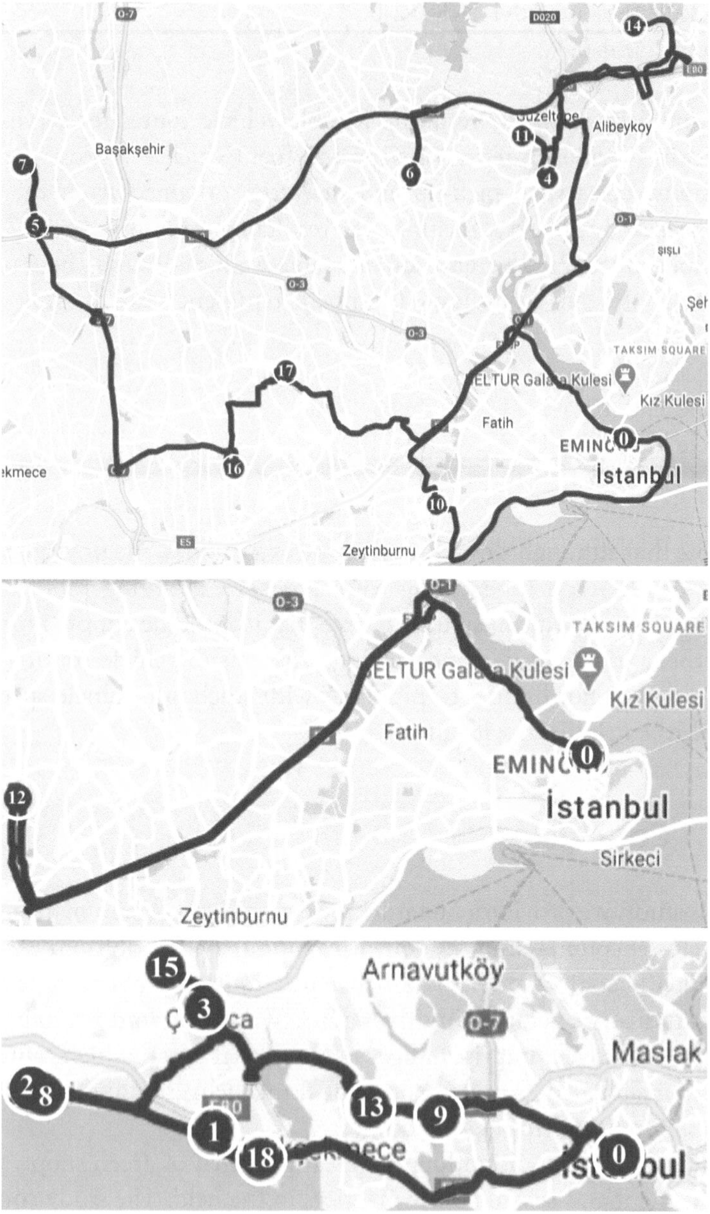

Fig. 5 The resulting three vehicle routes for urban travels with pickup and deliveries. (Source: Google Maps website)

theoretical contributions of this study are twofold: (1) to identify major research streams on soft-drinks supply chains and corresponding gaps and future research directions, some of which are related to sustainability issues; and (2) to show an example of how quantitative operations research techniques can aid decision makers in real life urban distribution planning instances.

The review on the topic reveals several gaps in the literature that could attract researchers' attention and progress the current academic knowledge. In terms of food logistics management, particularly in the soft-drinks industry, three prominent keywords for future research are food perishability, production process flexibility, and sustainability. An important note is that none of the accessed studies on soft-drink supply chains incorporate any urban-specific assumptions into their mathematical models, which reveals the need for future researches on the topic. We here acknowledge the fact that rather than providing a complete systematic literature review, we illustrate the research field with representative papers, the choice of which always remains subjective.

The proposed Approximate Dynamic Programming based algorithm can be regarded as an attempt that contributes to reducing carbon footprint and the relative climate impacts in a supply chain that involves both forward and backward flows. Our literature review reveals the need for researches on reverse logistics in soft-drink industry. The keyword analyses show that only a small fraction of the researches in the field have addressed backward flows of goods. The proposed algorithm respects backward flows of reusable bottles, and could aid decision makers in the soft-drinks industry as a decision-support tool while managing a closed-loop supply chain.

The provided model or its variants can be used by practitioners for sustaining logistics operations in an environmentally and socially responsible manner. The developed decision aid tool regards emissions from transportation operations, which has potential to reduce carbon footprints of logistics operations in urban areas. Generating carbon emissions in urban areas is not only harmful to global sustainability but also has a significant negative impact on social life in cities, causing various health problems.

The proposed Approximate Dynamic Programming algorithm has a number of possible parameters and settings selections of which may affect the performance of the heuristic. This study does not aim to find the best parameters and settings for the proposed algorithm but provides a promising decision aid tool for industrial decision makers. Therefore, a comprehensive analysis on parameters and settings of the algorithm is not performed, but an exemplary use of the algorithm with arbitrary settings of parameters (in line with the settings of Çimen & Soysal, 2017) is presented.

Appendix A: Distances Used for the Case Study

Table 8 Distances in kms between the soda-filling facility (node 0) and the crossdocking points (nodes 1–10)

	0	1	2	3	4	5	6	7	8	9	10
0	0	134.473	219.276	422.149	186.15	147.217	149.956	252.617	297.077	99.955	341.631
1	155.15	0	84.312	286.777	275.243	236.311	239.049	134.248	183.881	189.049	430.725
2	220.666	85.348	0	318.742	340.759	301.826	304.565	143.516	226.179	254.564	496.24
3	424.518	289.82	322.434	0	239.9	356.714	328.016	420.13	145.844	337.202	519.691
4	187.894	258.939	343.742	236.693	0	120.091	91.392	377.083	111.622	100.579	283.068
5	144.165	215.21	300.013	343.888	107.89	0	32.19	333.354	218.817	54.031	191.906
6	150.01	221.055	305.858	325.313	89.314	30.529	0	339.199	200.242	62.694	194.933
7	252.361	134.192	142.366	417.07	372.454	333.522	336.26	0	314.174	286.26	527.936
8	295.966	184.495	225.545	142.793	111.349	228.163	199.464	314.806	0	208.651	391.14
9	101.393	172.438	257.241	333.416	97.418	53.19	61.223	290.582	208.345	0	252.899
10	344.02	415.065	499.868	519.324	283.325	193.58	198.124	533.209	394.252	256.705	0

Source: Author

Table 9 Distances between the crossdocking point in İstanbul (node 0) and the dealers in İstanbul (nodes 1–18)

	0	1	2	3	4	5	6	7	8	9	10	11	12	13	14	15	16	17	18
0	0	56.311	73.148	57.62	10.71	20.293	19.406	21.506	72.307	25.837	10.453	10.628	18.987	31.21	13.727	51.836	19.355	14.272	48.965
1	55.39	0	24.096	19.218	53.921	40.52	48.938	42.95	23.255	33.335	52.935	54.127	47.926	32.608	57.144	13.434	47.704	45.741	10.211
2	72.596	22.613	0	36.425	71.128	57.727	66.145	60.157	1.251	50.541	70.142	71.334	65.133	49.815	74.35	30.641	64.911	62.948	27.073
3	57.46	20.862	28.03	0	55.992	42.591	51.009	45.021	27.189	35.405	55.006	56.198	49.997	34.679	59.214	9.831	49.775	47.811	23.479
4	9.486	55.759	72.597	57.069	0	19.741	5.008	20.954	71.756	25.286	11.253	0.206	16.476	30.658	7.708	51.284	17.091	14.258	48.761
5	21.241	43.293	60.13	44.603	19.772	0	14.789	4.456	59.29	9.238	18.786	19.978	10.196	18.192	22.995	38.818	9.974	11.592	36.295
6	18.865	51.49	68.327	52.799	5.148	15.472	0	16.685	67.486	21.016	19.127	5.067	22.878	26.389	11.782	47.015	22.657	15.614	44.492
7	21.726	42.718	59.555	44.027	20.257	5.266	15.275	0	58.714	8.273	19.272	20.463	14.585	17.617	23.48	38.243	14.364	12.077	35.72
8	71.726	21.743	1.158	35.555	70.258	56.857	65.275	59.287	0	49.671	69.272	70.464	64.263	48.945	73.48	29.771	64.041	62.077	26.203
9	26.42	34.045	50.882	35.355	24.952	9.346	19.969	9.551	50.042	0	23.966	25.158	17.944	8.944	28.174	29.57	17.722	16.772	27.047
10	10.94	53.739	70.576	55.048	15.336	17.721	16.834	18.934	69.735	23.265	0	12.251	9.446	28.638	16.83	49.264	10.06	8.003	39.112
11	9.692	55.759	72.596	57.068	0.206	19.741	4.784	20.953	71.755	25.285	11.459	0	16.682	30.657	7.708	51.283	17.297	19.883	48.761
12	16.216	49.406	66.243	50.715	21.043	9.65	20.902	14.601	65.402	18.932	8.984	17.957	0	24.305	29.107	44.931	0.311	4.375	33.667
13	32.056	27.485	44.322	28.794	30.588	17.187	25.605	19.617	43.482	10.001	29.602	30.794	24.593	0	33.81	23.01	24.371	22.407	23.613
14	18.227	60.173	77.01	61.482	10.413	24.155	13.484	25.367	76.169	29.699	20.202	10.619	25.425	35.071	0	55.697	31.339	24.297	53.175
15	49.862	13.264	30.101	9.577	48.394	34.993	43.411	37.423	29.26	27.807	47.408	48.6	42.399	27.081	51.616	0	42.177	40.213	15.881
16	16.173	41.199	66.423	50.895	21	9.83	21.082	14.781	65.582	19.113	8.941	17.915	0.196	24.485	22.494	45.111	0	4.555	33.624
17	14.129	47.955	64.792	49.264	18.594	11.937	13.612	13.15	63.951	17.481	7.701	18.8	4.141	22.854	21.817	43.48	3.92	0	40.957
18	46.883	8.533	26.066	24.58	49.463	36.061	44.48	38.492	25.225	28.876	40.33	49.669	36.007	25.526	52.685	18.796	35.885	41.282	0

Source: Author

Notes

1. http://www.dictionary.com/browse/sustainability, Online accessed: October 2017.
2. http://maps.google.com.tr, Online accessed: June 2018.

References

Accorsi, R., Cascini, A., Cholette, S., Manzini, R., & Mora, C. (2014). Economic and environmental assessment of reusable plastic containers: A food catering supply chain case study. *International Journal of Production Economics, 152*, 88–101.

Akkerman, R., Farahani, P., & Grunow, M. (2010). Quality, safety and sustainability in food distribution: A review of quantitative operations management approaches and challenges. *OR Spectrum, 32*(4), 863–904.

Akkerman, R., & van Donk, D. P. (2009). Product mix variability with correlated demand in two-stage food manufacturing with intermediate storage. *International Journal of Production Economics, 121*(2), 313–322.

Akkerman, R., Van Donk, D. P., & Gaalman, G. (2007). Influence of capacity- and time-constrained intermediate storage in two-stage food production systems. *International Journal of Production Research, 45*(13), 2955–2973.

Almeder, C., & Almada-Lobo, B. (2011). Synchronisation of scarce resources for a parallel machine lotsizing problem. *International Journal of Production Research, 49*(24), 7315–7335.

Almeida, C., Rodrigues, A. J. M., Agostinho, F., & Giannetti, B. F. (2017). Material selection for environmental responsibility: The case of soft-drinks packaging in Brazil. *Journal of Cleaner Production, 142*, 173–179.

Bard, J. F., & Nananukul, N. (2009). The integrated production–inventory–distribution–routing problem. *Journal of Scheduling, 12*(3), 257–280.

Bektaş, T., & Laporte, G. (2011). The pollution-routing problem. *Transportation Research Part B: Methodological, 45*(8), 1232–1250.

Bilgen, B. (2010). Application of fuzzy mathematical programming approach to the production allocation and distribution supply chain network problem. *Expert Systems with Applications, 37*(6), 4488–4495.

Bilgen, B., & Çelebi, Y. (2013). Integrated production scheduling and distribution planning in dairy supply chain by hybrid modelling. *Annals of Operations Research, 211*(1), 55–82.

Bilgen, B., & Günther, H. O. (2010). Integrated production and distribution planning in the fast moving consumer goods industry: A block planning application. *OR Spectrum, 32*(4), 927–955.

Brandenburg, M., & Rebs, T. (2015). Sustainable supply chain management: A modeling perspective. *Annals of Operations Research, 229*(1), 213–252.

Briskorn, D., Zeise, P., & Packowski, J. (2016). Quasi-fixed cyclic production schemes for multiple products with stochastic demand. *European Journal of Operational Research, 252*(1), 156–169.

Çatay, B. (2010). A new saving-based ant algorithm for the vehicle routing problem with simultaneous pickup and delivery. *Expert Systems with Applications, 37*(10), 6809–6817.

Cha, J. H., & Roh, M. I. (2010). Combined discrete event and discrete time simulation framework and its application to the block erection process in shipbuilding. *Advances in Engineering Software, 41*(4), 656–665.

Cha, J. H., Roh, M. I., & Lee, K. Y. (2010). Integrated simulation framework for the process planning of ships and offshore structures. *Robotics and Computer-Integrated Manufacturing, 26*(5), 430–453.

Chang, P. C., & Lin, Y. K. (2010). New challenges and opportunities in flexible and robust supply chain forecasting systems. *International Journal of Production Economics, 128*(2), 453–456.

Chen, H. K., Hsueh, C. F., & Chang, M. S. (2006). The real-time time-dependent vehicle routing problem. *Transportation Research Part E: Logistics and Transportation Review, 42*(5), 383–408.

Christopher, M. (2016). *Logistics & supply chain management*. Pearson, UK.

Çimen, M., & Kirkbride, C. (2017). Approximate Dynamic Programming algorithms for multidimensional flexible production-inventory problems. *International Journal of Production Research, 55*(7), 2034–2050.

Çimen, M., & Soysal, M. (2017). Time-dependent green vehicle routing problem with stochastic vehicle speeds: An Approximate Dynamic Programming algorithm. *Transportation Research Part D: Transport and Environment, 54*, 82–98.

Coelho, I. M., Munhoz, P. L. A., Ochi, L. S., Souza, M. J. F., Bentes, C., & Farias, R. (2016). An integrated CPU–GPU heuristic inspired on variable neighbourhood search for the single vehicle routing problem with deliveries and selective pickups. *International Journal of Production Research, 54*(4), 945–962.

Defra. (2007). *Guidelines to Defra's GHG conversion factors for company reporting*—Annexes updated June 2007. Technical Report, Department for Environment, Food and Rural Affairs.

Demir, E., Bektaş, T., & Laporte, G. (2011). A comparative analysis of several vehicle emission models for road freight transportation. *Transportation Research Part D: Transport and Environment, 16*(5), 347–357.

Dong, Y., Xu, K., & Evers, P. T. (2012). Transshipment incentive contracts in a multi-level supply chain. *European Journal of Operational Research, 223*(2), 430–440.

Ercin, A. E., Aldaya, M. M., & Hoekstra, A. Y. (2011). Corporate water footprint accounting and impact assessment: The case of the water footprint of a sugar-containing carbonated beverage. *Water Resources Management, 25*(2), 721–741.

Euchi, J., & Frifita, S. (2017). Hybrid metaheuristic to solve the 'one-to-many-to-one' problem: Case of distribution of soft-drink in Tunisia. *Management Decision, 55*(1), 136–155.

Ferreira, D., Clark, A. R., Almada-Lobo, B., & Morabito, R. (2012). Single-stage formulations for synchronised two-stage lot-sizing and scheduling in soft-drink production. *International Journal of Production Economics, 136*(2), 255–265.

Ferreira, D., Morabito, R., & Rangel, S. (2010). Relax and fix heuristics to solve one-stage one-machine lot-scheduling models for small-scale soft-drink plants. *Computers & Operations Research, 37*(4), 684–691.

Figliozzi, M. A. (2011). The impacts of congestion on time-definitive urban freight distribution networks CO_2 emission levels: Results from a case study in Portland, Oregon. *Transportation Research Part C: Emerging Technologies, 19*(5), 766–778.

Figliozzi, M. A. (2012). The time dependent vehicle routing problem with time windows: Benchmark problems, an efficient solution algorithm, and solution characteristics. *Transportation Research Part E: Logistics and Transportation Review, 48*(3), 616–636.

Franceschetti, A., Honhon, D., Van Woensel, T., Bektaş, T., & Laporte, G. (2013). The time-dependent pollution-routing problem. *Transportation Research Part B: Methodological, 56*, 265–293.

Gendreau, M., Ghiani, G., & Guerriero, E. (2015). Time-dependent routing problems: A review. *Computers & Operations Research, 64*, 189–197.

Hickman, J., Hassel, D., Joumard, R., Samaras, Z., & Sorenson, S. (1999). *Methodology for calculating transport emissions and energy consumption.* Technical Report, Transport Research Laboratory.

Iammarino, M., dell'Oro, D., Bortone, N., Mangiacotti, M., Damiano, R., & Chiaravalle, A. E. (2016). Radiostrontium levels in foodstuffs: 4-years control activity by Italian reference centre, as a contribution to risk assessment. *Food Chemistry, 210*, 344–354.

Ioannou, G. (2005). Streamlining the supply chain of the Hellenic sugar industry. *Journal of Food Engineering, 70*(3), 323–332.

Jabali, O., Van Woensel, T., & De Kok, A. G. (2012). Analysis of travel times and CO_2 emissions in time-dependent vehicle routing. *Production and Operations Management, 21*(6), 1060–1074.

Jordan, W. C., & Graves, S. C. (1995). Principles on the benefits of manufacturing process flexibility. *Management Science, 41*(4), 577–594.

Kok, A. L., Hans, E. W., & Schutten, J. M. (2012). Vehicle routing under time-dependent travel times: The impact of congestion avoidance. *Computers & Operations Research, 39*(5), 910–918.

Kuo, Y. (2010). Using simulated annealing to minimize fuel consumption for the time-dependent vehicle routing problem. *Computers & Industrial Engineering, 59*(1), 157–165.

Ligterink, N. E., Tavasszy, L. A., & de Lange, R. (2012). A velocity and payload dependent emission model for heavy-duty road freight transportation. *Transportation Research Part D: Transport and Environment, 17*(6), 487–491.

Lütke Entrup, M. (2005). *Advanced planning in fresh food industries: Integrating shelf life into production planning.* Springer Science & Business Media.

Luz, L. M., de Francisco, A. C., & Piekarski, C. M. (2015). Proposed model for assessing the contribution of the indicators obtained from the analysis of life-cycle inventory to the generation of industry innovation. *Journal of Cleaner Production, 96*, 339–348.

Malandraki, C., & Daskin, M. S. (1992). Time dependent vehicle routing problems: Formulations, properties and heuristic algorithms. *Transportation Science, 26*(3), 185–200.

Maldonado, M., Rangel, S., & Ferreira, D. (2014). A study of different subsequence elimination strategies for the soft-drink production planning. *Journal of Applied Research and Technology, 12*(4), 631–641.

Maravelias, C. T., & Sung, C. (2009). Integration of production planning and scheduling: Overview, challenges and opportunities. *Computers & Chemical Engineering, 33*(12), 1919–1930.

Meyr, H., & Mann, M. (2013). A decomposition approach for the general lot-sizing and scheduling problem for parallel production lines. *European Journal of Operational Research, 229*(3), 718–731.

Modak, N. M., Panda, S., & Sana, S. S. (2016). Two-echelon supply chain coordination among manufacturer and duopolies retailers with recycling facility. *International Journal of Advanced Manufacturing Technology, 87*(5–8), 1531–1546.

Mohamadghasemi, A., & Hadi-Vencheh, A. (2011). Determining the ordering policies of inventory items in class B using if-then rules base. *Expert Systems with Applications, 38*(4), 3891–3901.

Molina-Besch, K. (2016). Prioritization guidelines for green food packaging development. *British Food Journal, 118*(10), 2512–2533.

Nardi, P. C. C., da Silva, R. L. M., Ribeiro, E. M. S., & de Oliveira, S. (2017). Proposal for a methodology to monitor sustainability in the production of soft-drinks in ref PET. *Journal of Cleaner Production, 151*, 218–234.

Parragh, S. N., Doerner, K. F., & Hartl, R. F. (2008). A survey on pickup and delivery problems. *Journal für Betriebswirtschaft, 58*(1), 21–51.

Pureza, V., Morabito, R., & Reimann, M. (2012). Vehicle routing with multiple deliverymen: Modeling and heuristic approaches for the VRPTW. *European Journal of Operational Research, 218*(3), 636–647.

Ramanathan, U. (2012). Supply chain collaboration for improved forecast accuracy of promotional sales. *International Journal of Operations & Production Management, 32*(5–6), 676–695.

Ramanathan, U., & Muyldermans, L. (2010). Identifying demand factors for promotional planning and forecasting: A case of a soft-drink company in the UK. *International Journal of Production Economics, 128*(2), 538–545.

Ramanathan, U., & Muyldermans, L. (2011). Identifying the underlying structure of demand during promotions: A structural equation modelling approach. *Expert Systems with Applications, 38*(5), 5544–5552.

Santos, M. O., Massago, S., & Almada-Lobo, B. (2010). Infeasibility handling in genetic algorithm using nested domains for production planning. *Computers & Operations Research, 37*(6), 1113–1122.

Sel, C., & Bilgen, B. (2014). Hybrid simulation and MIP based heuristic algorithm for the production and distribution planning in the soft-drink industry. *Journal of Manufacturing Systems, 33*(3), 385–399.

Sel, Ç., & Bilgen, B. (2015). Quantitative models for supply chain management within dairy industry: A review and discussion. *European Journal of Industrial Engineering, 9*(5), 561–594.

Sel, Ç., Bilgen, B., & Bloemhof-Ruwaard, J. M. (2017). Planning and scheduling of the make-and-pack dairy production under lifetime uncertainty. *Applied Mathematical Modelling, 51*, 129–144.

Sel, C., Bilgen, B., Bloemhof-Ruwaard, J. M., & van der Vorst, J. (2015). Multi-bucket optimization for integrated planning and scheduling in the perishable dairy supply chain. *Computers & Chemical Engineering, 77*, 59–73.

Shahbazkhan, M. R., Shahriari, J. E., & Najafi, M. (2012). Identifying and evaluating effective factors on lean-agile supply chain. *Life Science Journal-Acta Zhengzhou University Overseas Edition, 9*(3), 1951–1961.

Silva, A. S., Medeiros, C. F., & Vieira, R. K. (2017). Cleaner production and PDCA cycle: Practical application for reducing the cans loss index in a beverage company. *Journal of Cleaner Production, 150*, 324–338.

Silvenius, F., Gronman, K., Katajajuuri, J. M., Soukka, R., Koivupuro, H. K., & Virtanen, Y. (2013). The role of household food waste in comparing environmental impacts of packaging alternatives. *Packaging Technology and Science, 27*(4), 277–292.

Simchi-Levi, D., & Wei, Y. (2015). Worst-case analysis of process flexibility designs. *Operations Research, 63*(1), 166–185.

Simon, B., Ben Amor, M., & Foldenyi, R. (2016). Life cycle impact assessment of beverage packaging systems: Focus on the collection of post-consumer bottles. *Journal of Cleaner Production, 112*, 238–248.

Soysal, M. (2016). Closed-loop inventory routing problem for returnable transport items. *Transportation Research Part D-Transport and Environment, 48*, 31–45.

Soysal, M., Bloemhof-Ruwaard, J. M., & Bektaş, T. (2015). The time-dependent two-echelon capacitated vehicle routing problem with environmental considerations. *International Journal of Production Economics, 164*, 366–378.

Soysal, M., Bloemhof-Ruwaard, J. M., Haijema, R., & van der Vorst, J. G. (2015). Modeling an inventory routing problem for perishable products with environmental considerations and demand uncertainty. *International Journal of Production Economics, 164*, 118–133.

Soysal, M., Bloemhof-Ruwaard, J. M., Meuwissen, M. P., & van der Vorst, J. G. (2012). A review on quantitative models for sustainable food logistics management. *International Journal on Food System Dynamics, 3*(2), 136–155.

Soysal, M., Bloemhof-Ruwaard, J. M., & Van der Vorst, J. G. A. J. (2014). Modelling food logistics networks with emission considerations: The case of an international beef supply chain. *International Journal of Production Economics, 152*, 57–70.

Soysal, M., & Çimen, M. (2017). A simulation based restricted dynamic programming approach for the green time dependent vehicle routing problem. *Computers & Operations Research, 88*, 297–305.

Temme, E. H. M., Toxopeus, I. B., Kramer, G. F. H., Brosens, M. C. C., Drijvers, J. M. M., Tyszler, M., et al. (2014). Greenhouse gas emission of diets in the Netherlands and associations with food, energy and macronutrient intakes. *Public Health Nutrition, 18*(13), 2433–2445.

Toledo, C. F. M., de Oliveira, L., Pereira, R. D., Franca, P. M., & Morabito, R. (2014). A genetic algorithm/mathematical programming approach to solve a two-level soft-drink production problem. *Computers & Operations Research, 48*, 40–52.

Toledo, C. F. M., Kimms, A., Franca, P. M., & Morabito, R. (2015). *The synchronized and integrated two-level lot-sizing and scheduling problem: Evaluating the generalized mathematical model.* Mathematical Problems in Engineering, 1. Research Article. Hindawi Publishing Corporation.

Villarreal, B., Sañudo, M., Duran, B., & Avila, L. (2009). A lean approach to vehicle routing. In *IIE Annual Conference Proceedings*, Institute of Industrial and Systems Engineers (IISE), 1096.

Vlontzos, G., & Pardalos, P. M. (2017). Data mining and optimisation issues in the food industry. *International Journal of Sustainable Agricultural Management and Informatics, 3*(1), 44–64.

Wan, X., & Dresner, M. E. (2015). Closing the loop: An empirical analysis of the dynamic decisions affecting product variety. *Decision Sciences, 46*(6), 1141–1164.

Wan, X., & Sanders, N. R. (2017). The negative impact of product variety: Forecast bias, inventory levels, and the role of vertical integration. *International Journal of Production Economics, 186*, 123–131.

Zeimpekis, V., Bloemhof-Ruwaard, J. M., & Bourlakis, M. (2014). Reverse logistics in food supply chains. In *Markets, business, and sustainability.* Bentham Science Publishers.

Zhou, Y. M., & Wan, X. (2017a). Product variety and vertical integration. *Strategic Management Journal, 38*(5), 1134–1150.

Zhou, Y. M., & Wan, X. (2017b). Product variety, sourcing complexity, and the bottleneck of coordination. *Strategic Management Journal, 38*(8), 1569–1587.

The Role of Informal and Semi-Formal Waste Recycling Activities in a Reverse Logistics Model of Alternative Food Networks

Luis Kluwe de Aguiar and Louise Manning

1 Introduction

Informal recycling is framed by the physical nature of the city e.g. population size, the socio-economic conditions, local urban policy, and available waste collection and management resources (Medina, 2000; Sembiring & Nitivattananon, 2010). The types of materials collected by waste pickers will vary from location to location but can include: aluminium and beverage cans, batteries, boxes, cardboard, construction and demolition waste, electronic and electrical items, fabric, glass, household items, newspaper, books and magazine waste, organic material, plastic and polythene, scrap metals, rubber including tyres, and wood (Nzeadibe & Anyadike, 2012; Suthar, Rayal, & Ahada, 2016). Labour intensive waste picking can in

L. K. de Aguiar
Harper Adams University, Newport, Shropshire, UK
e-mail: ldeaguiar@harper-adams.ac.uk

L. Manning (✉)
Royal Agricultural University, Stroud Road, Cirencester, Gloucestershire, UK
e-mail: Louise.Manning@rau.ac.uk

E. Aktas, M. Bourlakis (eds.), *Food Supply Chains in Cities*,
https://doi.org/10.1007/978-3-030-34065-0_5

some cities achieve high rates of recycling and reuse, increase the useful life of landfills, and provide low cost materials for local enterprises through the informal waste recycling process i.e. the human capital involved creates financial and physical capital, public infrastructure and social capital (Navarrete-Hernandez & Navarrete-Hernandez, 2018; Sembiring & Nitivattananon, 2010). Therefore, informal waste sectors on the one hand increase recycling rates and reduce environmental impact, and as a result capture economic value, whilst conversely being associated with socio-economic exclusion, child labour, poverty, and exploitation (Aparcana, 2017; Medina, 2000; Wilson, Velis, & Cheeseman, 2006).

Waste picking is also associated with "risk, unhygienic environments, criminal activities, homelessness, unemployment, poverty, and backwardness" (Sembiring & Nitivattananon, 2010, p. 802). Indeed, in their work in Mongolia, Uddin and Gutberlet (2018) highlight the precarious nature of the activity where two-thirds of the waste pickers they studied suffered from social marginalisation and experienced occupational health hazards including bone fractures, burns, cuts, stomach diseases, skin diseases, kidney and liver problems, and back pain. Waste pickers also have increased exposure to chemical hazards, traffic accidents, and economic exploitation (Hartmann, 2018); medical waste and needles, polluted water, or pests such as rodents or insects carrying disease (Sasaki, Araki, Tambunan, & Prasadja, 2014).

Multiple studies define the hierarchical nature of such informal/formal waste management systems. Typical structures include waste collectors, waste pickers, waste dealers, small stores and itinerant merchants, and then the more formal scrap traders and waste processing plants (Nandy et al., 2015). In the Philippines, the network is composed of waste pickers, waste collectors, itinerant waste buyers, junkshop dealers, and collection crews (Paul, Arce-Jaque, Ravena, & Villamor, 2012). Paul et al. (2012) argue that cooperative associations are possible for greater control at the waste picker level. In Hanoi, there is a complex network of waste collectors including city-based waste pickers, waste site pickers, and junk buyers working with intermediaries such as receivers, waste site depot operators, and roadside depot operators through to dealers (DiGregorio, 1997; Mitchell, 2008). A hierarchy of actors in the informal waste collection system is detailed in Fig. 1.

There are multiple barriers for individuals who wish to participate in urban waste management. Firstly, governance factors such as

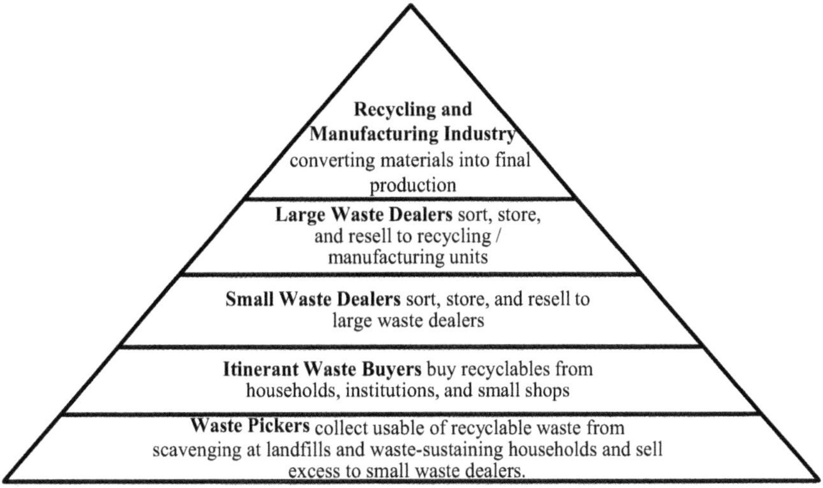

Fig. 1 Informal waste sector hierarchy. (Source: Adapted from Masood & Barlow, 2013; Sandhu, Burton, & Dedekorkut-Howes, 2017)

institutional, regulatory or legislative barriers; secondly, socio-economic or political barriers (including lack of advocacy, political power, limited access to assets, fear of violence, insecurity); thirdly, human capital barriers (personal confidence, education, knowledge, skills, or health); and finally natural and physical barriers i.e. lack of access to infrastructure, facilities, and transport (Nzeadibe & Anyadike, 2012). The formalisation and capital investment required in commercial organisations coming into the process to address MSWM systems threatens waste pickers' livelihoods, often refuting their role in valorising waste (Hartmann, 2018; Sandhu et al., 2017). Indeed Sandhu et al. (2017) note that the work of Samson (2009) shows that privatisation of MSWM in Cairo and Delhi reduced recycling rates due to less discrete separation of types of waste i.e. a lack of physical picking. Therefore, flexible formal/informal integration programmes for MSWM should focus on providing governance frameworks such as: legalising informal activities through waste picker co-operatives and public-private partnerships; developing knowledge and skills, creating social support programmes, providing advice and guidance and awarding contracts for collection, improving technical and management practices, and developing secondary material markets

(Ezeah, Fazakerley, & Roberts, 2013). In summary, cities can be considered as spaces of increasing consumption leading in the developing world to larger quantities of waste that in turn support informal waste collection activities (Mitchell, 2008). Thus, the process of informal urban waste collection is motivated by socio-economic pressures where there is demand for the recovered material and also a labour force that recognises that waste picking activities support their livelihoods (DiGregorio, 1994; Mitchell, 2008).

The aim of this research is to reflect on the practices developed in Brazil to address MSWM, particularly food waste management, as a case study to frame the social dilemma of simultaneous profligacy by some leading to municipal waste whilst others are in dire need and resort to picking through waste dumps for their livelihoods. Further, two business case studies are used to then develop a reverse logistics model to create alternative food networks that are of value in addressing both environmental pollution and also improving food security for poor urban communities.

2 Brazil as a Country of Study

Until the late nineteenth century, in Brazil, rubbish collection including night soil, transportation, and disposing of urban waste in the city dumps was a task exclusively of slave labour. In coastal urban areas, the approach was to dump rubbish near the seashore in marshes and mangroves as part of land reclamation, a practice that was in use until 1970. Usually urban waste collection in Brazil is by unskilled rural migrants looking for better work opportunities in the cities, many being illicit workers limiting employment opportunities because of the lack of skills and identification documentation. In Brazil today, most food waste collected from residences, factories, and commercial facilities is sent to landfill sites. Brazil is the 4th largest producer of food waste in the world, but despite this 1.6% of its population, some 3.6 million people, still experience daily hunger and the effects of malnutrition (Global Public Policy Network, 2017). Current legislation on solid wastes, their treatment, and destination is enacted in Brazil by the Decree 7404/2010 and comes into force through Law 12305/2010 that has evolved from previous legislation of 2000 and 2007, and environmental guidelines established by the Environmental Council such as CONAMA 385/2005 and CONAMA1/1998 (JusBrasil, 2010). The 2010 legislation establishes a policy for waste collection, treatment, and recycling, and sets out a need for companies and municipalities to develop a plan for recycling waste.

A more detailed analysis of solutions for the challenges of urban MSWM is hampered in Brazil by its continental dimensions. This means that distinct geographic areas present a variety of stages of economic development and associated cities reflect their regional differences, in terms of local culture and habits, which are, in turn, a result of different population dynamics including purchasing power and income. In low and middle-income countries, urban MSWM is mediated by environmental and socioeconomic issues, the systems used, and the development of informal waste management solutions (Aparcana, 2017). Waste collection is much less regulated in terms of what is sent to the city's landfill (Eigenheer, 2009). Cardboard, paper, metal, plastic, and sometimes wood is gathered around recycling centres or drop–off points in public areas and is segregated by sorting, then is baled, and sold. Yet, despite the subhuman conditions, the widespread presence of "legal" and illegal land sites in the urban periphery creates a work opportunity for large numbers of adults and children. They eke out a living by scavenging food to placate their immediate hunger as well as collecting any valuable material such as metal, paper, glass, wood, and plastic that they can exchange for cash. In the city centres, the homeless use improvised trolleys to search for discarded waste from offices, households, and industry. Food waste collected from residences, factories, and commercial facilities is then sent to landfill sites at the end of daily rounds because it has no value.

As a result of the Brazilian government's recognising rubbish picking as a profession, in recent years there has been a move towards the formalisation of the relationships in the sector between informal recyclers and the municipality rubbish collection service who take their materials to recycling centres or drop–off points in public areas. In these centres, different waste is sorted and, if possible, sold as a resource to be used by others. In 1985, in some neighbourhoods in Rio de Janeiro, the practice of household waste separation was introduced. Later, in 1988, the practice was adopted by the municipality of Curitiba (Eigenheer, 2009) and this has now progressed nationwide. Typical of less developed countries, the practice of dumping urban waste at open air landfills in the periphery of urban areas in Brazil could be justified by the lower relative price of the land utilised when it is compared to the cost of treating urban waste. With growing environmental concerns, the practice of incinerating rubbish for power generation became the established solution with mixed results due to air pollution from methane and the capital cost of incineration as a waste management option in Brazil (Leme et al., 2014; Oliveira & Rosa, 2003). So, can reverse logistics food supply models provide a solution?

2.1 Alternative Food Networks (AFNs)

Quite simply alternative food networks (AFNs) are alternative ways of provisioning food to what is standard and accepted practise (Maye & Kirwan, 2010). AFNs are defined by the socio-spatial proximity between source and consumers where the logistics channels employed often have community characteristics. AFNs are focused on sustainable food systems or in other words are structures that reconfigure, re-spatialise and re-socialise the systems of production, distribution, and consumption of food (Jarosz, 2008; Paül & McKenzie, 2013). In developed countries, AFNs are associated with higher welfare, higher quality, or local foods. The term AFNs is considered here in the context of urban and business MSWM as an alternative to existing food supply systems and practices that are in place. Indeed, Whatmore, Stassart, and Renting (2003) assert that AFNs underpin new market, state, and civic practices and objectives as can be demonstrated in the following example. AFNs associated with municipal solid waste are also developing in multiple urban areas in the developed world. In the United States (US) and Canada "dumpster diving" is the term generally used for obtaining items, in this case, food for consumption, from waste bins or dumpsters especially by low-income individuals to supplement their diet (Eikenberry & Smith, 2005; Miewald & McCann, 2014). Divers operate this survival strategy both as individuals and as groups (Rombach & Bitsch, 2015) often living outside of capital-based food systems (Mourad, 2016), either because there is no other option or alternatively as a lifestyle choice sometimes called freeganism (Nguyen, Chen, & Mukherjee, 2014). Thus, the AFNs associated with solid municipal waste identified in the developed world are also gaining ground in developing economies perhaps for different reasons. Whilst ensuring food and nutrition security is one driver for waste picking, alternatively the freegan movement is seen as a protest towards anti-consumption and excesses.

Sandhu et al. (2017) introduces the notion of an alternative informal business network that operates around MSWM from waste pickers to intermediary waste dealers, who then through other actors provide material for urban recycling units. This implies not only that the waste pickers live a precarious life exchanging waste for cash, but also the intermediaries base their business model on uncertain conditions as they could be evicted from their informal waste collection centres, by the local authorities or private site owners, at any time. Whilst the introduction of formal private sector solutions for MSWM has its advantages, the implementation of this approach can leave the informal waste sector threatened with displacement and livelihood loss (Sandhu et al., 2017). However, the authors argue that there are integration possibilities for the informal waste sector with more formal organisations and systems. This would involve waste pickers being involved in primary door-to-door collection so that the livelihoods based on recycling continue. They conclude: "Unless a viable policy is framed wherein the government can provide a feasible alternative to informal waste sector and waste picking in particular, no attempt should be made to fracture the informal waste sector." (Sandhu et al., 2017, p. 555). Thus, more should be done to develop inclusive reverse AFNs that utilise the social enterprise of waste pickers to drive effective waste management. So, what AFNs could develop in Brazil around waste picking?

When food waste is selectively separated and then collected it is classed as organic matter whose destination can be in the form of food redistribution, landfills, animal feed, or composting. Food waste produced from bars and restaurants is classified as commercial waste in Brazil and therefore it is the responsibility of the local municipality to collect, transport, and dispose of it. However, as food waste could be contaminated and thus, likely to cause harm to whoever ingests it, the current legislation does not encourage food waste redistribution as a recycling activity (GTZ, 2010). In recent years, direct contracts between informal sector organisations and local governments in Brazil have been established as part of a social programme to act as a catalyst for social development (GTZ, 2010). There are many examples of waste collectors organised under formal relationships with municipal governments. Waste recycling cooperatives have been set up to empower the waste collectors and enable them to relate to a legal and strong network of multiple stakeholders where municipal recycling partnership schemes has allowed for the regulation of contracts and legalisation of labour rights. Two examples of AFNs for food waste are introduced here, one focusing on food supply and the other on alternative fuel.

2.2 Reverse Logistics of Using Food Waste in Animal Feed

In the state of Rio Grande do Sul, (Fig. 2) rural migrants tend to move to the capital city searching for a better life. However, migrants who have generally little training and skills end up settling in the periphery of the urban area. Some decide to carry on what they know or what they do best, which is work in some agricultural related activity. Traditionally, migrants either produce fresh produce or engage in all kinds of livestock production, in this case, specifically, pig production. Nevertheless, the peri-urban farmers, unable to afford to buy processed animal feeds, tend to collect food waste from restaurants and food outlets around the capital as well as from established city dumps. In the process, usually the "farmsteads" become a small version of a waste dump, but more alarmingly they have little consideration for the quality of the collected waste food which is given to the animals (Waissman, 2002). Furthermore, there are also implications regarding the safety and quality of the meat produced especially if it is consumed in AFNs with limited governance and control.

In 1992, the Municipal Waste Collection Service (DMLU) of Porto Alegre decided to create a project: Reuse of Food Waste via Pig Production. Its aim was twofold: firstly, to reduce the quantity of food waste that went

Fig. 2 Map of Rio Grande do Sul. (Source: Google Maps, 2019)

directly into the city dumps (Thomas, 2013), and secondly, to reduce the number of clandestine waste dumps in peri-urban areas, and, finally, to improve the livestock's overall production conditions (Waissman, 2002). Feeding pigs with human food leftovers is an ancient practice. However, the practice has been discouraged since the outbreaks of Foot and Mouth disease (DEFRA, 2011) and other swine diseases such as Classic and African Swine Flu that can be transmitted via untreated meat in swill and have a devastating effect on commercial pig herds; hence it is banned in many countries (Dou, Toth, & Westendorf, 2017). Yet, according to the Organisation of International Epizooties (OIE) if swill is treated with a heat source in a sealed container at 30°C for 30 minutes it would suffice to the kill the Foot and Mouth virus (OIE, 2011).

In some countries, segregated food waste is recycled as feed, for example, Japan (35.9%) and South Korea (42.5%) (Salemdeeb, Zu Ermgassen, Kim, Balmford, & Al-Tabbaa, 2017). In 2010, due to the introduction of a more comprehensive waste management legislation in Brazil, the scheme became an important initiative regarding the recycling of solid organic matter waste.

The scheme came to fruition between 1995 and 1997 but for it to work, a pig producers' association with thirteen pig farmers was created. Such an institutional arrangement was needed as the association had to be responsible for heat-treating the food solids to be distributed amongst its members. As an incentive, the members of the association received technical support regarding animal husbandry, management as well as periodical animal health checks and screening for diseases.

Daily segregated collection of food waste is the responsibility of the DMLU, which initially collected food waste from thirteen different sources. In 1997, DMLU collected 2.7 tonnes of food waste daily, reaching 7.5 tonnes per day by 2002. By that time, the AFN included 38 collection points such as clubs, schools, businesses, hospitals, markets, and prisons. By 2013, there were already 73 restaurants, 51 companies, and 22 hospitals contributing to the AFN developed to collect and transport food waste for thermic treatment (Juffo, 2013; Thomas, 2013). Some 11 tonnes of food waste from meals was being collected in Porto Alegre (Juffo, 2013) which included DMLU's food solid waste from DMLU's own canteen. The municipal waste works is responsible for the collection and transportation of food waste to the treatment plant. The remaining twelve farmers could feed a considerably larger, herd, about 1200 pigs which is evidence of the economic benefit generated to small farmers. As a result of the initiative, the farmers who traditionally struggled to purchase commercial animal feed had access to a practically free feed source (DMLU, 2018; Juffo, 2013). Moreover, the farmers no longer acted within an informal economic system bereft of governance in theory and became fully integrated in a formal one where standards could be introduced thus having a direct impact on the safety associated with their agri-food system practices.

The pork obtained went to 22 community kitchens and a food bank. The pig producers' association, because of it becoming a formalised business entity and benefitting from an increased volume of meat produced, could then participate in the tenders for supplying 70 municipal schools (Thomas, 2013) as part of the local authorities' school meals programmes. This demonstrates the opportunity for AFNs with embedded circularity.

According to Waissman (2002), the cost of production to raise one pig with food waste up to 100Kg in 16 weeks was R$140.27 (approximately £29.03) whilst it would cost R$153.15 (approximately £31.25) for the same animal to reach 100Kg using a commercial feed. Yet, Juffo (2013) studied the quality of food waste used as pig feed. In Juffo's study food waste was collected from 14 restaurants within a shopping centre's food court in Porto Alegre. The results indicated that 54.5% of the waste collected had been incorrectly segregated. Consequently, it could have posed a risk to the health of the animals and by default to those consuming the meat. Typically, the food waste was contaminated with fruit skins such as pineapple and foreign bodies, to mention some, plastic cutlery and toothpicks, which can damage the gut of the animals. From all the samples collected and analysed for the purpose of nutrient content and presence of microorganisms, the food wasted was deemed clean from Salmonella and E. coli. As for the nutrient content, as part of Juffo's experiment, extra commercial feed was given to the pigs because of a slower growth rate. This shows that giving swill exclusively to pigs would take longer to fatten them up to 100Kg which is the standard weight of a pig to be sent to slaughter, compared to a commercial pig herd.

Since the beginning, the collection of food waste and delivery of treated material to farmers, this was part of a dedicated service by DMLU to bring food waste solids to small farmers. DMLU did not charge the pig producers' association any fee, but required the association to donate monthly a food basket to two nurseries/crèches located in Restinga, which is a very low-income neighbourhood (DMLU, 2018). The model in Fig. 3 shows at the top the interaction between different actors in the chain that has been described in this chapter with the waste pickers themselves acting as the reverse logistics process. The bottom reverse flow chain shows the governance bodies (yellow); the stages where value is added (green) and the overall reduction in environmental impact of food

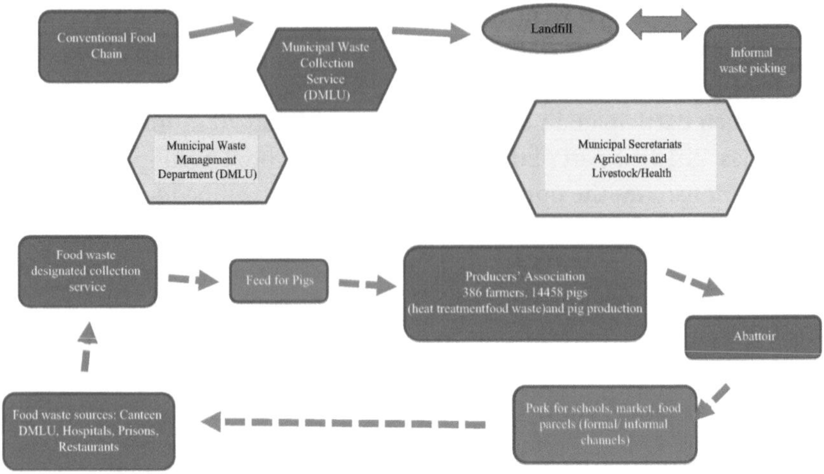

Fig. 3 AFN Model for Municipal Waste Collection Service (DMLU) of Porto Alegre via pig production. (Source: Authors' own elaboration)

waste. In the food waste reverse logistics, the environmental, economic and societal benefits are perhaps more indirect and felt in the long term. However, such systems need to be both resilient and economically sustainable. We now consider the second case study.

2.3 Reverse Logistics of Using Used Cooking Oil (Kato Oil and Life Oil)

Incorrect disposing of used cooking oil is of great concern regarding waste management. It can cause many environmental problems if it is poured into the drainage system, in domestic septic tanks, watercourses, and on soil. As a result, cooking oil will pollute aboveground and underground water sources and it will take a long time for them to recover to their natural state (DEFRA, 2001).

The problem of disposal of used cooking oil is ubiquitous thus affecting many countries. For many years in Brazil, saturated oils (used cooking oil) from bars and restaurants were dumped directly into the city's sewage system causing many problems with drainage during the rainy season as well as polluting the local water resources. At household level, it was common practice to throw old cooking oil down the drainpipe with similar detrimental consequences to the drainage system and local rivers and soil quality (da Silva César, Werderits, de Oliveira Saraiva, & da Silva Guabiroba, 2017). Household oil also ended up in landfill causing harmful environmental effects by contaminating the local water table. Household waste oil, after appropriate processing, can be used as an input to biodiesel production and this gave the incentive to develop a further reverse-logistics model for mitigating food waste.

In Brazil, a federal programme was introduced for the commercialisation of combining a new taxation model and social inclusion policies (da Silva César et al., 2017). The regulation of biodiesel is based on Law No. 11.097/2005 that became mandatory in Brazil in 2008. The development of a biodiesel sector is established by the National Program for the Production and Use of Biodiesel (called as PNPB—Programa Nacional de Produçao e Uso de Biodiesel) see da Silva Guabiroba, da Silva, da Silva César, and da Silva (2017). In Brazil, several Civil Society Organisations of Public Interest (CSOPIs) and some non-governmental organisations (NGOs) promoted oil collection and AFNs with businesses who have waste used oil, collection points, processors and companies that use the recycled oil for other activities (Ruiz, Oliveira, Struffaldi, Gabriel, & Bocatto, 2017). It is estimated that only 2.5% of the waste cooking oil produced in Brazil is recycled (da Silva César et al., 2017), offering significant opportunity to develop AFNs associated with recycling cooking oil.

Katu Oil is in Campo Grande, the capital city of the state of Mato Grosso (Fig. 4). Since used cooking oil has practically no commercial value, it is a low-cost raw material for further processing. Katu Oil as an "eco-innovator" local private company identified an opportunity in the market to collect and treat used cooking oil for the purpose of recycling it, hence avoiding it becoming a pollutant. Katu Oil decided to collect oil from bars and restaurants to feed an anaerobic digester that in turn generated electricity for the Brazilian national grid. To avoid door-to-door collection that could be costly and labour intensive as well as likely not to be efficient, a partnership with a regional food retailer, served as a centre for

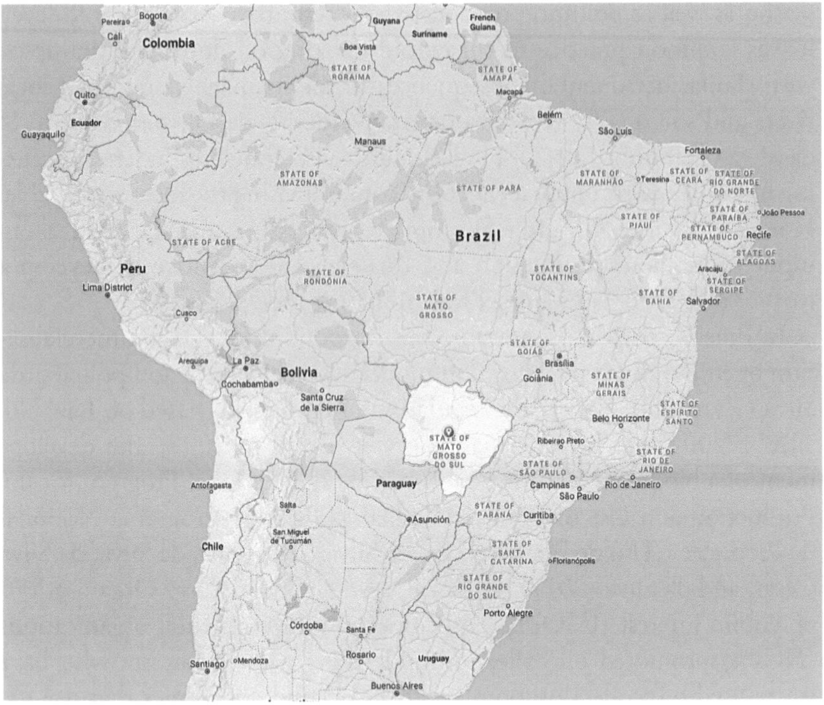

Fig. 4 Map of Brazil—Campo Grande. (Source: Google Maps, 2019)

the collection of used kitchen oil that was shoppers brought in when they visited the supermarket.

In 2010, the local government launched and initiative to collect used cooking oil: RECOL. At that time, two collection points were set up: the Municipal Market and the Central Fresh Produce Fair. That made sense since there were many fast food restaurants located in those areas. Nevertheless, after a while the initiative almost failed after the waning of the early awareness campaign as well as the limited number of collection points: only two (Bonifácio, 2017).

At the start in 2010, as an incentive to the public to avoid the used oil being dumped, the local government paid R$0.30/litre for the used oil. Bonifácio (2017) reports that the price paid is about R$0.50 (about £0.10/litre in 2017). In the meantime, Katu Oil developed a partnership with the local water and sewage works company because of shared interests. The water and sewage works company not only distributes water, collects and treats sewage, but also produces bottled mineral water from a spring. Therefore, preserving their aquifer from potential contamination was in their interest too. Both Katu Oil and the water company work together to educate school children about the importance of not polluting the environment; especially not disposing of used cooking oil down the drain.

As an eco-innovator, Katu Oil also recycles all its rainwater to be used in its process, therefore adding to their environmental claims. The initiative has also reached areas outside the capital city. Owners of restaurants have started to bring their oil to be either exchanged for cash or, as some prefer, in exchange for household cleaning products. In the beginning, all collected oil was put into a bio-digester to produce power to be fed to the national grid. On a weekly basis, about 20 thousand litres are now sent to another state, about 1200 km away, to be processed as an ingredient of animal feed. It has been estimated that Campo Grande has the capability to produce 60 tonnes of used cooking oil on a monthly basis. Yet, in 2017, about half of that quantity was actually being collected and recycled (Katu Oil, 2018). Katu Oil now collects used cooking oil from restaurants, schools and households of which 95% are commercial suppliers (restaurants and fast food outlets).

Life Oil is another eco-innovator company based in Campo Grande. They also collect used cooking oil and recycle it. However, differently from Katu Oil, the used cooking oil is processed into detergents or biodiesel. Feitosa (2014) reported that Life Oil's strategy was to collect used cooking oil from schools and residences. Taking advantage of the initial environmental awareness campaigns carried out in 2010 as well as education programmes with schools, the Managing Director of Life Oil believed the general public still lacked a deeper and embedded recycling culture. The public awareness initiatives need to be repeated to raise the general population concern for the problem despite the small incentive to recycle. Despite this, Life Oil has some 300 collection points for collections that takes place fortnightly or monthly depending on the need of the business or the households. In exchange for the used cooking oil it gives out toiletries, washing up liquids, and disinfectants. In the beginning, some local supermarket chains agreed to support the initiative and act as an eco-point generator. As time went on, the supermarket chain stated that the cost to maintain an employee to look after the process was too high and removed their involvement.

A shift in some consumer's practices might have also altered the way some households understood the source of used cooking oil as an opportunity. Feitosa (2014) stated that some households had reverted to some old practices of making their own soaps from the oil instead of exchanging it for some cash or cleaning products. That was more typical of residents of poorer neighbourhoods who still held onto old habits. Eco initiatives, particularly recycling, struggle to keep afloat. The cost to maintain a system (Fig. 5) that collects, transports, filters, and recycles

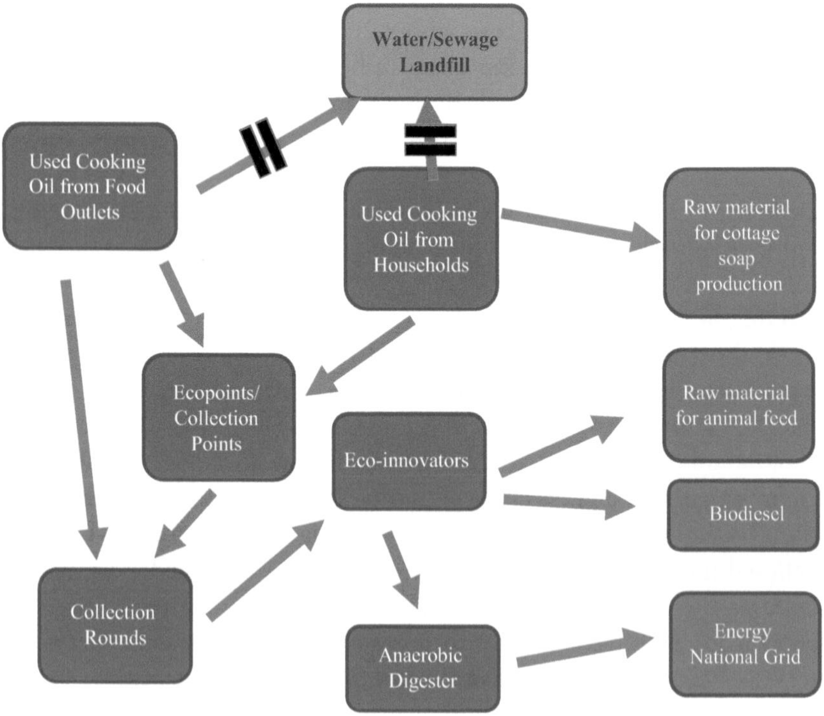

Fig. 5 AFN Model for waste oil collection and reuse. (Source: Authors' own elaboration)

used cooking oil works based on the oil being transformed into a product with some economic value such as biodiesel, or an ingredient for animal feed. In the case of Life Oil, the equivalent revenue from the sale of biodiesel is not enough to cover all its costs (Feitosa, 2014). The whole process is also labour-intensive. The oil drums need to be replaced often and the used cooking oil also requires pre-treatment as there is the need to filter all solids and water from the oil. Afterwards, the drums need to be labelled, cleaned and dropped back at the eco-point collection sites.

The summarised model (Fig. 5) seeks to contextualise the used cooking oil recycling initiative in Campo Grande. There are stages (green) that demonstrate the direct value added in the system. There are other indirect values in terms of reduced pollution and improved social capital, which are not shown in the figure. The viability of the process still depends on exogenous interaction by local government, with financial initiatives and promotional campaigns. In the process of recycling oil, value is created or re-created. However, the long-term valorisation in terms of environmental, social, and commercial value requires further investigation. The fact that this cooking oil can be transported for 1200 km and be used as an ingredient in animal feed and still at a competitive cost is remarkable. The long-term avoidance of the associated pollution of water resources impacts positively on the growing and thirsty urban population in the west of Brazil, but full econometric studies are required to consider this further.

The local authority, despite launching RECOL in 2010 has not expanded the collecting points, further creating an opportunity for eco-innovators to intervene. However, the eco-innovators also struggle to keep the initiative going. Thus, there is a need for more awareness on the part of the general public and the need for the local society to be more engaged as the lack of engagement is a barrier to driving pro-environmental behaviour. da Silva César et al. (2017, p. 251) suggest an integrated approach that includes: "*the creation of expanded awareness campaigns and encouraging the population; the creation of local/state laws that require legal entities and/or municipalities to collect the waste; and the establishment of cooperatives for the collection and plants for the transformation of waste into biodiesel.*"

3 Closing Thoughts

In the examples of reverse logistics networks shown here, the economics underpin the role of informal and semi-formal waste recycling activities in creating economic value. In a traditional food waste reverse logistics network, materials such as metal, cardboard/paper, glass, and plastic might derive a clear market value and thus, value creation can be demonstrated. The financial return might be more immediate for materials such as metal, for example, as metal has a clear market value. Social and environmental value creation is harder to quantify and depends on the timescale of value creation. da Silva Guabiroba et al. (2017) sought to apply value chain analysis to waste oil recycling in Brazil but only considered the economic aspect of sustainability. However, some studies have tried to broaden the analysis. Rutkowski and Rutkowski (2015) considered the value of moving waste pickers from landfill collection into co-operatives collecting door to door. Their data showed economic and environmental benefits in the net increase of tonnes of material recycled from 140 to 280 tonnes a month, but also social benefits of improved working conditions, occupational health and safety, and reduced child labour.

In the DMLU pork meat case study, livestock farmers who initially were not part of the formal economy, by adhering to the scheme became part of the formal economy. The association, by being a registered entity could then allow these farmers access to the municipal food tendering network (Personal communication: Thomas, 2018). It is clear from these cases the integration and the role of informal and semi-formal waste recycling activities in a reverse logistics network. The catalyst for enabling food waste recycling has been an initiative through the local government fostering a network of activities connecting supply and demand. Its positive impact was twofold: firstly it helped resolve the problem of disposing of hospital and commercial food waste as well as secondly, creating conditions for peri-urban living that are not connected to the formal economy that are also more responsible for animal welfare and the quality of the food that is produced. However, in the case of the pig producers' association, without local government support, the economics would be against the initiative and the subsequent social benefit would not have been realised.

Since 2013, the hospitals that supplied a large share of solid food waste have ceased to take part due to the risk of human to animal to human contamination. As a result, all food wasted from hospitals is being sent to the city dump. In some cases, the hospitals are setting up 'forced-air composting systems' to speed up the solid decomposing process (Personal communication: Thomas, 2018). This form of AFN is nothing new. Twenty years ago, it was highlighted that Havana's urban farms provided the population with 8500 tons of agricultural produce, 4 million dozens of flowers, 7.5 million eggs, and 3650 tons of meat (Companioni, 1996; Altieri et al., 1999). Linking this socio-economic benefit for urban communities with reducing the impact of urban solid waste has multiple benefits and provides opportunity for the development of sustainable food supply in urban locations.

The two cases on recycling used cooking oil considered in this chapter, despite the clear governmental intervention as a catalyst for enabling such an initiative, add a different dimension to AFNs especially with regard to the eco-innovation character of the enterprises. In the two cases described, the economic gains could be felt more immediately therefore levering the businesses' viability and multi-stakeholder engagement. The collecting, processing and selling of the used cooking oil take up the character of a tradable commodity whose monetary gain and return on initial investment is quicker. Conversely, the pork case has a longer economic cycle where payback is delayed. The efficiency of the feed conversion rate should also be considered as it is not advantageous for the pigs to take longer to finish. However, as soon as the oil is filtered and clean, it can be sold for animal feed or alternatively, used in a bio-digester to generate power. Moreover, both the used cooking oil cases and the pork one do not depend on the need for more sophisticated technology. Therefore, the examples can be replicated in many situations world-wide.

Reducing the environmental and social impact of MSWM has multiple benefits and provides economic opportunity for some of the poorest communities in the world. Why is this important in Brazil and other emerging economies? Well, it is estimated that in Brazil's large cities, in excess of half a million people survive by waste picking, reducing the waste in landfill, leading to cleaner cities and providing an economic benefit to the city and a livelihood for themselves (Fergutz, Dias, & Mitlin, 2011). These degrading conditions are ubiquitous in many other countries in Latin America, Africa and Asia. Some estimates suggest that there are 2 million waste pickers worldwide and between one and five percent of urban employment is associated with formal and informal solid waste management (Hoornweg & Bhada-Tata, 2012). Thus, the relevancy of the findings here will be of value in many cities around the developing world. A shift to formal systems of MSWM without consideration of the implications for poor communities eking a livelihood from informal processes is a danger to these marginal groups and so policy and governance structures need to recognise and mitigate any potential vulnerability that could occur. Food cultures do vary at both the regional and national level, so opportunities for reverse logistics approaches that support the nutritional security of the poor may be situational, but the case studies described here potentially have generalisability to other urban locations. In a world where the human population is rising and relocating to urban environments, how solid waste is managed and resources maximised and where possible recycled or reused in the city, where space can often be a premium, is of importance. As a result, as has been highlighted in this chapter, combining informal and formal MSWM networks has great merit to deliver in the longer term, to improving local environmental conditions by reducing land, water and air pollution whilst in the short to medium term economic return can be realised at local government and organisational level. However, most importantly this approach will improve the working conditions and nutrient security of some of the poorest and most vulnerable communities in the world.

References

Altieri, M. A., Companioni, N., Cañizares, K., Murphy, C., Rosset, P., Bourque, M., et al. (1999). The greening of the "barrios": Urban agriculture for food security in Cuba. *Agriculture and Human Values, 16*(2), 131–140.

Aparcana, S. (2017). Approaches to formalization of the informal waste sector into municipal solid waste management systems in low-and middle-income countries: Review of barriers and success factors. *Waste Management, 61*, 593–607.

Bonifácio, V. (2017, July). *Empresa compra óleo de cozinha usado para exportar.* [on-line]. Retrieved from http://www.diariodigital.com.br/geral/empresa-da-capital-compra-oleo-de-cozinha-para-exportar/160270/

Companioni, N. (1996). El Huerto Intensivo en la Agricultura Urbana de Cuba. In *Seminario Taller Regional "La Agricultura Urbana y el Desarrollo Rural Sostenible"* (pp. 39–48). FIDA-CIARA-MINGAG.

da Silva César, A., Werderits, D. E., de Oliveira Saraiva, G. L., & da Silva Guabiroba, R. C. (2017). The potential of waste cooking oil as supply for the Brazilian biodiesel chain. *Renewable and Sustainable Energy Reviews, 72*, 246–253.

da Silva Guabiroba, R. C., da Silva, R. M., da Silva César, A., & da Silva, M. A. V. (2017). Value chain analysis of waste cooking oil for biodiesel production: Study case of one oil collection company in Rio de Janeiro-Brazil. *Journal of Cleaner Production, 142*, 3928–3937.

DEFRA. (2001). *Guidance note for the Control of Pollution (Oil Storage) (England)* Regulations 2001. [on-line]. Retrieved from https://assets.publishing.service.gov.uk/government/uploads/system/uploads/attachment_data/file/69255/pb5765-oil-storage-011101.pdf

DEFRA. (2011). Supplying and using animal by-products as farm animal feed. [on-line]. Retrieved from https://www.gov.uk/guidance/supplying-and-using-animal-by-products-as-farm-animal-feed#abps-you-cant-use

DiGregorio, M. (1994). *Urban Harvest: Recycling as a Peasant Industry in Northern Vietnam.* Hawaii, East–West Center. pp. 1–12.

DiGregorio, M. (1997). *City and countryside in the Red River Delta: Notes on Hanoi's recycling industry* (pp. 2–17). Hawaii: East–West Center.

DMLU. (2018). *Suinocultura.* [on-line]. Retrieved from http://www2.portoalegre.rs.gov.br/dmlu/default.php?p_secao=115

Dou, Z., Toth, J. D., & Westendorf, M. L. (2017). Food waste for livestock feeding: Feasibility, safety, and sustainability implications. *Global Food Security, 17*, 154–161.

Eigenheer, E. M. (2009). *Lixo, limpeza urbana atraves dos tempos.* Pallotti. [on-line]. Retrieved from www.lixoeeducacao.uerj.br/images/pdf/ahistoriadolixo.pdf

Eikenberry, N., & Smith, C. (2005). Attitudes, beliefs, and prevalence of dumpster diving as a means to obtain food by Midwestern, low-income, urban dwellers. *Agriculture and Human Values, 22*(2), 187–202.

Ezeah, C., Fazakerley, J. A., & Roberts, C. L. (2013). Emerging trends in informal sector recycling in developing and transition countries. *Waste Management, 33*(11), 2509–2519.

Feitosa, L. (2014). *Coleta de óleo para reciclagem ainda se resume ações isoladas.* [on-line]. Retrieved from http://www.campograndenews.com.br/meio-ambiente/coleta-de-oleo-para-reciclagem-ainda-se-resume-a-acoes-isoladas

Fergutz, O., Dias, S., & Mitlin, D. (2011). Developing urban waste management in Brazil with waste picker organizations. *Environment and Urbanization, 23*(2), 597–608.

Global Public Policy Network. (2017). *Reducing food waste in Brazil: A Brazilian-Italian Legislative cooperation initiative.*

GTZ. (2010). *The waste experts: Enabling conditions for informal sector integration in solid waste management – Lessons learned from Brazil, Egypt and India.* [on-line]. Retrieved from www.giz.de/en/downloads/gtz2010-waste-experts-conditions-is integration.pdf

Hartmann, C. (2018). Waste picker livelihoods and inclusive neoliberal municipal solid waste management policies: The case of the La Chureca garbage dump site in Managua, Nicaragua. *Waste Management, 71*, 565–577.

Hoornweg, D., & Bhada-Tata, P. (2012). *What a Waste: A global review of Solid Waste Management.* Urban development series knowledge papers. March 2012. No 15. Washington, DC: World Bank.

Jarosz, L. (2008). The city in the country: Growing alternative food networks in Metropolitan areas. *Journal of Rural Studies, 24*(3), 231–244.

Juffo, E. L. D. (2013). *Residuos solidos organicos: da geracao em estabelecimentod de producao de alimentos em um shopping a destinacao final na alimentaçao de suinos.* MSc dissertation. PPGVet. UFRGS. [on-line]. Retrieved from http://www.lume.ufrgs.br/handle/10183/72041

JusBrasil. (2010). *Política Nacional de Residuos Solidos – Lei 12305/10.* [on-line]. Retrieved from https://presrepublica.jusbrasil.com.br/legislacao/1024358/politica-nacional-de-residuos-solidos-lei-12305-10 (Lei n° 12.305 – clique para ver).

Katu Oil. (2018). *Reciclagem.* [on-line]. Retrieved from http://www.katu-oil.com.br/

Leme, M. M. V., Rocha, M. H., Lora, E. E. S., Venturini, O. J., Lopes, B. M., & Ferreira, C. H. (2014). Techno-economic analysis and environmental impact assessment of energy recovery from Municipal Solid Waste (MSW) in Brazil. *Resources, Conservation and Recycling, 87*, 8–20.

Masood, M., & Barlow, C. Y. (2013). Framework for integration of informal waste management sector with the formal sector in Pakistan. *Waste Management & Research, 31*(10_suppl), 93–105.

Maye, D., & Kirwan, J. (2010). Alternative food networks. *Sociology of Agriculture and Food, 20*, 383–389.

Medina, M. (2000). Scavenger cooperatives in Asia and Latin America. *Resources Conservation Recycling, 31*, 51–89.

Miewald, C., & McCann, E. (2014). Foodscapes and the geographies of poverty: Sustenance, strategy, and politics in an urban neighborhood. *Antipode, 46*(2), 537–556.

Mitchell, C. L. (2008). Altered landscapes, altered livelihoods: The shifting experience of informal waste collecting during Hanoi's urban transition. *Geoforum, 39*(6), 2019–2029.

Mourad, M. (2016). Recycling, recovering and preventing "food waste": Competing solutions for food systems sustainability in the United States and France. *Journal of Cleaner Production, 126*, 461–477.

Nandy, B., Sharma, G., Garg, S., Kumari, S., George, T., Sunanda, Y., et al. (2015). Recovery of consumer waste in India – A mass flow analysis for paper, plastic and glass and the contribution of households and the informal sector. *Resources, Conservation and Recycling, 101*, 167–181.

Navarrete-Hernandez, P., & Navarrete-Hernandez, N. (2018). Unleashing Waste-Pickers' Potential: Supporting Recycling Cooperatives in Santiago de Chile. *World Development, 101*, 293–310.

Nguyen, H. P., Chen, S., & Mukherjee, S. (2014). Reverse stigma in the Freegan community. *Journal of Business Research, 67*(9), 1877–1884.

Nzeadibe, T. C., & Anyadike, R. N. (2012). Social participation in city governance and urban livelihoods: Constraints to the informal recycling economy in Aba, Nigeria. *City, Culture and Society, 3*(4), 313–325.

OIE. (2011). *Terrestrial Animal Health Code. Chapter 8.5 Foot and Mouth Disease.* [on-line]. Retrieved from http://www.oie.int/eng/A_FMD2012/docs/en_chapitre_1.8.5.pdf

Oliveira, L. B., & Rosa, L. P. (2003). Brazilian waste potential: Energy, environmental, social and economic benefits. *Energy Policy, 31*(14), 1481–1491.

Paul, J. G., Arce-Jaque, J., Ravena, N., & Villamor, S. P. (2012). Integration of the informal sector into municipal solid waste management in the Philippines – What does it need? *Waste Management, 32*(11), 2018–2028.

Paül, V., & McKenzie, F. H. (2013). Peri-urban farmland conservation and development of alternative food networks: Insights from a case-study area in metropolitan Barcelona (Catalonia, Spain). *Land Use Policy, 30*(1), 94–105.

Rombach, M., & Bitsch, V. (2015). Food movements in Germany: Slow food, food sharing, and dumpster diving. *International Food and Agribusiness Management Review, 18*(3), 1.

Ruiz, M. S., Oliveira, R. B. D., Struffaldi, A., Gabriel, M. L. D. D. S., & Bocatto, E. (2017). Cooking oil waste recycling experiences worldwide: A preliminary analysis of the emerging networks in the São Paulo Metropolitan Region, Brazil. *International Journal of Energy Technology and Policy, 13*(3), 189–206.

Rutkowski, J. E., & Rutkowski, E. W. (2015). Expanding worldwide urban solid waste recycling: The Brazilian social technology in waste pickers inclusion. *Waste Management & Research, 33*(12), 1084–1093.

Salemdeeb, R., Zu Ermgassen, E. K., Kim, M. H., Balmford, A., & Al-Tabbaa, A. (2017). Environmental and health impacts of using food waste as animal feed: A comparative analysis of food waste management options. *Journal of Cleaner Production, 140*, 871–880.

Samson, M. (2009). *Refusing to be cast aside: Waste pickers organising around the world*. Cambridge, MA: Women in Informal Employment: Globalizing and Organizing (WIEGO).

Sandhu, K., Burton, P., & Dedekorkut-Howes, A. (2017). Between hype and veracity; privatization of municipal solid waste management and its impacts on the informal waste sector. *Waste Management, 59*, 545–556.

Sasaki, S., Araki, T., Tambunan, A. H., & Prasadja, H. (2014). Household income, living and working conditions of dumpsite waste pickers in Bantar Gebang: Toward integrated waste management in Indonesia. *Resources, Conservation and Recycling, 89*, 11–21.

Sembiring, E., & Nitivattananon, V. (2010). Sustainable solid waste management toward an inclusive society: Integration of the informal sector. *Resources, Conservation and Recycling, 54*(11), 802–809.

Suthar, S., Rayal, P., & Ahada, C. P. (2016). Role of different stakeholders in trading of reusable/recyclable urban solid waste materials: A case study. *Sustainable Cities and Society, 22*, 104–115.

Thomas, A. (2013). *Soluçao conjunta para reaproveitamento de residuos organicos na alimentaçao de suinos*. Hospital de Clinicas de Porto Alegre. Retrieved from http://www.hospitaissaudaveis.org/arquivos/SHS%202013_Ana%20Lucia%20Thomas1.pdf

Uddin, S. M. N., & Gutberlet, J. (2018). Livelihoods and health status of informal recyclers in Mongolia. *Resources, Conservation and Recycling, 134*, 1–9.

Waissman, M. (2002). *Estudo da viabilidade economica do reaproveitamento de residuos organicos via suinocultura*. MSc Dissertation. PPGA UFRGS. [online]. Retrieved from http://www.lume.ufrgs.br/handle/10183/6908
Whatmore, S., Stassart, P., & Renting, H. (2003). What's alternative about alternative food networks? *Environment and Planning A, 2003*(35), 389–391.
Wilson, D., Velis, C., & Cheeseman, C. (2006). Role of informal sector recycling in waste management in developing countries. *Habitat International, 30*(4), 797–808.

Shortening the Supply Chain for Local Organic Food in Chinese Cities

Pingyang Liu and Neil Ravenscroft

1 Introduction

Since the 1990s, China has shown an increasing interest in transforming agricultural production away from its reliance on agri-chemicals, towards a new 'green' or sustainable approach which is part of what has been termed 'ecological citizenship' (Geall & Ely, 2015; Parr & Henry, 2016; Shi, 2002; UNEP, 2016; Weng, Dong, Wu, & Qin, 2015). In challenging not only Chinese economic policy, but also the capitalist imperative itself, ecological civilization has offered China the opportunity to re-evaluate its own ecological traditions and knowledge (Liu, Ravenscroft, Harder, & Dai, 2016; Wang, He, & Fan, 2014). Agriculture is central to this, not

P. Liu
Department of Environmental Science and Engineering, Fudan University, Shanghai, P.R. China

N. Ravenscroft (✉)
School of Real Estate and Land Management, Royal Agricultural University, Gloucestershire, UK
e-mail: Neil.Ravenscroft@rau.ac.uk

E. Aktas, M. Bourlakis (eds.), *Food Supply Chains in Cities*,
https://doi.org/10.1007/978-3-030-34065-0_6

only in reflecting how far China has moved from its traditional agrarian foundations (Liu et al., 2016; Liu & Ravenscroft, 2017), but also in offering the possibility of a new way forward, towards what might be termed a 'new ecological productivism' that operationalises the concept of 'Chinese Ecological Agriculture' (Lu, Ma, Zhang, Fu, & Gao, 2012; Shi, 2002), particularly in urban settings (Zhang, Zheng, & Wan, 2005).

At the core of this new approach to farming is a commitment to environmentally-benign low input or organic agriculture with short, green supply chains between producers and consumers (Chin, Tat, & Sulaimana, 2015; Geng, Mansouri, & Aktas, 2017; Geng, Mansouri, Aktas, & Yen, 2017; Krul & Ho, 2017). By improving the quality of their food and the trust that consumers can have in it, new ecological producers seek to improve the local environment by creating market conditions that support ecological civilization (Yan, 2012). However, few such farmers have yet achieved the full potential of this, with many of these farms on the margins of viability, constrained by the availability of land suitable to their needs, limited often to family and elderly labour, and characterised by immature and fragile routes to markets that are tainted by repeated food safety scandals (Ding, Liu, & Ravenscroft, 2018; EU SME Centre, 2015; Hao, Chen, Huang, & Liao, 2004; Sanders, 2006; Shi, Cheng, Lei, Wen, & Merrifield, 2011; Zhang et al., 2005). While this has led some commentators to argue for state intervention to support market development (see Sanders, 2006), in part by including food production within city planning regimes (Morgan, 2009), it has also created the conditions for new marketing and supply chain strategies that overcome both the physical and the social distance between ecological farmers and their customers (Ding et al., 2018; Jarosz, 2000; Shi et al., 2011; Thompson & Coskuner-Balli, 2007).

Central to this emerging opportunity is the growth in certified organic agriculture, with China now being second only to Australia in terms of the area subject to certification (Paull, 2007). The growth in organic farming has largely been a market-based phenomenon driven by a rapidly expanding urban middle-class seeking food that is both prestigious and safe to consume (Krul & Ho, 2017). In common with the West (McIver & Hale, 2015; Poulsen, 2017), this has created a market opportunity for small-scale ecological farmers to locate in urban and peri-urban locations where they are close to their customers, allowing them to combine traditional approaches to food production with—for China—new internet-based approaches to marketing and distributing the food. In many ways this is a reworking of Western approaches to Alternative Food Networks (AFNs), but very much with Chinese characteristics (Si, Schumilas, & Scott, 2015). Small scale organic farming in Chinese cities has therefore largely developed in parallel with the politics of ecological civilization, although there can be little doubt that they share the same developmental drivers associated with addressing environmental degradation. While most AFNs in China remain consumer-driven (Si et al., 2015), those who operate within them have been able to address the lack of trust that many Chinese people have with domestic food supply networks (Chen, 2015; Wang, Yu, Si, Scott, & Nam Ng, 2015), allowing them to improve the economic benefits that they receive from their farming (Li, Liu, & Min, 2011), largely by reducing supply chains to direct producer/consumer relationships that also promote circularity in terms of opening up new on-farm leisure and agri-tourism opportunities (Brinkley, 2017; Krul and Ho, 2007).

However, while benefitting from direct access to their market, these small-scale producers not only have to compete with large scale producers of organic produce, but they also face a range of institutional barriers such as the cost of organic certification (Jin, Li, & Li, 2017). In addressing these constraints, many of the more established enterprises have moved away from external verification of their produce, instead building up high levels of service that underpin customary *Guanxi* traditions of mutual trust with a network of customers (see Geng, Mansouri, Aktas, & Yen, 2017). In this way they are developing an element of reciprocity into their food supply chains, in encouraging customers to see their relationship with the producers as more than a mere economic exchange: that in return for high quality food delivered at their convenience from a known producer, the customers are making a personal and social, as well as financial, commitment which underpins the welfare of all those concerned in the production, distribution, and consumption of the food (Dedeurwaerdere et al., 2017; Forssell & Lankoski, 2015). While addressing the established relocalisation, or reterritorialisation, discourses about the value of 'food from here' (McMichael, 2009; Schermer, 2015;

Thompson & Coskuner-Balli, 2007), this also recognises that supply chains are often more than a unidirectional flow of goods (Brinkley, 2017; Forssell & Lankoski, 2015). Of course, this remains an asymmetrical relationship in which the majority of uncertainty and risk reside with the producers, but it does offer the possibility of building supply chains that are shorter and more sustainable in social and cultural, as well as financial, terms.

Our aim, in this chapter, therefore is to examine the components of these new AFNs with Chinese characteristics to determine how local food supply chains work in immature market settings and how far concepts such as reciprocity can be built into supply chain models as a way of boosting the sustainability of these supply chains. In so doing we will seek to examine how these supply chains differ from the more established versions in the West, and what lessons can be learned to improve both the circularity and the sustainability of small-scale supply chains that link ecological farmers with their markets. The chapter starts with a review of current literature on green supply chains within alternative food networks, then presents a case study of Shared Harvest Farm in Beijing (Fig. 1), which has established a new bespoke on-line direct sales platform involving 600 local customers, most of whom live in the city of Beijing, some 50km from the farm. Following this, the chapter concludes by offering insights into the development of green supply chains that support the growth of urban ecological agriculture.

2 Green Supply Chains for Urban Agricultural Produce

According to McMichael (2009), the passage of agriculture products between farmers and consumers is now entering its third phase, or regime. Originally, produce was consumed close to where it was grown, usually with limited intermediation via local produce markets. The second food regime was characterised by the opposite tendency, with food produced at a distance, geographically and socially, from consumers, often with a range of intermediaries processing, packing, and delivering food, as well

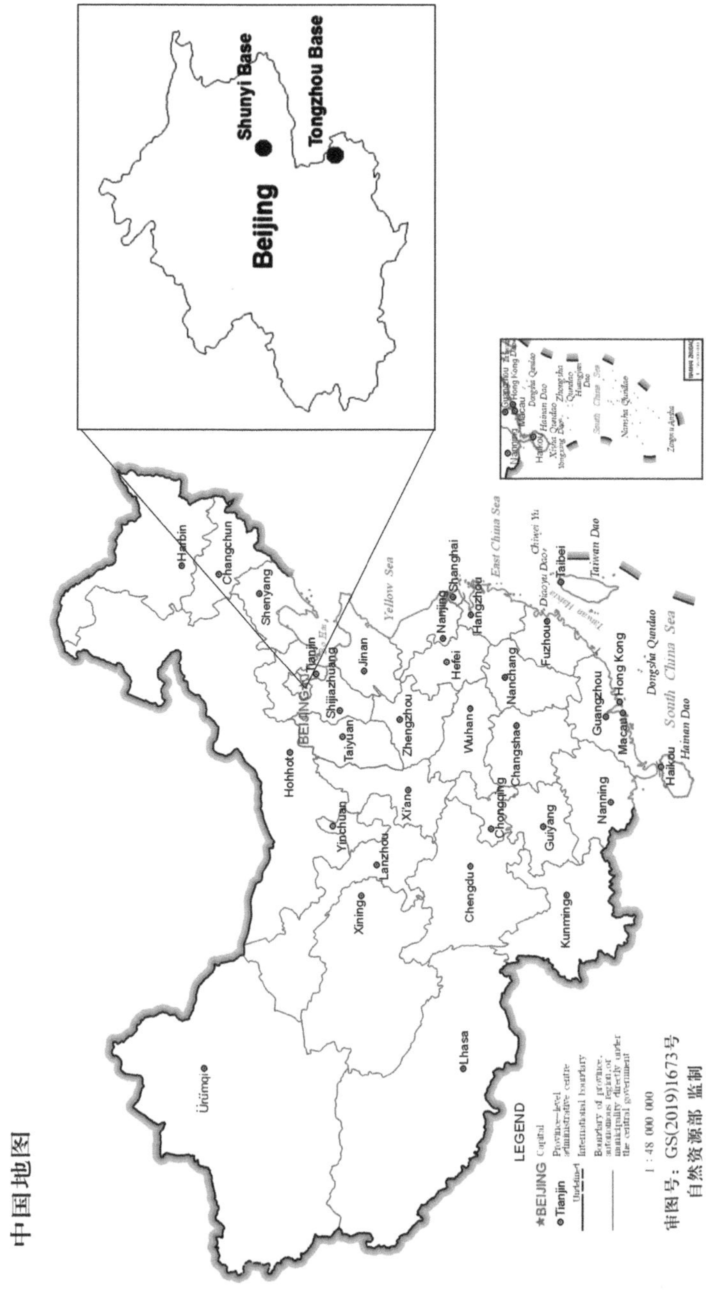

Fig. 1 Location of Shared Harvest Farm (Shunyi Base and Tongzhou base). (Source: Drawn by authors)

as eventually retailing it. This is now beginning to change again, as we enter what may well be the third food regime, in which there is once again a concentration on reducing the cultural and geographic distance between farmers and consumers—although this remains in the context of what McMichael (2009, p. 147) has termed '… the continuous transnationalization of hi-tech (biotechnology) agroindustries.' For Schermer (2015), this reterritorialising shift away from 'food from nowhere' to 'food from here' is characterised by local food cultures served by short food supply chains (Thompson & Coskuner-Balli, 2007). As in the first food regime, many of these chains are so short that there is little intermediation, with either a dyadic relationship between the farmer and the consumer, or a triadic one including market traders. In China, as Geng, Mansouri, Aktas, and Yen (2017) observe, these relationships are often formed around relational Guanxi between the actors that underpins mutual respect and trust, which is critical in addressing concerns over food safety. Indeed, as Ding et al. (2018) have found in Shanghai, even local markets are often distrusted by consumers, meaning that there is considerable pressure to develop strong dyadic farmer-consumer relationships. As Jin et al. (2017) have argued, such short supply chains are critical to those Chinese consumers who are concerned about the safety of their food, because there is a high degree of traceability, particularly where farms are selling only their own produce. It is also apparent that these short supply chains are about more than the transfer of food from producer to consumer: they are fundamental to fostering cultural change through social learning about sustainability (Dedeurwaerdere et al., 2017).

While it would seem that short supply chains that cut out intermediaries are ideal for maximising value for both producers and consumers (Kottila & Rönni, 2008; Flores & Villalobos, 2018), there are many warnings about the difficulties faced in making a success of this approach, particularly in achieving the 'last mile' segment (Fancello, Paddeu, & Fadda, 2017; Saskia, Mareï, & Blanquart, 2015). The recent widening of access to the internet in China, the emergence of e-commerce, and the presence of an established low-cost delivery infrastructure have provided new supply chain opportunities for small organic farms in many cities, including Shanghai (Ding et al., 2018). Indeed, Ding et al. (2018) report that these factors have provided a suitable environment for new farmers to enter the organic food market, with the number of such enterprises in Shanghai growing since 2010. Almost all of these new farms depend on direct sales, in many cases based initially on the social networks of the farmers, such as colleagues, ex-classmates, and business friends. These farmers understand the need to create and maintain a strong customer base, so spend lots of time promoting their produce via all kinds of social media, especially widely used platforms such as Wechat and Weibo (Fig. 2), where they can attract new customers.

While all home delivery schemes face a range of issues including uneven wholesale supply and relatively high costs associated with storing and assembling produce prior to delivery, the timeliness required for fresh food deliveries adds further complications (Jin et al., 2017; Kottila & Rönni, 2008; Zissis, Aktas, & Bourlakis, 2017). Yet there is little doubt that many consumers value home delivery, not least because it saves them time and energy (particularly when they do not own a private car or cannot park close to their home), as well as allowing them to order their shopping at their convenience, especially where there is an internet order platform (Durand & Gonzalez-Feliu, 2012; Saskia et al., 2015). In addition, for increasing numbers of people who are interested in organic food

Fig. 2 Weibo (L) and WeChat (R) of Shared Harvest Farm online platform. (Source: Mobile phone screen shot of the official Weibo site and the online wechat store of Shared Harvest, by authors)

as part of a lifestyle commitment to reducing their impact on the environment, home deliveries—including buying clubs—are associated with reducing the use of private transport (Si et al., 2015; Vieira, De Barcellos, Hoppe, & da Silva, 2013). Increasing interest in green logistics indicates that similar concerns are held by those responsible for supply chains (Aktas, Bloemhof, Fransoo, Gunther, & Jammernegg, 2017; Chin et al., 2015).

In addressing these market conditions, a range of new organisational forms have been developed under the broad umbrella term of 'Alternative Food Networks' (AFNs) (Forssell & Lankoski, 2015) and, more specifically, Community Supported Agriculture (CSA) schemes (Bloemmen, Bobulescu, Tuyen, & Vitari, 2015; Krul & Ho, 2017; Ravenscroft, Moore, Welch, & Church, 2012; Ravenscroft, Moore, Welch, & Hanney, 2013; Thompson & Coskuner-Balli, 2007). At their core, these approaches incorporate '... localized, and hence re-territorialized, market relationships [that] ... constitute a form of ethical consumerism' (Thompson & Coskuner-Balli, 2007, p. 217). While foregrounding the supply of fresh food, these producer/consumer relationships also emphasise attributes such as the direct sale of locally-grown food, predominantly small ecological farms and community engagement (Ding et al., 2018). Of course, positioning themselves as 'alternative' means that the status and the identity of these organisational forms are connected to the continuing dominance of the globalised and corporate food industry, with membership based on '... organising a protected space for learning and experimentation with lifestyle changes for sustainable food consumption and production practices' (Dedeurwaerdere et al., 2017, p. 123). From this perspective, the distribution of food within most AFNs continues to be market mediated, but in a more localised and co-operative context than is the case with the conventional food supply chain (Thompson & Coskuner-Balli, 2007, p. 277).

One of the ways in which many AFNs and CSAs differ from conventional supply chains is by creating a limited range of composite products, often known as food boxes or fresh food portfolios in which producers pre-select 'packages' or 'bundles' of goods, depending on what is in-season and available on the farm:

> These portfolios of fresh agricultural products … usually comprise a combination of fresh produce to meet the demands of a family with three to five members for one week. Consumers order the combination of fresh produce online and the firm delivers the fresh produce to the consumer's door in specific delivery areas. Fresh Produce Portfolios (FPPs) provide a good solution to the problems encountered during fresh produce e-commerce circulation because the provision of bulk deliveries reduces the unit delivery cost and serving specific delivery areas (usually within a city) means that fresh produce e-commerce firms can ensure the freshness of the produce they deliver. (Jin et al., 2017, p. 818)

In moving beyond a simple, arm's length transaction, many of those who use fresh food portfolios or other home delivery schemes seek to supplement the financial transaction with a social relationship that creates commonality and reciprocity between producer and consumer (Jarosz, 2000). This is highly important to farmers in retaining a stable customer base in which the consumers feel that they have a personal vested interest in the success of the farm and its delivery system. Indeed, as Brinkley (2017) reports, for many farms the supply of food is only one of a range of services that farms offer, with other services often based on provision of leisure and tourism activities. Thus, effectively, home delivery and other forms of direct marketing producers sell a locally-embedded social connection between the farmer and the consumer (Hinrichs, 2000, p. 299). Yet, certainly in China, it is not yet clear whether those promoting fresh food portfolios and short supply chains are really moving beyond neo-liberal logic, with attempts to shorten the social as well as geographic distance between producers and consumers often being as much a commercial as a social practice:

> Despite some evidence of these elements of alternativeness, our fieldwork shows that the degree of their alternativeness is open to question. As many of the CSA farms in China are founded by market-oriented entrepreneurs, operating within rather than beyond the neoliberal market logics, it is hard for them to escape the circle of profit-motivated commodity production. Some of the elements of alternativeness may thus be subdued in order to cater to consumer needs. For example, although "eating seasonally" has been widely praised by CSA farmers, we still observed an online debate on

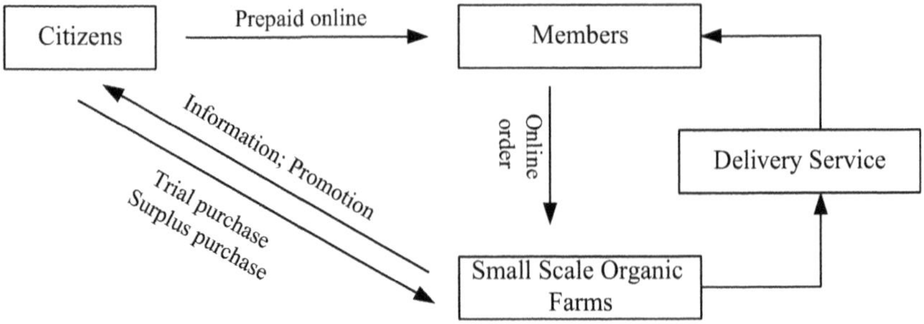

Fig. 3 Supply chains for fresh produce in Shanghai. (Source: Drawn by authors)

> microblogs between some CSA farms on whether it was appropriate to grow vegetables in greenhouses, thereby violating the principle of "eating seasonally". (Si et al., 2015, p. 305)

Based on research by Ding et al. (2018), Fig. 3 describes the typical supply chains for fresh organic produce in Shanghai. Even though most of the farms in the survey claim to use a CSA model, it is apparent that they mainly operate in a virtual online community—a social media group—based on dyadic, one-to-one, connections, rather than the usual CSA model of one-to-all connections. However, even where membership schemes have been established, they remain fragile due to a residual lack of trust, even where food is certified organic. This is largely because most of the members/customers do not know the farmers well and have insufficient knowledge to be able to tell the difference between organic and non-organic produce. Thus, members/customers are easily affected by negative episodes, such as stale vegetables, rumours about the origin of the produce, or unsatisfied expectations. Many customers also remain price sensitive, even when they know that organic produce needs to command a premium over conventional foods. Consequently, the direct sales customer base varies a lot for most farms.

In addition to these types of concerns, a strong argument remains that while appearing to be more community-oriented and, thus, democratic, than conventional approaches, many green short supply chains continue to mask discrepancies in relative power between those involved. In particular, as Si et al. (2015) have argued, such value chains are mainly characterised by relatively wealthy middle-class producers and consumers who have the cultural capital to communicate effectively, whether in person or via the Internet, meaning that those involved in producing the food are often marginalised or excluded (see also Ding et al., 2018). Equally, those who do not have access to fast Internet connections, or who live in parts of cities that are isolated from the areas served by food deliveries, also tend to be excluded (Ilbery, Courtney, Kirwan, & Maye, 2010; Saskia et al., 2015).

Thus, while there is some evidence in the fresh food sector that short supply chains can promote forms of loyalty and reciprocity in which those who purchase the food have a connection with those who produce it, there remain a number of gaps in knowledge that need to be addressed. Some of these are technical, in developing better web-based platforms for exchanging information about the produce as well as allowing customers to exercise choice in what they purchase (Ding et al., 2018). Others are logistical, in developing effective systems for convenient, cost-effective, and environmentally-friendly 'last mile' delivery of fresh produce. There are also cultural barriers to be addressed, in terms of ensuring that all of those involved in the supply chain have a stake in its success, understood as the equitable distribution of value, in terms of produce, finance, and labour (Shi et al., 2011).

In starting to address these gaps, Si et al. (2015, p. 305) have highlighted the initiative being undertaken at Shared Harvest Farm in Beijing, where a group of 'food activists' have set out to ' ... experiment with value redistribution through the model of working with, rather than hiring as labour, small peasants, and "sharing more harvest" with them.' This approach, described by Yu (2015) as 'trailblazing', has involved redesigning most elements of the fresh food supply chain so that it takes on a genuinely communicative form in which there is collaboration in all stages of the growing, marketing, distribution, and consumption cycle. Using a single case study of Shared Harvest Farm, the next section of the chapter assesses the extent to which the gaps in knowledge set out above have been addressed, and what potential this offers for the viability of other urban ecological farms, in China and beyond.

3 Research Design and Methods

The data for this chapter have been generated from a single case study of Shared Harvest Farm in Beijing. As Flyvbjerg (2006) and Seawright and Gerring (2008) have argued, single case approaches are suitable where the generation of evidence is focussed on observation and deduction and a means of initial theory development. While there is debate about the nature and generalisability of this evidence (Bennett & Elman, 2006), Flyvbjerg (2006, pp. 227–228) argues strongly that single cases are particularly well suited to environments in which the parameters of knowledge are unclear or at an early stage of development. This approach does not dismiss the value of large-scale data-driven studies, but suggests that single cases can be a good way of identifying and framing issues and problems. This is very much the purpose of this chapter, which seeks to address the multiple barriers facing small scale urban ecological farms and, through the case of Shared Harvest farm, generate a body of evidence that begins to identify the current boundary of knowledge and practice in China. In addition, its owner, Shi Yan, is well-known and highly influential within both the Chinese and the global Community Supported Agriculture movement, making access to comprehensive information possible. The case study is based on observation from site visits over a two-year period (2015–2017), with special attention paid to the information below:

- The changing scales of membership, including the numbers and types of people joining and leaving the customer base;
- The types and areas of land obtained and farmed, the number of employees and workers, the range of outputs, and the management of production;
- The approaches taken to improving the production capacities of the farms, such as the diversity of vegetables, and also the development of new service activities (for example, education and lodging), new products and new technology application;
- Emerging problems and their solutions, such as imbalanced membership demand and farm supply, the issues associated with managing

non-production costs, and the risk of declining consumer-producer trust.

Five semi-structured interviews—conducted by one of the co-authors of this paper—were undertaken between Nov. 7th and 12th, 2016. For Shi Yan, the farm owner, the questions were mainly focused on: the history of the farm (cross-checked with the available information from Internet); how she has developed the operational processes on the farm; her approach to managing the farmers and other staff; and her plan for the Farm future. Shi Yan's husband, Cheng Cunwang, is also the developer of the Good Farm app. He was interviewed mainly about the efficiency and cost of the farm operations, as well the development, application and promotion of the app. More detailed information on the farm management and operation were obtained and cross checked with 2 department managers, one from the membership department (Manager M), and the other from the education department (Manager E). These interviews were structured according to the development of their departments, the past and current problems that they have encountered, and their views on the farm. The final interview was with the farmer in charge of the vegetable production. The questions were mainly about his cooperation with Shi Yan, and how he managed to shift from conventional farming to organic farming. Information of the development of the farm, growing production capacities, monitoring and guarantee of organic farming were cross checked with other interviews and information from the Internet.

In addition, published and web-based information on the farm, as well as information from social media platforms such as Wechat and Weibo, was cross-checked with the primary data to build up a comprehensive study of the farm (Fig. 4). All materials collated in the case study were recorded and, where necessary, transcribed into Mandarin text form. These texts were then coded in Mandarin, with the main themes drawn out of the data in an open coding approach involving 'deep reading' of the transcripts (Basit, 2003). The findings were translated into English, as were the quotes used in the text.

While this approach to generating data provided a rich textual source from which to draw out observations and tentative arguments, it must be

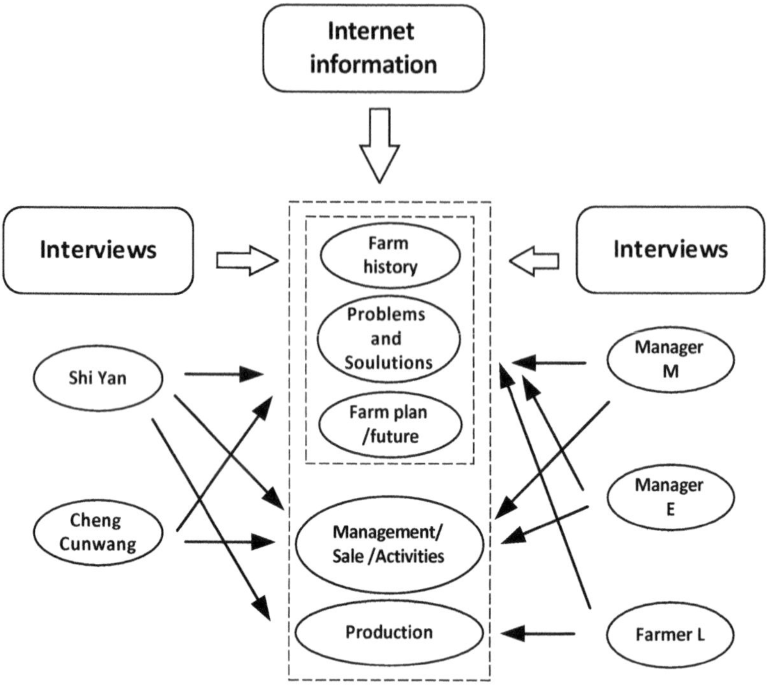

Fig. 4 Sources of information. (Source: Drawn by authors)

recognised that it did not generate a similar body of quantitative data relating to sales, distribution, profitability and so on. Much of this information undoubtedly exists, but it is commercially sensitive and any attempt to obtain or use it would have entailed so much generalisation and anonymisation of the qualitative as well as the quantitative data that it well have rendered the case study meaningless. As a result, the case study is necessarily limited to less sensitive observations, but equally to observations and conclusions that can inform others in the sector, if not precisely through generalisability then certainly though comparison and association of ideas.

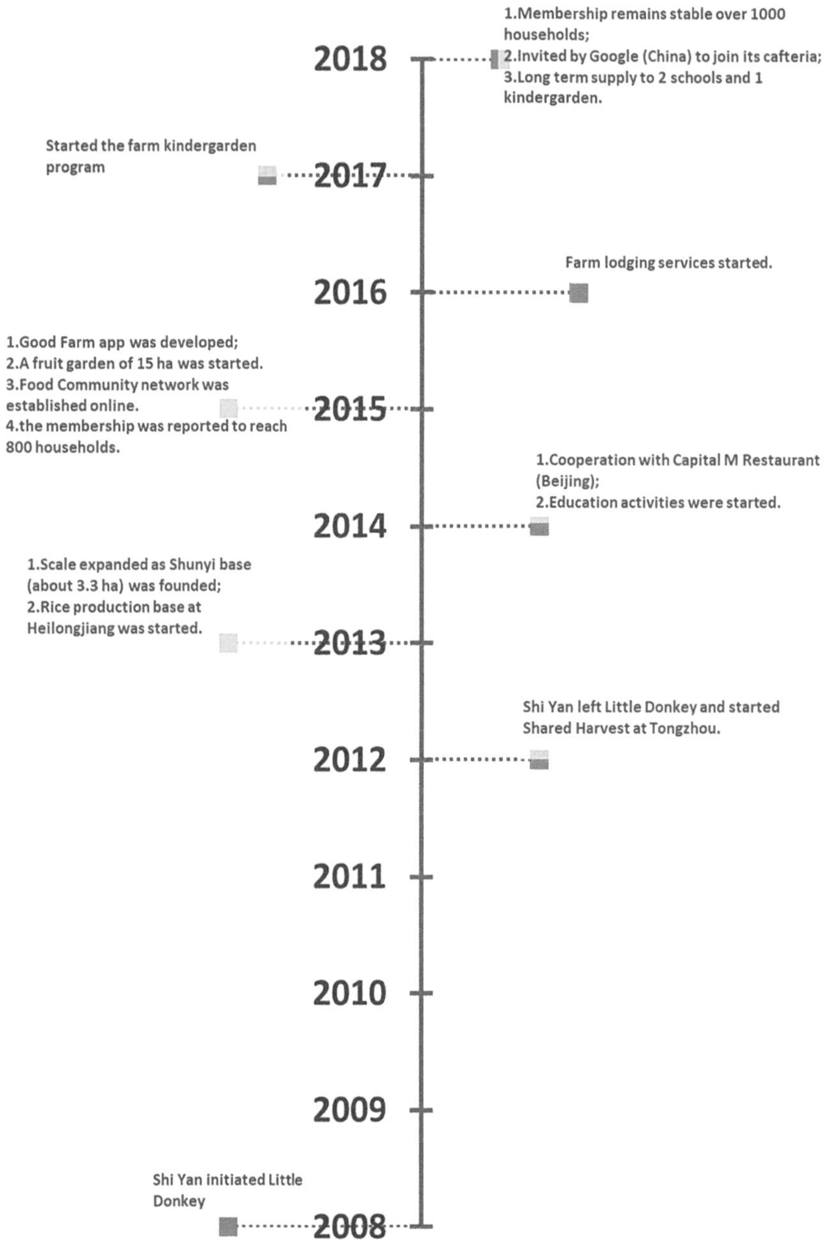

Fig. 5 Timeline of Shared Harvest Farm. (Source: Drawn by authors)

4 Findings

Shared Harvest Farm was established in 2012, in Shunyi District, a suburb of Beijing (Fig. 5). The founder, Shi Yan, had previously been involved with the pioneering Little Donkey Farm, also in Beijing, as well as having undertaken an internship at Earthrise Farm in the USA (Yu, 2015). Shared Harvest Farm was initiated when Shi Yan persuaded retired local farmers to start farming again, with a shift to organic production. This established a small production base for Shared Harvest. In the beginning, it was difficult to produce an even flow of quality produce, which meant that it was hard to build and maintain a stable CSA membership. In addressing this the business expanded in 2013 by taking on another farm (Shunyi Base) with greenhouses for vegetable production and its own production team. This allowed the business to maintain a more consistent output with improved quality, which underpinned a steady growth in membership and sales. In 2015, another production unit was added, when Shared Harvest rented a substantial fruit farm close to the main production base.

Even though the team had little experience of fruit production, Shi Yan wanted to diversify the food choices available to members while, in addition, also creating some on-farm accommodation for agri-tourism. As She Yan explained:

> (The expansion of the land area) was mainly driven by a need to address a problem. We were seriously short of supply in the winter of 2012 because the Tongzhou Base did not have enough greenhouses. … If we failed to deliver to the members, we did not earn money but still had to pay salaries for the whole group. The overall cost to the business would have increased sharply. Thus, we had to look for lands with infrastructure and we found this place (Shunyi Base). … Shunyi Base had too much infrastructure and was not a good place for holding visitor activities. It's also inconvenient to hold activities at the Tongzhou Base as it's too far from our central services. We also wanted to diversify our home delivery service by adding fruits … so last year we managed to rent a fruit farm nearby. (She Yan, interviewed in 2016)

This fostered considerable interest in the farm, in terms of on-line orders of produce as well as visits and other activities. Basic online orders have now reached more than 1000 per week, with specialist and one-off orders in addition. This encouraged the farm to develop its visitor business, which now includes rural tourism, pick-your-own vegetables, nature education for children, and day visits by members. No data are available to help quantify the visitor business, but it is currently small-scale compared to the on-line sale and distribution of produce.

Thus, the development of Shared Harvest Farm is typical of that found by Ding et al. (2018), in their research on small ecological farms in Shanghai. In particular, successful ecological farms in China tend to base their business model on direct sales and home delivery, which often accounts for around 70% of sales. One of the key factors in the success of Shared Harvest Farm—and one that distinguishes it from most other such farms—was the early development of its direct sales supply chain, which was put in place before the farm commenced operation. This was achieved by establishing a core of supporters who were willing to prepay their 3-year initial membership. Ten supporters provided about 300 thousand RMB (USD 44K) for investment in the farm.[1] As She Yan emphasised on many occasions: 'You have to find 200 memberships first, and then you can initiate the farm. Not vice versa' (Shi, 2016 interview). The farm now also operates its own internet Wechat and Taobao shops, selling its own products and also products from other farms that have passed the farm's certification process (Fig. 6). The farm also sells through the Beijing farmers' market, has a community buying group and is working to expand its cooperation with local kindergartens and primary schools.

The business is emblematic of the global AFN/CSA movement: there are direct sales from the farm to people who identify as members and who express support for the farm; there are several small organic farms within the overall business model that are viable because of the retail premium that they receive for their produce; and there is an organisational structure in which the risks and responsibilities of farming and food production are shared between the farmers and the consumers (Liu, Gilchrist, Taylor, & Ravenscroft, 2017; Ravenscroft et al., 2012, 2013).

Fig. 6 Online products from Shared Harvest Farm (left) and other farms (right, usually marked by fair trade 公平贸易). (Source: Mobile phone screen shot of the online wechat store of Shared Harvest, by authors)

Consistent with the work of Lievegoed (1980), the farm is organised on the principle of a four-leafed clover (see Fig. 7) in which the central management team oversees a circular chain of processes, from membership and marketing, to a number of production units (farms) to the support services required to run the organisation, to the information and communication management function which brings the farmers into dyadic relationships with the members. In this way, each department has a single area of responsibility and is expected to operate simultaneously in an autonomous and a collaborative way that promotes operational efficiency at the individual and collective levels. This was explained by the Manager of the Membership Department:

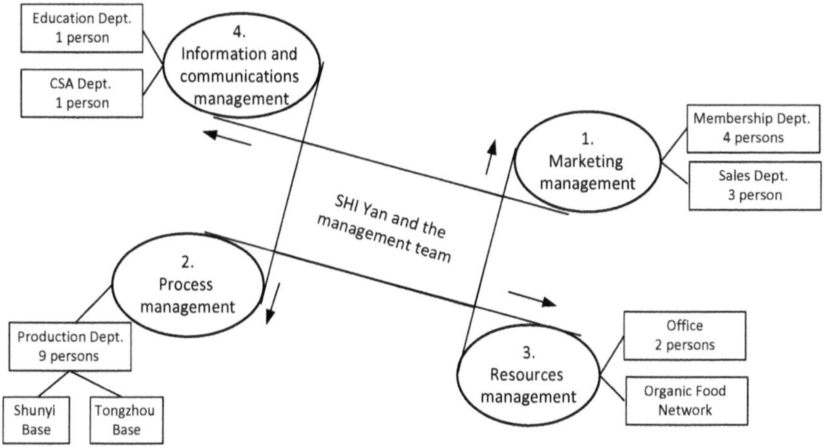

Fig. 7 Integrated organization structure at Shared Harvest Farm. (Source: Liu, Ravenscroft, Ding, & Li, 2019)

> The division of each department was based on our past practices ... such as the membership department is in charge of the orders, delivery, customer service and feedback, and member-only activities, while the sales department is responsible for online shopping, farmers' market and group purchase. ... But these are not fixed. It's common to help each other whenever necessary. (Manager of Membership Department, interviewed November, 2016)

This operational structure is designed to reduce management costs to the minimum by freeing each worker and work team to act on their own skills and initiative, but within a single overall framework. This approach to integration is important to Shi Yan:

> The regular work is done by each department. Sometimes we communicate over WeChat. As a group, we are not like a company. For example, many problems are negotiated and solved over lunch together. Some of the [managers] live on the farm, and they can negotiate and solve many problems together. For me it's the same. What I want to know, and they want to ask, can also be done at the lunch table. (Shi Yan, interviewed November, 2016)

This approach also informs the way that key decisions, such as what to grow and in what quantities, are made. Although not described as such, the decision process is broadly dialectic in the sense that evidence and argument are used to set up a proposition, which is either agreed on, or more discussion and reflection needs to take place, as described by Shi Yan:

> We make plans for the next year before the Chinese Spring Festival. Usually my husband and I, the Production Department, M (responsible for the Shunyi Base) and L (responsible for the Tongzhou Base) will attend the meeting. The (production) plan is made according to our estimate of the membership for the coming season. … We also do some consumer interviews to better understand what they need. We don't vote on the final decision. Voting is superficial. We discuss and persuade first, and we then make the final decision. (Shi Yan, interview November, 2016)

While it is apparent that Shi Yan, as the owner and driving force of the business, is in charge of all aspects of the business, the division of responsibilities is also apparent in the production planning meetings:

> Local experiences matter in this process—and the farmers have their own experience and expertise. For example, Lang (at Tongzhou Base) does much better than Ma (at Shunyi Base) in vegetable growing; … Ma is good at growing tomatoes. …So, when we make the production plan, we demand few tomatoes from Lang and few vegetables from Ma, and their own advantages can be complementary and overall maximised. (Shi Yan, interview November, 2016)

This observation about the relative abilities of the farmers is accepted by the farmers themselves, who understand that their job is to grow agreed volumes of high-quality produce without having the normal farmer concerns about whether they can sell the produce and at what price:

> Shi Yan is responsible for sales and I am in charge of the (vegetable) production. … At first, she made big demands of me (on vegetable quantity, quality, and diversity). Since then I have managed to grow up to 20 kinds of vegetables, and she is satisfied with it. The whole production plan is

always in my head—and Shi Yan trusts me. (Lang, farmer at Tongzhou Base, interviewed November, 2016)

At the centre of the Shared Harvest Farm business model is a bespoke Internet platform called the Good Farm App, which was developed in 2015 by Cheng Cunwang, Shi Yan's husband. According to information from the Weibo webpage for Good Farm,[2] the App was not only designed for communicating with members and customers, but also to provide a range of purchase options for members that include the ability to order regular food boxes and to supplement regular orders as required. One-off orders can also be made. Delivery—by private courier—is available through the app, while customers can also elect to visit the farm to collect their order. Customers and members register on the App for free, and then order the produce that they want or buy different kinds of membership and delivery options. Once registered, customers will also receive reports and updates from the farm. This allows the farmers to explain what is happening on the farm and what produce is available each week, and it allows the members to make informed choices about their purchases. Not only does this help build long term trust between the farm and its members, but according to Shi Yan, it has reduced considerably the complaints made by members prior to the launch of the App. In addition to listing its own produce, the Good Farm app also lists produce from nearby farms that have passed their certification.

Thus, for the farm, the App is designed to be more than a sales medium. Rather, it is designed to support management efficiency, which is widely regarded as one of the biggest challenges for the survival of small-scale organic farms (Ding et al., 2018). For example, Shared Harvest has over 600 members and offers more than 30 kinds of vegetables, so the daily orders can have considerable variety, as well as financial value. Working with the App has reportedly reduced mistakes in packing and dispatching the orders, as well as helping the farm to understand what its members demand most, as well as selling all kinds of products, including personalised processing, by gathering order data. Although there are no metrics available about the magnitude of the efficiency gains, the app designer, Cheng Cunwang, explained in an interview and on his personal Weibo page that prior to the introduction of the App, the farm required five people to handle the orders, with mistakes made in 5–8% of the orders processed. Since the introduction of the App, fewer staff are required, even though the orders have increased, with few mistakes in processing the orders. Cheng Cunwang claims that this has saved the business 200–300 thousand yuan (£20,000–£30,000) per year.

The App has now been made available to other CSA farms in China. There is no up-front charge for using the app, but instead, farms are charged a percentage of the sales completed through the app. Although precise details remain confidential, Cheng reported that the app has been widely adopted by over 250 farms in 30 provinces of China, with each farm paying a fee of up to 10% of the value of their on-line sales. In the future Cheng intended to add additional functions to the App, including details of organic certification and monitoring data, as well as supporting systems for production planning.

Working at the core of Shared Harvest, Shi Yan has a unique understanding of the farm, which she likens to a 'food community'. While growing most of the produce sold through the Good Farm App, the farm now sells organic food from all over China, as long as it passes third party testing:

> We have two types of cooperation: One is incubated by us, and the other is certified by us and we help sell their products. The rice from Heilongjiang Province is the first type. We needed rice and found a household there, and we helped with their production locally at the beginning. Now we do the monitoring and quality testing annually, taking records, advocating the products, designing and packing the rice, as well as selling it. (Shi Yan, interviewed November, 2016)

Another example of the incubator approach has been Shi Yan's desire to start a bakery:

> Now we also mill flour to sell online. The [development of a] bakery is a good direction, but the most important thing is to find an appropriate person to do it. It's far more than making the bread, and includes many management tasks, such as advertising and sales. … We have made such mistakes before. For example, we used to process cabbage boxes, but they

> all rotted because of the poor store management at that time. So, we don't want to increase the products if we don't have a complete management plan. … the cost of management matters. Sometimes it even surpasses the value added by the products.
>
> The best choice for the bakery is that there is someone who is really interested in it. We can help them fulfil their dream as a career. We can help the sales through our membership, with the baker getting a percentage from the sales. We can also offer a basic salary. By doing this, the baker has a career and guaranteed income, and we have new products and profits. It's a win-win model. If we hire someone to do it, they are just an employee, with no incentive, which means only that our management costs will increase. (Shi Yan, interviewed November, 2016)

In contrast to this, Shared Harvest also works with established local businesses (this is the second type of co-operation outlined above by Shi Yan). These include locally-grown walnuts and locally-produced noodles. In both cases these businesses were successful before they worked with Shared Harvest, but elected to go through the selection, review, and certification process so that they could add their products with their own brand to the Shared Harvest list. This provided them more sales security which allowed them to concentrate on improving the efficiency of their production processes and the quality of their products.

5 Discussion and Conclusions

As the case study has illustrated, the supply chains for organic food in China are developing almost exclusively outside of the mainstream models that link farms to retailers. This applies not only to McMichael's (2009) second food regime of long supply chains linking industrial farming to processing and to large retail outlets, but also to more traditional models in which there has often been small scale localised wholesale and retail intermediation between farmer and consumer. In place of these models, the increasingly dominant form of supply chain for fresh organic produce is short and direct, featuring on-line purchasing and home delivery from farmer to customer, sometimes supplemented by sales through farmers' markets. Two-way communication is built into these supply chains, through features such as CSA membership, newsletters, farm visits and a range of educational services. Thus, although in many ways the Chinese market for local organic and ecological food is immature and often fragmented, it is also increasingly sophisticated in its use of on-line technology and social media-based communication.

At the core of these new supply chains are: an emphasis on the provenance of the food, reflecting public concern about food safety and the readiness of the emerging middle class to pay a premium for produce that they can trust—within a culture in which most farmers and agribusinesses are not trusted; supply chains that are dominated by consumer convenience, in terms of the range of food available, web-based ordering and practically 24/7 delivery; and, crucially, a two-way communication between the farmer and consumer, often linked to forms of scheme membership and pre-payment. As such, it is apparent that while these short supply chains resemble their counterparts in other countries, there are also some significant Chinese characteristics emerging, particularly associated with the mix of contemporary consumerist Internet platform and more traditional mutual respect (Guanxi) between producer and consumer. While Guanxi itself is fracturing, between traditional neighbour and clan relationships and newer formations intermediated by money (Du, Liu, Ravenscroft, & Su, 2020), it is clear that increasing numbers of consumers do recognise their reliance on, and increasing interest in, those who produce the food—and where the food is produced—while the producers understand full well that they rely on these customers to keep them in business. This means that more and more customers are beginning to see themselves as part of the endeavour—quasi-members of the enterprise—who visit the farms and associate themselves with the responsibility for producing and consuming locally grown organic food. This is very much akin to Wendell Berry's American argument about the need for eating to be understood as an agricultural practice (Berry, 1989). Thus, this emerging culture—in China and elsewhere—of mutual respect and reliance is foundational to creating more sustainable approaches to food production and consumption that can underpin the vitality of both the supply chains and those involved in growing and distributing the food.

5.1 Future Research

This is clearly an immature area of business development, with few examples of successful operations available to research, and little peer-reviewed literature yet in the public domain. This has meant that data generation has been limited to a single case study approach which has yielded a range of observational and narrative data. What these data indicate, by way of an initial theory about the development of short supply chains for fresh food, is that business success is predicated on trust, loyalty, ease of ordering, comprehensiveness of the offer and, once the business is established, a range of value-added services such as rural tourism. These are not unique to fresh produce, nor China. However, they do address a key gap in current knowledge, which is about the significance of internet-enabled tools as the preferred form of intermediation for many customers, and the power that this offers to those who are able to develop and implement such tools. With these tools in place, small farms can concentrate on developing the quality and range of produce that they are able to supply, within a framework in which trust is fostered and customers become increasingly loyal. Clearly, this is a tentative theory that will require additional research—particularly using more quantitative data—before it can be verified and refined. to be verified. There is also a need for research on the extent to which the findings can be translated to other scenarios (in China or elsewhere) or are scalable—either upwards or downwards. Nevertheless, as this chapter has revealed, even in emerging markets with few data, qualitative methods such as the single case study can be useful in generating an initial understanding of the parameters and dynamics in which new businesses develop.

Notes

1. Source: 'Learn CSA at USA and Work as a Farmer when returned', published at China Youth Daily, 2017-04-26. Also confirmed by the onsite interview in 2016.
2. Source: 'From CSA farmer to IT CEO', Official Weibo of Good Farm, 2015-10-24.

References

Aktas, E., Bloemhof, J. M., Fransoo, J. C., Gunther, H., & Jammernegg, W. (2017). Green logistics solutions. *Flexible Services and Manufacturing Journal*. https://doi.org/10.1007/s10696-017-9301-y

Basit, T. (2003). Manual or electronic? The role of coding in qualitative data analysis. *Educational Research, 45*(2), 143–154.

Bennett, A., & Elman, C. (2006). Qualitative research: Recent developments in case study methods. *Annual Review of Political Science, 9*, 455–476.

Berry, W. (1989). The pleasure of eating. In Berry, W. (2017). *The World-ending fire: The essential Wendell Berry*, selected and introduced by P. Kingsnorth. London: Random House UK.

Bloemmen, M., Bobulescu, R., Tuyen, N., & Vitari, C. (2015). Microeconomic degrowth: The case of Community Supported Agriculture. *Ecological Economics, 112*, 110–115.

Brinkley, C. (2017). Visualizing the social and geographical embeddedness of local food systems. *Journal of Rural Studies, 54*, 314–325.

Chen, W. (2015). *The revolution of the vegetable basket – Typical cases of CSA farms in China*. Beijing: Economic Science Press.

Chin, T. A., Tat, H. H., & Sulaimana, Z. (2015). Green supply chain management, environmental collaboration and sustainability performance. *Procedia CIRP, 26*, 695–699.

Dedeurwaerdere, T., De Schutter, O., Hudon, M., Mathijs, E., Annaert, B., Avermaete, T., et al. (2017). The governance features of social enterprise and social network activities of collective food buying groups. *Ecological Economics, 140*, 123–135.

Ding, D., Liu, P., & Ravenscroft, N. (2018). The new urban agricultural geography of Shanghai. *Geoforum, 90*, 74–83.

Du, Y., Liu, P., Ravenscroft, N., & Su, S. (2020). Changing community relations in southeast China: The role of Guanxi in rural environmental governance. *Agriculture and Human Values*. https://doi.org/10.1007/s10460-019-10013-8.

Durand, B., & Gonzalez-Feliu, J. (2012). Urban logistics and e-grocery: Have proximity delivery services a positive impact on shopping trips? *Procedia – Social and Behavioral Sciences, 39*, 510–520.

EU SME Centre. (2015). *The food and beverage market in China*. Beijing: EU SME Centre in partnership with the China-British Business Council.

Fancello, G., Paddeu, D., & Fadda, P. (2017). Investigating last food mile deliveries: A case study approach to identify needs of food delivery demand. *Research in Transportation Economics, 65*, 56–66.

Flores, H., & Villalobos, J. R. (2018). A modeling framework for the strategic design of local fresh-food systems. *Agricultural Systems, 161*, 1–15.

Flyvbjerg, B. (2006). Five misunderstandings about case-study research. *Qualitative Inquiry, 12*(2), 219–245.

Forssell, S., & Lankoski, L. (2015). The sustainability promise of alternative food networks: An examination through "alternative" characteristics. *Agriculture and Human Values, 32*, 63–75.

Geall, S., & Ely, A. (2015). *Innovation for sustainability in a changing China: Exploring narratives and pathways*. STEPS Working Paper 86. Brighton: STEPS Centre.

Geng, R., Mansouri, S. A., & Aktas, E. (2017). The relationship between green supply chain management and performance: A meta-analysis of empirical evidences in Asian emerging economies. *International Journal of Production Economics, 183*, 245–258. https://doi.org/10.1016/j.ijpe.2016.10.008

Geng, R., Mansouri, S. A., Aktas, E., & Yen, D. A. (2017). The role of Guanxi in green supply chain management in Asia's emerging economies: A conceptual framework. *Industrial Marketing Management, 63*, 1–17.

Hao, Z., Chen, L., Huang, Y., & Liao, T. (2004). Research on the development of ecological urban agriculture in Chongqing. *Contemporary Agriculture Research, 25*(1), 64–67.

Hinrichs, C. C. (2000). Embeddedness and local food systems: notes on two types of direct agricultural market. *Journal of Rural Studies, 16*, 295–303.

Ilbery, B., Courtney, P., Kirwan, J., & Maye, D. (2010). Marketing concentration and geographical dispersion: A survey of organic farms in England and Wales. *British Food Journal, 112*(9), 962–975. https://doi.org/10.1108/00070701011074345

Jarosz, L. (2000). Understanding agri-food networks as social relations. *Agriculture and Human Values, 17*, 279–283.

Jin, S., Li, H., & Li, Y. (2017). Preferences of Chinese consumers for the attributes of fresh produce portfolios in an e-commerce environment. *British Food Journal, 119*(4), 817–829. https://doi.org/10.1108/BFJ-09-2016-0424

Kottila, M.-R., & Rönni, P. (2008). Collaboration and trust in two organic food chains. *British Food Journal, 110*(4/5), 376–394. https://doi.org/10.1108/00070700810868915

Krul, K., & Ho, P. (2017). Alternative approaches to food: Community supported agriculture in urban China. *Sustainability, 9*(5), 844.

Li, W., Liu, M., & Min, Q. (2011). China's ecological agriculture: Progress and perspectives. *Journal of Resources and Ecology, 2*(1), 1–7.

Lievegoed, B. C. J. (1980). *The developing organization*. Berkeley, CA: Celestial Arts.

Liu, P., Gilchrist, P., Taylor, B., & Ravenscroft, N. (2017). The spaces and times of community farming. *Agriculture and Human Values, 34*, 363–375.

Liu, P., & Ravenscroft, N. (2017). Collective action in implementing top-down land policy: The case of Chengdu, China. *Land Use Policy, 65*, 45–52.

Liu, P., Ravenscroft, N., Ding, D., & Li, D. (2019). From pioneering to organised business: The development of ecological farming in China. *Local Environment, 24*(6), 539–553.

Liu, P., Ravenscroft, N., Harder, M. K., & Dai, X. (2016). The knowledge cultures of changing farming practices in a water town of the Southern Yangtze Valley, China. *Agriculture and Human Values, 33*(2), 291–304. https://doi.org/10.1007/s10460-015-9607-x

Lu, Y., Ma, Z., Zhang, L., Fu, B., & Gao, G. (2012). Redlines for the greening of China. *Environmental Science & Policy, 33*, 346–353.

McIver, D., & Hale, J. (2015). Urban agriculture and the prospects for deep democracy. *Agriculture and Human Values, 32*(4), 727–741.

McMichael, P. (2009). A food regime genealogy. *Journal of Peasant Studies, 36*(1), 139–169.

Morgan, K. (2009). Feeding the city: The challenge of urban food planning. *International Planning Studies, 14*(4), 341–348.

Parr, B., & Henry, D. (2016). *A new starting point: China's eco-civilisation and climate action post Paris*. Briefing Paper 6. Melbourne, VIC: Melbourne Sustainable Society Institute, University of Melbourne.

Paull, J. (2007). China's organic revolution. *Journal of Organic Systems, 2*(1), 1–11.

Poulsen, M. N. (2017). Cultivating citizenship, equity, and social inclusion? Putting civic agriculture into practice through urban farming. *Agriculture and Human Values, 34*, 135–148.

Ravenscroft, N., Moore, N., Welch, E., & Church, A. (2012). *Connecting communities through food: The theoretical foundations of community supported agriculture in the UK*. Working Paper No. 115. CRESC Working Paper Series. Centre for Research on Socio-Cultural Change, Open University and University of Manchester.

Ravenscroft, N., Moore, N., Welch, E., & Hanney, R. (2013). Beyond agriculture: The counter-hegemony of community farming. *Agriculture and Human Values, 30*(4), 629–639.
Sanders, R. (2006). A market road to sustainable development? Ecological agriculture, green food and organic agriculture in China. *Development & Change, 37*(1), 201–226.
Saskia, S., Mareï, N., & Blanquart, C. (2015). Innovations in e-grocery and logistics solutions for cities. *Transportation Research Procedia, 12*, 825–835.
Schermer, M. (2015). From 'food from nowhere' to 'food from here:' Changing producer-consumer relations in Austria. *Agriculture and Human Values, 32*, 121–132.
Seawright, J., & Gerring, J. (2008). Case selection techniques in case study research: A menu of qualitative and quantitative options. *Political Research Quarterly, 61*(2), 294–308.
Shi, T. (2002). Ecological agriculture in China: Bridging the gap between rhetoric and practice of sustainability. *Ecological Economics, 42*, 359–368.
Shi, Y., Cheng, C., Lei, P., Wen, T., & Merrifield, C. (2011). Safe food, green food, good food: Chinese Community Supported Agriculture and the rising middle class. *International Journal of Agricultural Sustainability, 9*(4), 551–558.
Si, Z., Schumilas, T., & Scott, S. (2015). Characterizing alternative food networks in China. *Agriculture and Human Values, 32*(2), 299–313.
Thompson, C. J., & Coskuner-Balli, G. (2007). Enchanting ethical consumerism. The case of Community Supported Agriculture. *Journal of Consumer Culture, 7*(3), 275–303.
UNEP. (2016). *Green is gold: The strategy and actions of China's ecological civilization*. Nairobi: United Nations Environment Programme.
Vieira, L. M., De Barcellos, M. D., Hoppe, A., & da Silva, S. B. (2013). An analysis of value in an organic food supply chain. *British Food Journal, 115*(10), 1454–1472. https://doi.org/10.1108/BFJ-06-2011-0160
Wang, R., Yu, R., Si, Z., Scott, S., & Nam Ng, C. (2015). The transformation of trust in China's Alternative Food Networks: Disruption, reconstruction, and development. *Ecology and Society, 20*(2), 19.
Wang, Z., He, H., & Fan, M. (2014). The ecological civilization debate in China. *Monthly Review, 66*, 6. Retrieved from http://monthlyreview.org/2014/11/01/the-ecological-civilization-debate-in-china/
Weng, X., Dong, Z., Wu, Q., & Qin, Y. (2015). *China's path to a green economy. Decoding China's green economy concepts and policies*. IIED Country Report. London: International Institute for Environment and Development.

Yan, Y. (2012). Food safety and social risk in contemporary China. *The Journal of Asian Studies, 71*(3), 705–729.
Yu, K. (2015, November 25). Meet the woman leading China's new organic farming army. *Aljazeera*. Retrieved from https://www.aljazeera.com/indepth/features/2015/11/woman-leading-china-organic-farming-army-beijing-151123140338900.html
Zhang, L., Zheng, Y., & Wan, Z. (2005). A summary of research on urban agriculture in China. *Guangdong Agricultural Science, 3*, 85–87.
Zissis, D., Aktas, E., & Bourlakis, M. (2017). A new process model for urban transport of food in the UK. *Transportation Research Procedia, 22*, 588–597.

How Digital Business Platforms Can Reduce Food Losses and Waste?

Luciana Marques Vieira
and Daniele Eckert Matzembacher

1 Introduction

Inefficient agricultural activities are among the major causes of environmental degradation (Foley et al., 2011). Companies are increasingly worried about circumstances in which food is produced and offered, seeking solutions that encompass three dimensions of sustainability (Beske, Land, & Seuring, 2014). In this context, food loss and waste are receiving special attention (Aschemann-Witzel, de Hooge, Amani, Bech-Larsen, & Oostindjer, 2015), since due to the complexity of coordinating the members in the food supply chain, food waste has increased over the past few years (Govindan, 2018).

L. M. Vieira (✉)
Industrial Production and Operations Administration Department,
Sao Paulo, Brazil
e-mail: luciana.vieira@fgv.br

D. E. Matzembacher
Graduate School of Business Management, Federal University of Rio Grande do Sul (UFRGS), Porto Alegre, Brazil

E. Aktas, M. Bourlakis (eds.), *Food Supply Chains in Cities*,
https://doi.org/10.1007/978-3-030-34065-0_7

According to the Food and Agriculture Organisation of the United Nations (2013), one-third of the food produced for human consumption is lost or wasted globally. Two thirds of this amount (about 1 billion tons) are wasted every year along the supply chain (Zhong et al., 2017). In this sense, definitions are not universal worldwide (Buzby & Hyman, 2012), but in most of the academic publications it is possible to distinguish two relevant concepts according to the stage of the supply chain in which the phenomenon occurs.

Food loss is a decrease in mass or nutritional value of food that was originally intended for human consumption. Inefficiencies in the food supply chain mainly cause these losses. Some problems may relate to poor infrastructure and logistics, lack of technology, insufficient skills, knowledge and management capacity of supply chain actors and lack of access to markets. They also relate to natural disasters (FAO, 2013). Food losses take place at production, postharvest, distribution and processing stages in the food supply chain (Parfitt, Mark, & Sarah, 2010).

Food waste refers to food that is appropriate for human consumption but discarded in latter stages of the supply chain. It encompasses also food left to spoil that after it is kept beyond its expiry date. Some of the reasons relate to oversupply due to markets and individual consumer shopping/eating habits (FAO, 2013). It relates to retailers and consumers' stages in the food supply chain (Parfitt et al., 2010).

The causes of food loss and food waste in medium/high-income countries mainly relate to consumer behaviour, as well as to a lack of coordination between different agents in the supply chain. At consumer level, lack of purchase planning, expiring 'best-before-dates' in combination with the careless attitude of those consumers who can afford to waste food are the main causes. The non-acceptance of non-standard products are also important causes of food waste, i.e., products that do not have perfect appearance, with some small deformations in shape or size, despite having the same nutritional quality. Since consumers do not accept these products, retailers reject them justifying it as a non-standard compliance and send them back to the producer. In low-income countries, the causes are mainly connected to financial, managerial and technical limitations in harvesting techniques, storage and cooling facilities in difficult climatic conditions, infrastructure, packaging and marketing systems.

Other factor that may also contribute to this problem are quality standards, which reject food items that are not perfect in shape or appearance—non-standard products (Gustavsson, Cederberg, Sonesson, Van Otterdijk, & Meybeck, 2011).

Food loss and waste impact on sustainability in the environmental, social, and economic dimensions (Alexander et al., 2017). Therefore, minimising food losses and waste should lead to economic resources savings, cost reduction, and alleviation of social and environmental negative impacts (Pullman & Wikoff, 2017; Thyberg & Tonjes, 2016).

The most advantageous way to deal with food waste is by prevention initiatives (Göbel, Langen, Blumenthal, Teitscheid, & Ritter, 2015; Papargyropoulou, Lozano, Steinberger, Wright, & bin Ujang, 2014). It can be achieved by surplus reduction and avoiding edible food to be discarded through the entire food supply chain. Moreover, food waste solutions require a combination of multiple actors (Aschemann-Witzel et al., 2015), such as institutions, supply chain actors, and consumers (Stuart, 2009). There is an urgent call for new modes of food supply chain involving coordination mechanisms and more collaboration between farmers, food processors, distributors and retailers (Zhu et al., 2018). Therefore, to achieve sustainable food systems the food supply chain needs to be better coordinated (Govindan, 2018).

In this sense, Brazil presents some challenges related to coordination in food supply chains that may be related to food loss and waste generation. The country is one of the largest food producers in the world. It is responsible for, on one hand, mass scale production and export of commodities such as coffee, soybeans, and beef, which represented 22% of the GDP in 2015. On the other hand, Brazil faces relevant challenges in the internal food distribution. Managing fresh food supply in large urban cities, such as Sao Paulo (the largest financial and economic centre of the country and South America) is not a trivial task. According to Kawano, de Vargas Mores, da Silva, and Cugnasca (2012) it involves aspects of larges distances, perishability and supply chain coordination. When the result is not satisfactory, it can generate losses in prices and food available.

The aim of this chapter is to discuss how digital business platforms (DBP) can be an alternative to reduce food losses and waste by increasing coordination across supply chain in a large metropolis. The relevance is to describe how a new stakeholder in the chain, by using information and communication technology, can make food demand and supply closer, and minimise losses and waste. This seems a feasible alternative to be expanded and adopted by other chains in different countries and contexts. Theoretically, it proposes a new alternative form of food supply chain coordination with high social and environmental positive impact.

In fact, the way food is produced, processed, transported and consumed has a great impact on whether sustainability is achieved throughout the whole food supply chain (Govindan, 2018). Novel digital technologies are creating major opportunities for the food industry (Zhu et al., 2018). In this sense, the use of information and communication technology, business/supply chain strategy, and collaboration are identified as highly relevant for future sustainable supply chain management (Sauer & Rebs, 2018) and new technologies might be a valuable source when enabling a more sustainable business environment. Therefore, DBP are expected to be able to bring alternatives for reducing food waste by increasing coordination among the links in the supply chain.

Despite the popularity and importance of digital platforms, there is a lack of research in this area (Holland & Gutiérrez-Leefmans, 2018). As DBP are disruptive and have potential to revolutionise traditional sectors such as food (Moazed & Johnson, 2016), investigating the relation between food waste solutions, coordination mechanisms, and DBP can bring valuable contributions to more sustainable food systems. This chapter investigates two research questions: 1. How can digital business platforms contribute to food waste solutions? 2. How do they improve supply chain coordination?

2 Theoretical background

The emergence and growth of sharing economy happened as influential trend for the practice and theorization of management: Digital Business Platforms (DBP) not follow the configuration of traditional companies but rather comprise value networks, which are associated with the firm's ability to mediate exchanges between groups of users (Stabell & Fjeldstad,

1998). As a consequence, the buyer-supplier relationship in a supply chain is not organized as the traditional model but via mediation services.

As digital business platforms gain market share from traditional businesses, they are influencing the practices within the food industry and, more importantly, whether they address the problems of food waste. Preventing waste in food chains—in particular food waste at retail or household (Brancoli, Rousta, & Bolton, 2017)—is central for the sustainability of value chains. However, agri-food chains in Brazil are fragmented, and levels of education and technological literacy vary widely, which results in major food losses throughout the supply chain and households. Among the factors influencing the optimization of food systems, coordination is central. According to Nooteboom (1999), rather than signifying control only, it expresses multiple interests.

Hence, balancing interests and power constitutes one of the greatest challenges. Besides economic and technological issues, we need to consider moderating factors of interorganizational relationships, such as trust, relational risks, (Cao & Lumineau, 2015), regulatory structures, sustainability issues, political factors, among others (MacCarthy, Blome, Olhager, Srai, & Zhao, 2016). In the specific case of digital business platforms, supply chain coordination can be embedded in the architecture of the business, as a structural way of changing behaviours. The two main topics, DBPs and supply chain coordination, are discussed below.

2.1 Supply Chain and Coordination Mechanisms

Coordination refers to methods by which the relationships between organizations are managed, i.e., interactions among partners to adjust their goals by allocating functions and cooperating effectively to accomplish joint and individual tasks (Grandori, 1997; Gulati, Wohlgezogen, & Zhelyazkov, 2012). Coordination mechanisms are typically divided into formal and informal (Alvarez, Pilbeam, & Wilding, 2010; Lumineau & Henderson, 2012). The formal mechanisms are related to control and communication systems as more structured command structures or legal contracts (Alvarez et al., 2010; Gulati & Singh, 1998; Pilbeam, Alvarez, & Wilson, 2012). The informal mechanisms (Giannakis, Doran, &

Chen, 2012) include trust, commitment and communication, influencing knowledge-sharing routines and collaboration among the actors (Alvarez et al., 2010).

Collaboration is an organizational practice that seeks information sharing, strategic alliance, performance, cost reduction, and inventory management (Chen et al., 2017). Collaborative activities start with network creation, then followed by practicing communication, knowledge and resource sharing, and developing a shared sustainability strategy with the stakeholders along the whole value chain (Dangelico, Pontrandolfo, & Pujari, 2013; Sancha, Gimenez, Sierra, & Kazeminia, 2015; Vachon & Klassen, 2008).

Collaboration is considered as the fundamental aspect to promote a better-quality interaction among stakeholders (Dania, Xing, & Amer, 2018). It also focuses on how companies in the supply chain are technically and logistically integrated (Vachon & Klassen, 2008). In fact, collaboration efforts correspond to one of the frequently mentioned aspects in the literature as barriers to the implementation of sustainable supply chains (Seuring & Müller, 2008).

Most of the supply chain coordination focuses on buyer—supplier (dyadic) relationships as described by Chen and Paulraj (2004). However, empirically size asymmetries and geographical distances make these two supply agents far from communicating to each other. This can lead to coordination problems.

Based on the previous literature, Neutzling, Land, Seuring, and do Nascimento (2018) provide a framework with elements to analyse inter-organisational relationships addressing solutions to food waste: collaboration, and coordination mechanisms (formal and informal mechanisms). They are presented in Table 1, which provides the analysis categories that is intended to be used for data collection and analysis. The proposition is that by combining formal and informal coordination mechanisms companies are able to develop new and sustainable strategies through inter-organisational relationships. At the same time, by using efficient mechanisms they are able to better manage interactions in the supply chain and to promote collaboration.

Table 1 Framework to analyze inter-organisational relationships providing solutions to food waste

Categories of analysis	Elements	Literature
Collaboration	Communication and information sharing; Investment in specific relationships; Joint development projects (cross-functional and technology integration); Planning and logistics integration; Resource sharing for specific objectives; Development of joint practices in sustainability programs; Values and social norms sharing; and Social ties (trust, commitment)	Beske (2012); Beske et al. (2014); Chen and Paulraj (2004); Rodríguez-Díaz and Espino-Rodríguez (2006); Zacharia, Nix, and Lusch (2011); Touboulic and Walker (2015);
Coordination mechanisms (formal and informal mechanisms)	(a) Formal mechanisms: Control and communication systems; Command structures; Incentive systems; Standardized procedures of operations; Troubleshooting; Legal contracts. (b) Informal mechanisms: Self-regulations (rules, conventions, or standards); Requirements of suppliers' self-evaluation processes; Informal social ties or social norms; Information sharing; Value systems; and Schema and culture	Alvarez et al. (2010); Dekker (2004); Gulati and Singh (1998); Pilbeam et al. (2012)

Source: Authors, based on Neutzling et al. (2018)

2.2 Digital Business Platforms (DBPs)

DBP can be labelled as either internal or external, according to the level of analysis. Internal platforms focus on manufacturing companies and the overall development of new products within a single entity (Gawer, 2009). The primary goal of this kind of DBP is to fashion production in line with a more flexible and dynamic response to market demands and trends (Thomas et al., 2015). A classic example would be the automobile industry, where different 12 automobile models can share and co-exist on the same platform. Academic researchers are interested in understanding how the development and management processes occur in internal platforms (Gawer & Cusumano, 2014).

External platforms operate within multiple companies in the same supply chain, which facilitates the production and development of new products amongst all the partners (Gawer, 2009). The goals are similar to those of internal platforms, specifically flexibility, innovation, and cost savings through the reuse of components. The most prominent difference refers to the platform's ownership, or "a firm that owns a core element of the technological systems that defines the forward evolution" (Gawer & Henderson, 2007, p. 4).

At the supply chain level, the platform's property can be shared by several companies rather than owned by a single firm, as it is with internal platforms. External platforms can also be represented by industry platforms. They are associated companies that develop complementary applications, technologies or services, but do not necessarily have transactional relationships with each other, as occurs in supply chain platforms (Gawer, 2009). Industry platforms are responsible for conceiving an ecosystem that focuses on the interdependence between businesses. These BP must meet two prerequisites: firstly, the resolution of obstacles in a sector; and secondly the ease of interaction between current users and new participants (Gawer & Cusumano, 2008).

Industry platforms can assume the roles of both "maker" and "exchange" (Moazed & Johnson, 2016). Industry maker platforms offer a diverse spectrum of tools, technologies, and infrastructure for content production, mainly in those content platforms that provide articles (Wikipedia), videos (YouTube), and so on. Industry exchange platforms, in turn, support the interaction between consumers and producers, such as Airbnb, eBay, and Paypal. In these cases, platforms can generate transaction leverage, since they eliminate the need for intermediaries for clinching deals (Thomas et al., 2015).

Given the different types of platform available, the research project will focus on digital BP that operate at the industry level, whose primary function is the exchange between three groups of actors: the platform owner, responsible for offering the technological solution; the end users, who demand the service; and the service providers (Ruutu, Casey, & Kotovirta, 2017; Zhu & Furr, 2016). The following topic presents to what extent DBP have brought an innovative approach to doing business when compared to traditional business models.

DBP offer new technologies that can originate innovative business models. They began as electronic marketplaces (Chu, Leung, Van Hui, & Cheung, 2007), connecting buyers and sellers, to facilitate the exchange of information, goods, services, and payments in market transactions (Bakos, 1998).

DBP are connecting food producers and consumers in many new ways (Zhu et al., 2018). They approximate distinct groups and catalyse a virtuous cycle: more demand from one side requests increased supply at the other (Eisenmann, Parker, & Van Alstyne, 2006). They also act as coordination mechanisms to promote innovation systems that foster effective co-evolution, i.e., mutual interactions and adaptation over time between the technological, social and institutional components of an innovation (Kilelu, Klerkx, & Leeuwis, 2013). Perhaps these DBP also have coordination mechanisms that can help reduce food waste and increase sustainability in food systems.

In fact, as digital business increases, many data-driven systems will be the future of food supply chain management, enhancing its sustainability (Zhong et al., 2017). It also appears in various organisational forms, such as across and within supply chains (Gawer, 2014). For instance, farmers who cannot supply their food products that are considered out of retail standards (non-standard appearance due to size, colour, texture) can connect, in business platforms, to end consumers that demand more sustainable food initiatives.

Therefore, DBP may represent an alternative coordination mechanism that might prevent food losses and waste. Pagell and Wu (2009) suggest that business models and environmental and social elements of sustainability need to be aligned to lead to more sustainable supply chains. In this sense, focus must be both on coordination and flow (Ahi & Searcy, 2013). A study conducted by Kilelu et al. (2013) identified that the key role of platforms is in connecting the institutional change to other components of innovation by establishing effective patterns of interactions.

As Aghalaya and Verma (2018) propose, there is a need for focal players and coordinating mechanisms to ensure that players in the supply chain are setting and achieving sustainability targets for a coordinated supply chain. It the case of DBP it is expected that these organisations have some of the formal or informal coordination mechanisms and collaboration aspects previous explained and listed in Table 1 in their interactions with agents in their supply chain in a way that positively impacts food waste reduction. It is also expected that they can act as a kind of bridge between production and consumption and even other stages of supply chain.

Figure 1 illustrates this proposition:

Figure 1 is a symbolic representation of a generic supply chain in the food industry. Each agent has its associated food losses or waste causes. It is proposed that DBPs can promote new offers and demands (such as for non-standard products) and make the origin of the food more visible changing consumers′ behaviour towards food. By doing so, it acts as a kind of bridge between suppliers and buyers, avoiding disposal of food throughout supply chain.

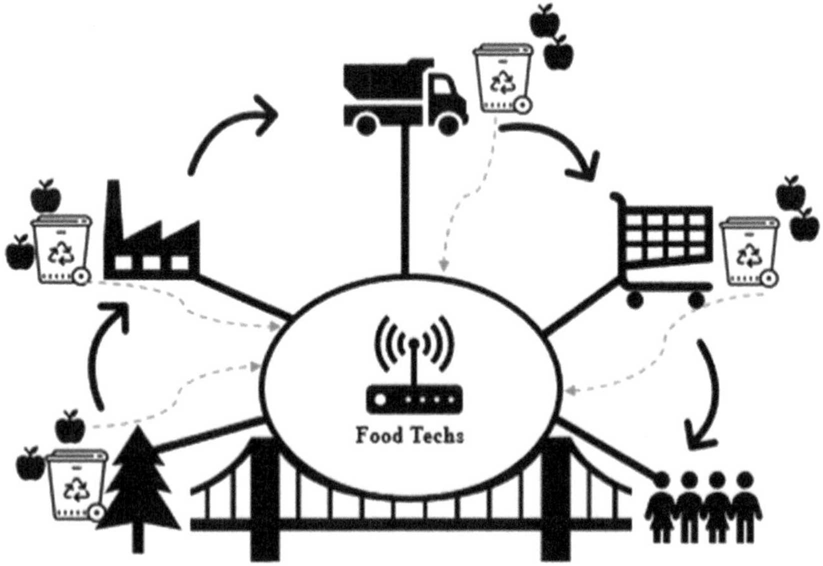

Fig. 1 DBP as promoters of coordination mechanisms to reduce food waste. (Source: Authors' own elaboration)

3 Methods

This research uses a qualitative approach and was carried out in two phases, based on primary data collection, through observation visits and interviews, as well as secondary data. This investigation is part of a larger project, called "Prospects for the Reduction of Losses and Waste in Agri-Food Chains". The objective of the project was to analyze the relationships in the agrifood chain with regard to losses and waste of ten types of food products in the city of São Paulo, Brazil. In this paper the focus was to understand the role of DBP in providing solutions for food waste.

3.1 Data Collection

The first phase investigated producers, associations/cooperatives, and government support agencies. The objective of this phase was to understand difficulties related to coordination mechanisms faced in food supply chains in Brazil regarding food waste. The process of data collection followed the case study protocol proposed by Yin (2017) in relation to cases selection criteria, approach to organizations, preparation for data collection, and conduct of interviews and observation. The criterion that determined the choice of the interviewees was: a) located in a relevant production area in the State of São Paulo; and b) to have availability for an observation visit and to an in-depth interview.

Observations visits were made to all these stakeholders. A total of 16 interviews were held between August and November of 2017, talking with a total of 28 people. Producers and associations/cooperatives interviewed were of small and medium size, the majority selling their products to intermediaries, in supply distribution centres, or, in a minority, with supply contracts for supermarkets.

The questionnaire applied related to production and commercialization process, causes of food loss and waste related to these processes; and the relationships with the main clients in terms of collaboration and formal and informal coordination mechanisms—elements presented in Table 1. As the interviewed were Brazilians, the interviews were conducted in the native language. The data was analyzed in Brazilian

Portuguese, seeking to preserve the integrity of the information, and after it was translated into English.

The second phase was based on primary and secondary data collection from three case studies about food DBP. The objective of this phase was to understand if they overcome the difficulties related to coordination mechanisms and collaboration found in the first phase of the study and, if so, to investigate how they have overcome them. As proposed by Silvestre (2015), this phase was carried out with the focal companies, since they are central agents to encourage more sustainable behaviour of all members of the supply chain.

The process of data collection in this phase also followed the case study protocol proposed by Yin (2017). The criteria that determined the choice of the interviewees was: (a) to be a DBP of food related products; (b) to have availability for an observation visit and to an in-depth interview. Visits to the physical locations of platforms and interviews were conducted with the owners of each platforms, between September and October 2017. The secondary data collected were the websites, reports and news made with these platforms.

The questionnaire applied was the same in terms of relationships with the main clients, i.e., collaboration and formal and informal coordination mechanisms. The difference is that the questions were related to their suppliers, since they are in a different position in the supply chain. The questions about commercialization process, food loss and waste related to these processes were the same. Also, the procedures related to Brazilian Portuguese and English language were the same as in the previous stage.

The Table 2 synthetizes the information regarding data collection:

3.2 Data Analysis

All interviews were taped, transcribed, and analysed along with the field notes and photographs during the visits. The gathered data were analysed through content analysis, with the support of NVivo 11 software. Two different groups of researchers analysed the data to ensure greater reliability/validity: one group made visits and carried out the interviews and

Table 2 Interviews in data collection

Phase	Interviewed	Interviewed	Interview period
1	Agents of the supply chain related to the productive process	16 producers 7 agronomists 1 cooperative/ association coordinator 3 cooperative/ association presidents 1 cooperative/ association director	August and November of 2017
2	Digital platform A	1 Owner/founder	Sep. 2017
2	Digital platform B	1 Owner/founder	Sep. 2017
2	Digital platform C	1 Owner/founder	Oct. 2017

Source: Authors' own elaboration

the second group had access to interviews' recordings and transcriptions to make the coding. Both groups came to similar results.

The data and results obtained were validated during a workshop held in April 2018. The objective was to present and validate with the supply chain agents and other stakeholders in the food sector the initial results found in the research, as well to discuss possible collaborative solutions. In this stage, 19 people were present, among academic researchers studying food waste, fresh food suppliers, government agencies such as representatives of the Brazilian Agricultural Research Corporation, food bank managers, and owners of platforms A and B. Due to the distance and daily routine, horticultural producers couldn´t join the workshop.

4 Findings

The results are presented according to the phases of the research. Phase 1 identifies the coordination problems that impact on food waste. Phase 2 presents the results on how DBP help to overcome the difficulties related to coordination mechanisms found in the first phase of the study and promote better collaboration with food producers.

4.1 Coordination's Problems related to Food Waste

Through the research it was identified that in Sao Paulo the producers are based in the outskirts of the city and main buyers are large wholesalers and retailers. The transactions have low asset specificity and occur in the spot market. Due to the high perishability of the product, the uncertainty of sale, and the eventual rejection of some food products by retail, producers face relevant food waste issues.

Many producers face commercialization problems in relation to products that have a nonstandard aesthetic appearance, whether due to their shape or size. Retail rejects these products. There is a settled agreement that these products are not accepted. These foods are often returned to producers and end up being disposal. Sometimes these foods can be sold to retail at lower prices, which do not even offset the cost of the fuel during transportation. For this reason, many times farmers throw away nonstandard products after the harvest, although they have the same nutritional value as foods that fit the standard defined by the market as being ideal.

This problem affects both large and small producers, regardless of whether they use high or low mechanisation. In such cases the food waste varies according to the producer, the region in which this producer markets its products, the product itself, and the time of year.

Producers that participated in this study do not have exact quantification regarding this issue, but they report it is a usual problem faced that has a relevant impact on the amount of food thrown away. In Figs. 2 and 3, it is possible to observe potatoes rejected by retail for not attending aesthetic standards. The producer did not know what to do with these products, explaining that it would probably be discarded.

A second problem identified that affects mainly small producers is the waste of food due to the lack of marketing channels. They usually have manual and low-volume production, selling their food to supply chain intermediaries, food fairs, distribution centres, chefs, or even direct to consumers. Fig. 4 demonstrates a productive area in which there was no market for that food. Therefore, it was not be harvest. Natural resources, financial resources and labour were lost in this process.

Fig. 2 Retail-rejected food returned to producer for failing to meet aesthetic standards. (Source: Authors' own elaboration)

Fig. 3 Potatoes reject by retail. (Source: Authors' own elaboration)

Fig. 4 Disposal of food on land due to the lack of marketing channels. (Source: Authors' own elaboration)

These producers do not have planning activities for their production process, nor do they have articulation with other links in the supply chain, in order to estimate the demand. As consequence, they generate surpluses, in addition to the high concentration of the same products in the same region. When asked about the impact of their food losses and waste, these producers consider it as an inherent risk of production, with low awareness about the negative effects.

4.2 Digital Business Platforms

Platform A is a small company founded in 2016 that aims to reduce food waste by raising awareness around conscious consumption. It is an intermediary between farmers and consumers, offering a signature delivery service of baskets with food that would be discarded by producers, either because they do not fit the aesthetic standard required, or because the food production exceeds the market demand. This problem was affecting other stages of the supply chain and impacting on losses for farmers that was up to 20% of the production. They used the 'problem' as a market opportunity. Currently, the company has 850 subscribers.

Regarding their motivation the interviewee explained that:

> My wife and I have always wanted to be entrepreneurs and we like food [...] In 2014 I started to do a post graduate degree in business management, with a focus on sustainability. So, there I started to see some cool things about companies that did not just care about [...] profit, they worried about how to develop the supply chain [...]. Then we decided to see how to help this chain that is behind us, in the fruits and vegetables, the producers are the most suffering, they have many intermediates, there are many food losses.

The business itself was developed in collaboration with producers, as the two founders, who had a different initial idea of work, began to visit farmers to understand what their biggest difficulties was. Based on the exchange of knowledge with these producers, who showed them the amount of food wasted because of the same difficulties found in phase 1, specially for non-standard appearance products, the owners developed the idea of Platform A venture. The problem was most affecting small producers, mainly non-organic foods grown by family farms. Knowing the problem, they did research and were inspired by some international initiatives. From this they have entered into a supply agreement. They would purchase from these producers the foods that would be thrown away because they did not have commercialisation in the market due to the aesthetic standards or harvest larger than the demand.

The owner reports that as soon as the platform began to operate, they sought to fit a transport to go to the producers or even try to economically make the products come to São Paulo. As most of these producers send other products to the São Paulo supply centre, they combined this place as a meeting point to facilitate the transportation and make the price viable for both. On the coordination of the activity, the interviewee comments that in the beginning they went from producer to producer questioning about those products that would be wasted:

> We asked to producers: 'Do you have any problem with the product [related to food waste]? Are you going to throw away something?' And then we combined a price that justified sending the product. So, the relationship

began, and then we started to appear in the media a lot, so a lot of producers came to look for us. [...] We have 40 producers now, who already know us.

There is no formal specification on the products, what exists is an informal agreement in which the Platform guides producers frequently. The product that cannot be accepted is only those that are visibly crumpled, as they would spoil before they reach the customer. But the different appearance of the market patterns is accepted or even wanted by the platform, since the marketing of these products are part of the business strategy.

The performance on reducing food waste is positive. They estimate that they have so far prevented 300 tons of fruits and vegetables from being wasted over a period of 24 months. Positive feedback has also been received from producers: "We always talk to them, they say, 'Actually, this time 10% of the production was going to be lost, but I sold it to you.'" As for consumers, the business is much more focused on the awareness of food waste. They focus a lot on awareness talks at schools, companies and events.

At the same time, there is a process of education of producers. The interviewee tells the case that they combined the purchase of non-standard lettuce and when they went out on the farmer the other colleagues had passed with the tractor on top, as they were accustomed to this product not being sold. Many producers need to be re-educated in this sense, to know that there are alternatives to losing produce:

After we started to show this project [to producers] and they said 'it's really, that's the product that we consume here at home, it's good the same as the other' from there they began to see... but there is still that thought of 'but it has always been like that, the supermarket never bought, no one has ever bought this product from us. Will it really sell? Is not it a negative advertisement for my business? 'There are times until we try to talk' there's going to be a reporting, someone wants to talk to you" and they say 'but I'm going to talk about the bad products out of here?" but it is not a problem, it is a normal product, and when they start to see that they have a market for it, they really began to have a better relationship. But at the beginning everyone was asking: 'but are you going to be able to sell it anyway?' 'Do we have a market for it?'

The concept that producers can sell their products directly to end-customers through internet is not a new idea. The main difference that Platform A brings as DBP is in relation to the product sold, since it is a "waste/disposal" in a traditional supply chain. In this specific case, DBP provides solutions for food waste related to the two main problems identified in phase 1: a second market for nonstandard aesthetic appearance food (problem presented in Fig. 3 in Phase 1), and for products without marketing channels (problem presented in Fig. 4 in Phase 1). Through the collaborative process instituted by this DBP products out of the market gain a new space and a new appreciation.

Platform B is a small company founded in 2017 that aims to deliver fresh and regional produce direct from the producer to the consumer. The customer makes the shopping list every week and receives products at home. It offers a signature delivery service of local and seasonal food. The business proposition is to favour local development and to promote the traceability of products.

Four people work on the platform B, including two agronomist engineers and two information and communication technology managers. The idea of the business came from the perception that people are more interested to consume local food. Considering their professions, they were motivated to set up a platform that would provide local products, fresh and of good nutritional quality. Customers place the order on a weekly basis. The platform requests the producers to separate the food, which will be withdrawn directly from the producers. Then the delivery is made to the consumers using the Platform B's own assets. One of the objectives of Platform B is to educate consumers about how to store and how to handle food to improve its preservation, avoiding waste.

The relationship with the producers is "very good". There are five fixed producers. Most contacts take place either in person or through a mobile communication application, especially via "WhatsApp". They seek cooperation in the sense of also helping producers, once they have technical knowledge:

> Once a month I also try to go on their property to talk, by my training too, to see if they are in need of any help. That's the kind of thing we're trying to do, too. When we go to rural properties, we take a look, talk, discuss

something they have some doubt, express some of our opinions too, sometimes even come home with doubts to solve for them.

According to the interviewee's report, the relationship with the producers is based on trust. He explains that they do not select the aesthetic of food they will receive, it is up to the producer. The only criterion they ask producers is to maintain the quality of the product to use the products in the crop that are released by the regulation and that they are careful not to exceed the stipulated quantity. In addition to this criterion, they report that they do not make other demands on producers, especially considering that the partnership has worked well.

Also, in relation to food waste, a proposal of them is to instruct the producer on how much he has to plant due to demand planning, since they have witnessed that one of the farmers is without demand for one of his products. In this sense, Platform B address food waste solutions for products without marketing channels (problem presented in Fig. 4 in Phase 1), as well as educating the farmer to prevent this problem from occurring. Therefore, through the collaborative process instituted by this DBP products out of the market gain a new space and a new appreciation.

Internally, they have worked to minimise waste, which is virtually nil because of the way the business is organised, especially in relation to seeking to deliver local products that require less time for storage and transportation. As they say: "You lose less, if you deliver straight [to consumer]; you do not have this loss of the product in the market. Also, as we always deliver fresh product well, you do not lose in the hour that will clean this product". Despite this, due to the use of packaging, they generate packaging waste. The problem is indicated by themselves, and they have been looking for packaging and carton solutions that will reduce the impact on the product during handling.

Platform C was found in 2012 and sells oranges directly from the farm to consumers. The business began after the owner observed that a relevant part of the fruits of the farm had been discarded and served as fertilizer, since the juice producing industries were with excess of supply. The business operates through a system of signatures. This is because the rule is that customers do not buy only once, but usually receive the products to help. This helps to predict demand. The customer receives a package of oranges of the chosen weight once a week. There is also the possibility of making single purchases. Clients are both individuals, who can select up to monthly delivery option, such as legal entities, for example, restaurants.

On motivation, the interviewee explains that her family has worked exclusively with oranges for about 40 years. The motivation to begin the platform:

> It started on a weekend that I was on the [family's] farm, the industry had rejected the orange because they were surplus of product [...] The orange, if it has no consumption, it is harvested, ends up throwing on the ground, and serves as natural fertiliser. Being able to avoid, if this orange that had been thrown could have been consumed of another form, I as a person would have felt happier.

Initially, she began to sell oranges on Facebook and gradually the business evolved and grew. Currently, there are ten people working exclusively with the platform. All products harvested are sent to customers, there are no appearance criteria, as long as they are fit for consumption. Although not rejecting nonstandard aesthetic appearance, this is not the great contribution of Platform C. The mains contribution to solve the problems identified in phase 1 by Platform C is related to the fact that it provides food waste solutions for products without marketing channels (problem presented in Fig. 4 in Phase 1). Therefore, through the collaborative process instituted by this DBP products out of the market gain a new space and a new appreciation.

As a rule, the relationship is with a single supplier, who is your family, so coordination occurs very informally. In fact, producer and platform "mix in this case, it is a group of people divided, but it is a family working for the same purpose." Only when there is an off season, if necessary, look for some partner to provide oranges. The orange is harvested and boxed by the producer. From there it is placed in a truck that goes to the central of supply in São Paulo, along with other deliveries. From there, the platform makes distribution to all customers. Since oranges are delivered directly to customers three days after harvest and the distribution is local, in addition to the business of selling the food that would be wasted, minimises storage and transportation damage.

Table 3 presents a summary of the results in phases 1 and 2:

Table 3 Summary of the results in phases 1 and 2

Category	Food waste problems identified in phase 1	Food waste solutions identified in Phase 2		
		Digital platform A	Digital platform B	Digital platform C
Collaboration	• Low level of communication • Low exchange of information • No formation of social ties	• Hight level of communication • High exchange of information • Social ties (trust, commitment) resulting from personal encounters with the producer • Planning and logistics integration (in relation to the delivery point)	• Hight level of communication • High exchange of information • Social ties (trust, commitment) resulting from personal encounters with the producer • Producers educations and help related to plan production according to demand	• Hight level of communication • High exchange of information • Social ties (trust, commitment) resulting from personal encounters with the producer
Coordination	• Mostly formal mechanisms (legal contracts) • Command structures very focused on retail • High degree of standardisation	• Informal mechanisms (mainly self-regulations and informal social ties) • Command structure focused on DBPs with less degree of standardisation • Information sharing	• Informal mechanisms (mainly self-regulations and informal social ties) • Command structure focused on DBP with less degree of standardisation • Information sharing	• Informal mechanisms (mainly informal social ties) • Command structure focused on DBP with less degree of standardisation • Information sharing

Food waste reduction	• Food waste due to the lack of marketing channels (there is no way to plan demand in advance) • Food Waste generated by aesthetic requirements, which discard products that are good for consumption	• The model of signatures allows the forecast of the demand, being able to scale in advance the quantity that will be necessary. • New market for products that would be wasted for aesthetic reasons or demand • Promotion of consumer awareness related to the acceptance of non-standard products	• The model of signatures allows the forecast of the demand, being able to scale in advance the quantity that will be necessary. • Consumer education to avoid waste of food at household level	• The model of signatures allows the forecast of the demand, being able to scale in advance the quantity that will be necessary. • New market for products that would be wasted for aesthetic reasons or lack of demand

Source: Authors' own elaboration

5 Discussion

Preliminary findings of the cross-case analysis indicate five forms that Digital Food Platforms contribute to food waste solutions improving supply chain coordination:

First, they establish collaboration to promote a new market for products that would be discarded due to aesthetic standards set by retail or surplus demand. This new market is characterised by less degree of standardisation which in these cases rely on informal coordination mechanisms. This is in line with propositions established by the literature, since according to Gulati and Singh (1998) pooled interdependence are likely to have structures with less hierarchical control than those with interdependent alliances. The collaboration was an important factor in determining the initial structure of the supply chain and the success of the relationship (Alvarez et al., 2010).

DBP promote a short supply chain with a positive result on food waste reduction. First because they decrease transportation distances avoiding losses inherent to this stage. Second, as proposed by Alvarez et al. (2010), few actors in this supply chain led to intense and frequent communication among all parties. Since communications schedule were not formally stipulated, coordination mechanisms are mostly informal, relying on norms rather than on formal contracts (Alvarez et al., 2010). Also, communication has been found to improve buyer-supplier relationships and it represents an important enabler to capture value from the supply chain. In fact, logistics integration and enhanced communication are collaborative sustainable supply chain practices (Beske et al., 2014). Therefore, DBP supply chains are related to both logistics and cooperation aspects with positive results on food waste reduction.

The delivery system facilitates aspects of planning and management. It improves the problem faced by producers of food waste due to the lack of marketing channels. Seasonal fluctuations in sales faced by producers is a problem identified in the literature (Kilelu et al., 2013) that DBP seem to help reduce.

The DBPs also encourage local markets, acting as a bridge between producers and consumers. As proposed by Holland and Gutiérrez-Leefmans (2018), e-commerce platforms are designed specifically to help small businesses. DBP engage mostly small family farms in a collaborative network that can provide better quality of life to them. This is an important aspect, since although small farmers play a significant role in sustainable food supply chains, they have weak market power compared to their competitors and are often neglected in published research studies that optimise production and supply (Zhu et al., 2018).

Local distribution is also related to food waste reduction. All these benefits enhance transparency, reputation, trust, and image of supply chain agents. Reputation is some kind of informal social control mechanisms related to trust (Dekker, 2004). Trust was found to be an important element in defining the initial conditions and governance mechanisms (Alvarez et al., 2010), since it can reduce the likelihood of hierarchical controls on alliances (Dekker, 2004; Gulati & Singh, 1998). Trust acts as an informal mechanism that contributes to better relationship quality and facilitates knowledge sharing (Touboulic & Walker, 2015).

Finally, some of the DBP also promote consumer awareness regarding non-standard products. It is a relevant aspect, because consumers need to perceive value in this new market created by DBP. This is in line with Touboulic and Walker (2015) who identified that investments in relation-specific assets, such as substantial knowledge exchange or learning process between customers and suppliers in the sense of having a common goal and structuring solutions are some of the enablers of collaboration in supply chain. This bridges the gap between production and consumption.

Positive results are visible, since the reduction of waste is significant. For example, Platform A prevented a total of 300 tons of food from being lost and wasted in the last two years, resulting in various environmental and social benefits. Also, horticultural producers receive higher incomes and consumers pay lower prices. Returning to the initial proposition, our findings indicate that DBP contribute to food waste solutions improving supply chain coordination mostly by informal mechanisms, increasing collaboration aspects between supply chain agents. DBPs act as a bridge between production and consumption, creating a new market for food that would be discarded in many stages of the supply chain either by non-standard conditions or by the lack of market demand, but they also create the consumer demand through awareness and subscription purchase mechanisms.

6 Final Remarks

As a theoretical contribution, this study illustrates the interface between innovative business models and sustainable supply chains in a megalopolis such as Sao Paulo. It also highlights the importance of digital business platforms as an alternative coordination mechanism in the supply chain and the business opportunities that arise from such practices. DBPs act as a bridge between horticultural producers located in peri-urban areas and urban consumers who are more aware of sustainability. The practical implications of this research are that it presents opportunities to reduce food losses and waste through digital business platforms, able of positively affecting food supply chain sustainability addressing all three dimensions of sustainability. It reinforces recent trend in supply chain management literature of looking not only to vertical linkages but to other stakeholders that have a social driver and are influencing changes on behaviour (see, for example, Johnson, Dooley, Hyatt, & Hutson, 2018)

The most important social impact refers to the fact that food losses and waste theme is part of the new agenda for integrating efforts to increase global sustainability, specifically food and nutrition security (2030 Sustainable Development Agenda). A better coordination of food supply chain proposes an active role of private sector to reduce FLW and consequently improve food and nutrition security in an emerging country that has a double burden (undernutrition and overweight/obesity).

A research limitation is that this research only considers data from three digital business platforms located in Sao Paulo and their main stakeholders. These initiatives are recent and represent a small share of fresh food supply chain. Future studies should explore different contexts and conduct longitudinal studies, for example, returning to these platforms in a few years. Future research may also explore the coordination mechanisms of DBP with other supply chain agents, since this study was carried out with short supply chains.

Finally, as recommendations for the survival of these BPDs it is first suggested to promote consumer awareness in order to stimulate the demand for these products and increase their gains in scale. In terms of coordination, it is suggested to approach with producers through mechanisms of informal coordination, seeking to expand aspects of trust and valuations. Observations throughout this research indicate that, for producers, personal contact, the valorization of their products and less formality are essential aspects to expand collaboration and coordination activities.

References

Aghalaya, S. N., & Verma, P. (2018). Identifying sustainability metrics for a coordinated value chain using QFD. The 5th international EurOMA Sustainable Operations and Supply Chains Forum, University of Kassel, Germany, March 2018.

Ahi, P., & Searcy, C. (2013). A comparative literature analysis of definitions for green and sustainable supply chain management. *Journal of Cleaner Production, 52*, 329–341.

Alexander, P., Brown, C., Arneth, A., Finnigan, J., Moran, D., & Rounsevell, M. D. A. (2017). Losses, inefficiencies and waste in the global food system. *Agricultural Systems, 153*, 190–200. https://doi.org/10.1016/j.agsy.2017.01.014

Alvarez, G., Pilbeam, C., & Wilding, R. (2010). Nestlé Nespresso AAA sustainable quality program: An investigation into the governance dynamics in a multi stakeholder supply chain network. *Supply Chain Management: An International Journal, 15*(2), 165–182.

Aschemann-Witzel, J., de Hooge, I., Amani, P., Bech-Larsen, T., & Oostindjer, M. (2015). Consumer-related food waste: Causes and potential for action. *Sustainability, 7*(6), 6457–6477.

Bakos, Y. (1998). The emerging role of electronic marketplaces on the Internet. *Communications of the ACM, 41*(8), 35–42.

Beske, P. (2012). Dynamic capabilities and sustainable supply chain management. *International Journal of Physical Distribution & Logistics Management, 42*(4), 372–387.

Beske, P., Land, A., & Seuring, S. (2014). Sustainable supply chain management practices and dynamic capabilities in the food industry: A critical analysis of the literature. *International Journal of Production Economics, 152*, 131–143.

Brancoli, P., Rousta, K., & Bolton, K. (2017). Life cycle assessment of supermarket food waste. *Resources, Conservation and Recycling, 118*, 39–46.

Buzby, J. C., & Hyman, J. (2012). Total and per capita value of food loss in the United States. *Food Policy, 37*(5), 561–570.

Cao, Z., & Lumineau, F. (2015). Revisiting the interplay between contractual and relational governance: A qualitative and meta-analytic investigation. *Journal of Operations Management, 33–34*, 15–42.

Chen, I. J., & Paulraj, A. (2004). Towards a theory of supply chain management: The constructs and measurements. *Journal of Operations Management, 22*(2), 119–150.

Chen, L., Zhao, X., Tang, O., Price, L., Zhang, S., & Zhu, W. (2017). Supply chain collaboration for sustainability: A literature review and future research agenda. *International Journal of Production Economics, 194*, 73–87.

Chu, S. C., Leung, L. C., Van Hui, Y., & Cheung, W. (2007). Evolution of e-commerce Web sites: A conceptual framework and a longitudinal study. *Information & Management, 44*(2), 154–164.

Dangelico, R. M., Pontrandolfo, P., & Pujari, D. (2013). Developing sustainable new products in the textile and upholstered furniture industries: Role of external integrative capabilities. *Journal of Product Innovation Management, 30*(4), 642–658.

Dania, W. A. P., Xing, K., & Amer, Y. (2018). Collaboration behavioural factors for sustainable agri-food supply chains: A systematic review. *Journal of Cleaner Production, 186*, 851–864.

Dekker, H. C. (2004). Control of inter-organizational relationships: Evidence on appropriation concerns and coordination requirements. *Accounting, Organizations and Society, 29*(1), 27–49.

Eisenmann, T., Parker, G., & Van Alstyne, M. W. (2006). Strategies for two-sided markets. *Harvard Business Review, 84*(10), 92.

FAO. (2013). Food wastage footprint: Impacts on natural resources – Summary report. Retrieved April 13, 2017, from http://www.fao.org/

Foley, J. A., Ramankutty, N., Brauman, K. A., Cassidy, E. S., Gerber, J. S., Johnston, M., et al. (2011). Solutions for a cultivated planet. *Nature, 478*(7369), 337–342.

Gawer, A. (2009). Platform dynamics and strategies: From products to services. *Platforms, Markets and Innovation, 45*, 57.

Gawer, A. (2014). Bridging differing perspectives on technological platforms: Toward an integrative framework. *Research Policy, 43*(7), 1239–1249.

Gawer, A., & Cusumano, M. A. (2008). Platform leaders. *MIT Sloan Management Review*. 68–75. Boston, MA: MIT Sloan School of Management.

Gawer, A., & Cusumano, M. A. (2014). Industry platforms and ecosystem innovation. *Journal of Product Innovation Management, 31*(3), 417–433.

Gawer, A., & Henderson, R. (2007). Platform owner entry and innovation in complementary markets: Evidence from Intel. *Journal of Economics & Management Strategy, 16*(1), 1–34.

Giannakis, M., Doran, D., & Chen, S. (2012). The Chinese paradigm of global supplier relationships: Social control, formal interactions and the mediating role of culture. *Industrial Marketing Management, 41*(5), 831–840.

Göbel, C., Langen, N., Blumenthal, A., Teitscheid, P., & Ritter, G. (2015). Cutting food waste through cooperation along the food supply chain. *Sustainability, 7*(2), 1429–1445.

Govindan, K. (2018). Sustainable consumption and production in the food supply chain: A conceptual framework. *International Journal of Production Economics, 195*, 419–431.

Grandori, A. (1997). Governance structures, coordination mechanisms and cognitive models. *Journal of Management & Governance, 1*(1), 29–47.

Gulati, R., & Singh, H. (1998). The architecture of cooperation: Managing coordination costs and appropriation concerns in strategic alliances. *Administrative Science Quarterly, 43*, 781–814.

Gulati, R., Wohlgezogen, F., & Zhelyazkov, P. (2012). The two facets of collaboration: Cooperation and coordination in strategic alliances. *Academy of Management Annals, 6*, 531–583.

Gustavsson, J., Cederberg, C., Sonesson, U., Van Otterdijk, R., & Meybeck, A. (2011). *Global food losses and food waste: Extent, causes and prevention.* Rome: FAO.

Holland, C. P., & Gutiérrez-Leefmans, M. (2018). A taxonomy of SME e-commerce platforms derived from a market-level analysis. *International Journal of Electronic Commerce, 22*(2), 161–201.

Johnson, J. L., Dooley, K. J., Hyatt, D. G., & Hutson, A. M. (2018). Emerging discourse incubator: Cross-sector relations in global supply chains: A social capital perspective. *Journal of Supply Chain Management, 54*, 21–33.

Kawano, B. R., Mores, G. D. V., Silva, R. F. D., & Cugnasca, C. E. (2012). Estratégias para resolução dos principais desafios da logística de produtos agrícolas exportados pelo Brasil. Revista de Economia e Agronegócio/Brazilian Review of Economics and Agribusiness, 10(822-2016-54244), 71–88.

Kilelu, C. W., Klerkx, L., & Leeuwis, C. (2013). Unravelling the role of innovation platforms in supporting co-evolution of innovation: Contributions and tensions in a smallholder dairy development programme. *Agricultural Systems, 118*, 65–77.

Lumineau, F., & Henderson, J. E. (2012). The influence of relational experience and contractual governance on the negotiation strategy in buyer-supplier disputes. *Journal of Operations Management, 30*(5), 382–395.
MacCarthy, B. L., Blome, C., Olhager, J., Srai, J. S., & Zhao, X. (2016). Supply chain evolution–theory, concepts and science. *International Journal of Operations & Production Management, 36*(12), 1696–1718.
Moazed, A., & Johnson, N. L. (2016). *Modern monopolies: What it takes to dominate the 21st-century economy*. New York: St. Martin's Press.
Neutzling, D. M., Land, A., Seuring, S., & do Nascimento, L. F. M. (2018). Linking sustainability-oriented innovation to supply chain relationship integration. *Journal of Cleaner Production, 172*, 3448–3458.
Nooteboom, B. (1999). Innovation and inter-fi rm linkages: New implications for policy. *Research Policy, 28*, 793–805.
Pagell, M., & Wu, Z. (2009). Building a more complete theory of sustainable supply chain management using case studies of 10 exemplars. *Journal of Supply Chain Management, 45*(2), 37.
Papargyropoulou, E., Lozano, R., Steinberger, J. K., Wright, N., & bin Ujang, Z. (2014). The food waste hierarchy as a framework for the management of food surplus and food waste. *Journal of Cleaner Production, 76*, 106–115.
Parfitt, J., Mark, B., & Sarah, M. (2010). Food waste within food supply chains: Quantification and potential for change to 2050. *Philosophical Transactions of the Royal Society B: Biological Sciences, 365*(1554), 3065–3081.
Pilbeam, C., Alvarez, G., & Wilson, H. (2012). The governance of supply networks: A systematic literature review. *Supply Chain Management: An International Journal, 17*(4), 358–376.
Pullman, M., & Wikoff, R. (2017). Institutional sustainable purchasing priorities: Stakeholder perceptions vs environmental reality. *International Journal of Operations & Production Management, 37*(2), 162–181.
Rodríguez-Díaz, M., & Espino-Rodríguez, T. F. (2006). Redesigning the supply chain: Reengineering, outsourcing, and relational capabilities. *Business Process Management Journal, 12*(4), 483–502.
Ruutu, S., Casey, T., & Kotovirta, V. (2017). Development and competition of digital service platforms: A system dynamics approach. *Technological Forecasting and Social Change, 117*, 119–130.
Sancha, C., Gimenez, C., Sierra, V., & Kazeminia, A. (2015). Does implementing social supplier development practices pay off? *Supply Chain Management: An International Journal, 20*(4), 389–403.

Sauer, P. C., & Rebs, T. (2018). Identifying theories, methods, and applications in qualitative and quantitative SSCM research – A Delphi Study. The 5th international EurOMA Sustainable Operations and Supply Chains Forum, University of Kassel, Germany, March 2018.

Seuring, S., & Müller, M. (2008). From a literature review to a conceptual framework for sustainable supply chain management. *Journal of Cleaner Production, 16*(15), 1699–1710.

Silvestre, B. S. (2015). A hard nut to crack! Implementing supply chain sustainability in an emerging economy. *Journal of Cleaner Production, 96*, 171–181.

Stabell, C. B., & Fjeldstad, Ø. D. (1998). Configuring value for competitive advantage: On chains, shops, and networks. *Strategic Management Journal, 19*(5), 413–437.

Stuart, T. (2009). *Waste: Uncovering the global waste scandal.* London, UK: Penguin.

Thomas, L. D., Autio, E., & Gann, D. M. (2015). Architectural leverage: Putting platforms in context. *Academy of Management Perspectives, 28*(2), 198–219.

Thyberg, K. L., & Tonjes, D. J. (2016). Drivers of food waste and their implications for sustainable policy development. *Resources, Conservation and Recycling, 106*, 110–123.

Touboulic, A., & Walker, H. (2015). Love me, love me not: A nuanced view on collaboration in sustainable supply chains. *Journal of Purchasing and Supply Management, 21*(3), 178–191.

Vachon, S., & Klassen, R. D. (2008). Environmental management and manufacturing performance: The role of collaboration in the supply chain. *International Journal of Production Economics, 111*(2), 299–315.

Yin, R. K. (2017). *Case study research and applications: Design and methods.* Los Angeles: Sage Publications.

Zacharia, Z. G., Nix, N. W., & Lusch, R. F. (2011). Capabilities that enhance outcomes of an episodic supply chain collaboration. *Journal of Operations Management, 29*(6), 591–603.

Zhong, R., Xu, X., & Wang, L. (2017). Food supply chain management: Systems, implementations, and future research. *Industrial Management & Data Systems, 117*(9), 2085–2114.

Zhu, F., & Furr, N. (2016). Products to platforms: Making the leap. *Harvard Business Review, 94*(4), 72–78.

Zhu, Z., Chu, F., Dolgui, A., Chu, C., Zhou, W., & Piramuthu, S. (2018). Recent advances and opportunities in sustainable food supply chain: A model-oriented review. *International Journal of Production Research, 56*, 1–23.

The Role of Food Hubs in Enabling Local Sourcing for School Canteens

Laura Palacios-Argüello, Ivan Sanchez-Diaz, Jesus Gonzalez-Feliu, and Natacha Gondran

1 Introduction

Food supply chains face multiple challenges related to traceability, transparency, sustainability, trust and flexible legislation (De Fazio, 2016). These challenges have increased consumer mistrust in globalized agri-food systems, and have led to a growing interest in, and calls for, more environmentally sustainable, local and organic food (Aubry & Kebir, 2013). Politicians and public authorities have noticed these concerns, and have responded by requesting food industries to re-examine the

L. Palacios-Argüello • J. Gonzalez-Feliu • N. Gondran
Institut Henri Fayol, École des Mines de Saint-Étienne, UMR CNRS 5600 EVS, Saint-Étienne, France
e-mail: laura.palacios@emse.fr; jesus.gonzalez-feliu@emse.fr; natacha.gondran@emse.fr

I. Sanchez-Diaz (✉)
Technology Management and Economics, Chalmers University Technology, Gothenburg, Sweden
e-mail: ivan.sanchez@chalmers.se

E. Aktas, M. Bourlakis (eds.), *Food Supply Chains in Cities*,
https://doi.org/10.1007/978-3-030-34065-0_8

sustainability of their supply chains, and by increasing policies and strategies aimed at increasing the competitiveness of local and organic food (De Fazio, 2016; Nicholson, Gómez, & Gao, 2011).

Such initiatives as direct sales from producers to consumers, producers' shops, urban farmers' markets and e-commerce for local quality products, continue to emerge in this context (Bosona & Gebresenbet, 2011; Holguín-Veras, Sánchez-Díaz, & Browne, 2015; Sánchez-Díaz, 2018). While these producer-focused initiatives are attractive to consumers because of the sustainability of the products offered, they face common challenges related to physical distribution costs. The main strategies to reduce costs are to decrease the number of intermediaries who do not add value to the final product, and design distribution systems that allow local rural products to reach consumers in a cost-efficient manner. Distribution systems with these characteristics are referred to in the literature as Short Food Supply Chains (SFSC) (De Fazio, 2016). In addition to the obvious benefits for local producers, SFSC can also benefit society by reducing overall negative environmental externalities (De Fazio, 2016; Migliore, Schifani, & Cembalo, 2015).

In 2009, the French Ministry of Agriculture adopted an official national definition of SFSC: "*the supply chain is short when it has at most one intermediary between the agricultural producer and the consumer*" (Aubry & Kebir, 2013). SFSC stimulate food relocation by reducing intermediaries and distance by creating proximity between producers and consumers (Aubry & Kebir, 2013). According to FAO, SFSC are considered an invaluable contribution to the fight against hunger and malnutrition, and helpful in alleviating social, economic, and environmental sustainability issues (Rodriguez Reyes, 2012).

Policymakers have identified the market segment of institutional catering serving public entities as one with great potential to increase the share of local and organic food because public authorities can establish strict specifications for products and suppliers. However, they also recognise the need to redesign the physical distribution system to make SFSC more competitive and sustainable. As Fredriksson and Liljestrand (2015) and Nordmark, Ljungberg, Gebresenbet, Bosona, and Jüriado (2012) mention, this requires an analysis and redesign of current logistics systems to identify issues and potential solutions. One key strategy to enable a successful implementation of short, sustainable and efficient food supply chains is the introduction of Food Hubs (FH). FH are specialized logistics platforms where supplies from multiple local vendors are gathered and consolidated before being distributed either to local kitchens for further preparation (i.e., cooking), or directly to consumers locations (Morganti, 2011), which correspond to canteens in institutional catering (Morganti & Gonzalez-Feliu, 2015).

This chapter will identify the functions and characteristics of FH used for institutional catering, particularly school canteens, and analyse how these FH should be designed to support the objective of increasing local and organic products for the distribution of meals in these settings.

2 Background and Context

2.1 Institutional Catering

Food deliveries generate a significant amount of urban freight traffic, and food customers often demand customized delivery services, which lead to various logistics and organisational constraints (Behrends, 2016). To ensure that food remains fresh when delivered, this sector must find ways to cope with unpredictability, such as reducing order sizes and increasing the frequency of deliveries. Practices such as just-in-time deliveries have become common in this market.

Catering services is a retail activity located at the very end of food supply chains (Eriksson, Osowski, Malefors, Björkman, & Eriksson, 2017). A catering service can be classified depending on the customer and type of contract. Catering services can be divided between (1) commercial catering and (2) institutional catering, depending on whether the food traded is part of the customer's commercial activity, or the customer purchases catering services as a secondary service to feed regular consumers at an institution. Commercial catering includes food deliveries to cafeterias, food self-service, and fast-food restaurants; it represents 51% of out-of-home meals in France. Institutional catering provides regular meals to an institution under an agreement or a contract made with a private or public-sector entity; it represents 49% of out-of-home meals in France (ADEME, 2016). Catering management can be either internal or outsourced, via lease and concession contracts. Overall, catering services represent about 12% of food consumed in France, corresponding to 7.3 billion meals per year (Lessirard, Patier, Perret, & Richard, 2017).

Lessirard et al. (2017) classify institutional catering based on the type of consumer:

- School catering: serves primary school, public and private high schools, colleges and university canteens.
- Enterprise or corporate catering: serves restaurants for private companies and restaurants for local authorities.
- Health catering: serves hospitals canteens (private or public) and social establishments, such as nursing homes.

Institutional catering represents a stable market for all food chains with a growth of 0.4% to 0.5% per year. The direct management and delegated management of institutional catering represents 60% and 40%, respectively (ADEME, 2016).

In institutional catering, the customer defines the meal specifications in terms of menu quality and food quantity based on the consumer's health recommendations. The catering company then analyses the customer's food requirements and coordinates with its suppliers, who may be wholesalers or food distributors, food cooperatives, and/or primary producers. Figure 1 explains the flow of information between the stakeholders, which operates in the opposite direction of the flow of food.

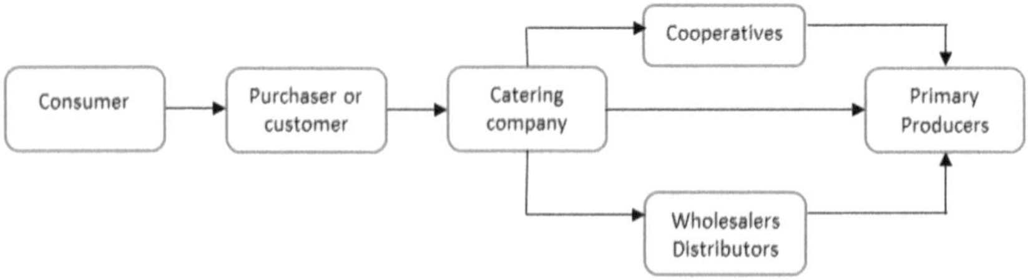

Fig. 1 The flow of information among institutional catering stakeholders. (Adapted from Lessirard et al., 2017). (a) The consumer could be the parent or the student who eats at the school canteen; the patient in the hospital; or the employee in the enterprise, among others. (b) The customer or purchaser could be a territorial authority, public institution, or private enterprise

In institutional-concession catering, local institutions (customer/purchaser) can identify preferred producers types and encourage agreements that prioritize local products. This is achieved by establishing annual quantities to be purchased, negotiating the price directly with producer cooperatives, investing in the development of new products such as organic local products, and sharing transportation among producers. In the French market, institutional-concession catering is dominated by three international groups, which represent 75% of the turnover of concession catering (Lessirard et al., 2017).

According to Lessirard et al. (2017), on average, 70% of French public purchasers demand that the food served in their canteens comes from the local department (i.e. administrative subdivision used in France) or from neighbouring departments. In fact, during 2017, 100% of the calls for tenders showed that all local authorities introduced a requirement for short circuits and local products in institutional catering. In institutional catering in Paris, local products represent 20–30% of the institutional catering food supply, but 80% are local fresh products and 90% are local fruits and vegetables (Lessirard et al., 2017). Some strategies implemented in French institutional-concession catering aimed at fostering SFSC and increasing the share of local products include: (1) developing regional FH to increase consolidation and decrease kilometres driven per ton of products transported between producers and customers; (2) using central kitchens to combine and transform local fresh vegetables from different producers in a facility that also serves as a consolidation centre; and (3) designing menus composed exclusively of seasonal products.

2.2 FH for Organic and Local Products

FH Definition

Multiple definitions of FH have emerged, see E. Morganti (2019) and L. P. Palacios-Argüello and Gonzalez-Feliu (2016). Research on FH focuses on the role of specialized logistics platforms for food product collection and distribution (Morganti & Gonzalez-Feliu, 2015). This type of distribution system structure often helps to improve service, and the adaptability to demand and lower transport costs through consolidation, which encourages sustainable food transportation (Morganti, 2011). FH seek to coordinate the distribution of food products from the same origin to conventional or hybrid markets through agreements between the food chain stakeholders (Cunha & Elmes, 2015; Matson & Thayer, 2013; Morganti, 2011; Morley, Morgan, & Morgan, 2008).

The United States Department of Agriculture (USDA), defines the concept of a Food Hub as "*An organization that actively manages the aggregation, distribution, and marketing of source-identified food products primarily from local and regional*" to strengthen local capacities to meet the demands of wholesalers customers, retailers, and end customers (Barham et al., 2012).

Eleonora Morganti (2011) extends the classical notion of FH (Morley et al., 2008) to urban logistics. The author distinguishes the following stakeholders as the main actors in urban FH:

- *Wholesalers, and in a minor proportion, producers* produce and store food wholesale products, and can be considered commercial and logistical nodes in the food supply chain.
- *Urban transport and logistics agents* perform the last-mile transport and delivery services in urban areas. They can be suppliers, manufacturers, carriers, and logistics service suppliers (3PLs).
- *Retailers and food services* are the final receptors of food products. This type of stakeholder includes traditional outlets, distribution marketplaces, Ho.Re.Ca[1] stakeholders, and institutional catering, such as, school canteen systems.

L. P. Palacios-Argüello and Gonzalez-Feliu (2016) considers a slightly different classification by connecting stakeholders in urban FH to the traditional stakeholder classification in supply chain management:

- *Load Generators* (LG), *or shippers*, are the stakeholders that generate the logistics flows, i.e. those that are at the origin of food shipping.
- *Load Beneficiaries* (LB) are the final destinations of foods, i.e. those who benefit from the cargo.

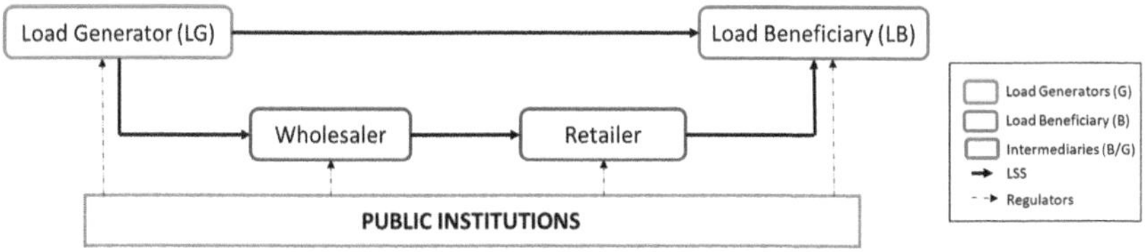

Fig. 2 Food Hub Stakeholders. (Adapted from Palacios-Argüello & Gonzalez-Feliu, 2016)

- *Logistics Services Supplier* (LSS) are the carriers in charge of transporting the food between generators and beneficiaries.
- *Public Institutions* regulate the role of each stakeholder.

Several stakeholders can have several roles at the same time. Figure 2 shows the relationship between these stakeholders.

Palacios-Argüello, Morganti, and Gonzalez-Feliu (2017) define a Food Hub in its broadest sense as a collaborative system between producer, distributor, and trader, eliminating middlemen to shorten the food supply chain. Its main function is to strengthen the supply of agro-industrial products by adding value to the final product. The characteristics of FH can be related to their commercial or logistical interests. This is the case of FH aimed at increasing the organic local products offered for catering. The authors also propose a typology of FH based on an analysis of the literature summarized also in L. P. Palacios-Argüello and Gonzalez-Feliu (2016), and presented in Table 1.

3 Research Design and Methods

The aim of this research is to identify the functions and characteristics of FH used for institutional catering, and analyse how these FH should be designed to support the objective of increasing local and organic products. In so doing, a mixed approach that combines quantitative and qualitative data production is proposed. The method designed is based on the principles of case study research (Bryman & Bell, 2015; Eisenhardt, 1989; Maxwell, 2013; Yin, 1994).

Table 1 Considerations of FH

FH elements	Description	
Stakeholder	• Focused on the retailer • Focused on the producer • Focused on the consumers • Hybrid models • Led by the wholesale and food services	
Structure	• Non-profit organization: Developed from community-based initiatives. • External private structure: Limited Liability Company, or other corporate structure. • Cooperative owned either by producers and/or consumers • Led by the public sector	
Functions	*Commercial functions* • Processing of convenience, i.e. activities that add value to the product, like washing, peeling, and cutting food, among others. • Aggregation, classification packaging. • Marketing, promotion, sales. • Training activities for accounting or commercial purposes • Payment and financial services.	*Logistical functions* • Food consolidation and distribution services (first and/or last mile). • Urban logistics areas and aggregation of logistics activities. • Warehousing and storage services. • Inventory management. • Transport organization and planning.

Adapted from L. Palacios-Argüello et al. (2017)

The Auvergne-Rhône-Alpes region in France was selected as the case study given: (1) its goal of using exclusively local and organic food in the public-school canteens system; and (2) the canteens' suppliers have successfully implemented FH to enable short supply chains.

The method is developed in two phases, the first seeking to understand the role of FH in short supply chains, and the second seeking to assess quantitatively different designs of a Food Hub-based distribution system.

3.1 First Phase

This phase uses a qualitative research strategy where data is collected via interviews with key private and public-sector stakeholders involved in the food distribution system. A descriptive and interpretive approach is used to analyse these data and to understand the role of FH in short supply chains.

The data collection includes primary and secondary data. The secondary data uses a documentary analysis of scientific and technical/legislation documents to describe cases and context. The data is complemented with primary data collected through semi-structured interviews based on open-ended and theory-driven questions. These interviews were performed with stakeholders of the institutional catering supply chains in the Auvergne-Rhône-Alpes region of France. The sample was designed following the snowball sampling method (Bryman & Bell, 2015). The first interview was held with the head of the school meals program at Saint-Etienne, who referred us to the regional manager of organic local offer for institutional catering in Auvergne-Rhône-Alpes. At the end of the interviews the regional managers pointed identified other regional managers who could be contacted to ensure a complete overview of the system.

To perform the semi-structured interviews and ensure high-quality research, two documents were created: an interview protocol and an interview guide. The protocol was used to explain the purpose of the interview: the study aim, time, type of questions, main topics to consider, feedback, and use of information. The interview guide was used to make the questions asked as effective and efficient as possible, without forgetting any important topic, and to document all impressions and experiences from each interview.

Each potential stakeholder was contacted via e-mail. The e-mails included the purpose of the study, a description of the research area, and an invitation to be interviewed. Potential respondents were informed of the study aims and the structure of the case studies. In total, nine semi-structured interviews were conducted. The duration of each interview was between 30 and 90 minutes. All interviews were conducted in French. Table 2 shows the position of the stakeholder interviewee, and the date of the first interview. These interviews were followed up with a validation process, and secondary interviews were performed from 2017 to 2018.

Additional discussions were held with researchers to verify the quality of the research approach, and to ensure that a sufficient number of stakeholders were interviewed. The interviews were analysed using secondary data for comparison. To increase trustworthiness, the secondary data of

Table 2 Information about interviewees

Stakeholder	Position of the interviewee(s)	Date of the first interview
Public authority	Head of the school meals program at Saint-Etienne	25/05/2016
Meal contractor	Manager of meal distribution for school canteens at Saint-Etienne	18/11/2016
Public authority	Manager of organic local offer for institutional catering in Rhône and Loire departments.	22/11/2017
Public authority	Regional manager of organic local offer for institutional catering in Auvergne-Rhône-Alpes	23/10/2017
FH1	Marketing manager, in charge of the local producers, transformers and institutional catering customers' communication.	11/06/2018
FH2	Supply manager, in charge of local producers and transformers	15/06/2018
FH3	Sales Manager, in charge of central kitchens, corporate restaurants and institutional catering	11/11/2017
FH4	Supply manager, in charge of local producers and transformers	18/06/2018
FH5	Sales manager, in charge of central kitchens, corporate restaurants and institutional catering	17/07/2018

Source: Authors' own elaboration

each FH and school canteen system were read in advance of meetings. This was done to reduce any possible misunderstandings between interviewers and interviewees.

3.2 Second Phase

This phase seeks to assess quantitatively different designs of a Food Hub-based distribution system. To this effect, secondary-data provided by the stakeholders interviewed was analysed using a Vehicle Routing Problem.

The data collected from Regional Organic Agriculture Observatory encompasses data from 152 producers offering organic products (e.g. vegetables and fruits). A Vehicle Routing Problem (VRP) Spreadsheet

Solver developed by Erdoğan (2017) was then applied to these data to assess different designs of a Food Hub-based distribution system.

The alternative designs of a Food Hub-based distribution system can be:

1. Allocation of producers to FH for organic local products according to geographical proximity.
2. Allocation of producers to FH for organic local products according to the administrative subdivision (i.e., the Department, which is the political division in France).

The VRP Spreadsheet Solver is an open-source unified platform for representing, solving, and visualizing the results of vehicle routing problem (VRP). It unifies Excel, public GIS and metaheuristics. Three functions of the public GIS were used: *geocoding, which* converts addresses into the latitude/longitude values; *direction, which* is the function that returns the distance and driving time between two points; and *static maps, which* return image files defined by their centre point, zoom level, and size. In the VRP Spreadsheet Solver the optimization algorithm is of the same nature and relevance as those of route construction and fleet management software tools, and thus can be used as a reference for obtaining realistic food transport routes (Erdoğan, 2017).

4 Findings

The Auvergne-Rhône-Alpes has the second highest number of organic farms among regions of France, with about 4800 organic farms by the end of 2016. Sixty percent of organic farms in this region sell at least part of their production directly through short circuits, or through only one intermediary between producers and consumers. There are FH that have been created specifically to have a robust offer for the institutional catering market as an alternative to solve dysfunctions in the aggregation and distribution of locally grown food.

Those FH have been supported by the Regional Federation of Organic Agriculture of Auvergne-Rhône-Alpes (FRAB AuRA), entity adhered to the FNAB (National Federation of Organic Agriculture), which protects the interests of organic producers at the national level. FRAB AuRA represents the collective interests of organic farmers in Auvergne- Rhône-Alpes, with regional organizations and communities. It promotes different actions for the development of organic agriculture in the region. This entity is one of the most important interlocutors of farmers and stakeholders in the agriculture sector, representing them with public authorities.

4.1 The Role of Food Hubs (FH) in Short Supply Chains

Food Hubs (FH) Description

The common goal of these five FH is to develop fair local agricultural market practices for producers; to make the consumption of fresh, quality and seasonal products available to all; and to develop the region's economy and agricultural jobs. Table 3 summarizes the description of FH.

FH1 is an association that aims to sustainably link local agricultural production with residents and professionals of the territory by setting up supply solutions adapted to the constraints of stakeholders in the distribution. This platform was created as a project co-built by regional producers and consumers Its philosophy is to make these products accessible throughout the territory by putting logistics at the service of consumer needs. This FH is looking to be certified as a cooperative society with non-profit status.

FH2 is a cooperative-type company that brings together producers and processors from the region with the aim of developing profitable local agriculture for the producers. FH3 is a regional Food Hub for organic products created by ten local-organic producers motivated to work with institutional catering. They contacted other FH, and they work together to strategize the producers' network, producers' governance and cooperative work with other platforms. In 2011, FH3 became a non-profit entity.

FH4 is a regional Food Hub for organic products created to facilitate the regular and gradual introduction of quality foods and more proximity in the catering and commercial markets. It has been a non-profit since 2012. FH4 prioritizes fair trade, and allows food service providers to

Table 3 FHs characteristics

FH's characteristics	FH1	FH2	FH3	FH4	FH5
Year of creation	2005	2007	2006	2011	2005
Number of employees	5	4	6	11	11
Products	95% organic producers and 5% in organic conversion or non-organic	100% organic Meat, dairy products, fruits, vegetables, cereals	100% organic	100% organic	• 85% organic and local products • 15% foreign organic products
Number of producers	80–100	30	• 80 producers • Partnership with Biocoop Restauration to offer complements of organic range, and to offer products not available locally	• 30 local and organic producers. • 10 producers with a national offer and foreign producers.	• 140 producers • Partnership with Biocoop Restauration to offer complements of organic range, and to offer products not available locally

(continued)

Table 3 (continued)

FH's characteristics	FH1	FH2	FH3	FH4	FH5
Customers	70% represents customers from institutional catering, 15% retailers and convenience stores and 15% individual customers.	100% institutional catering: • Corporate and administrative restaurants • Commercial restaurants • School canteens.	90% institutional catering: • Corporate and administrative restaurants • School canteens. 10% Commercial restaurants	90% institutional catering: • Corporate and administrative restaurants • Holiday resorts. • Hospitals • Elderly care centres. 10% Commercial catering	90% institutional catering: • Corporate and administrative restaurants • School canteens. 10% Commercial restaurants and convenience stores. In total 140 customers
Infrastructure Logistic organization	1 platform • Upstream: Coordinate collection of the products from the FH and/or direct deliveries from producer-customer. • Downstream: 2 freight vehicles (3.5 tons and refrigerated) for delivering the products.	1 platform • Upstream: Coordinate collection of the products from the FH, 100% of the product flow passes through this platform. • Downstream: External carriers.	1 platform • Upstream: Coordinate collection of the products from the FH and/or direct deliveries from producer-customer. • Downstream: External carrier.	1 platform • Upstream: Coordinate collection of the products from the FH, 80% of the product flow passes through this platform. • Downstream: 80% External carriers, and 20% directly from producers to customers.	1 platform • Upstream: Coordinate collection of the products from the FH, 65% of the product flow passes through this platform. • Downstream: 65% external carriers, and 35% directly from producers to customers.

Source: Authors' own elaboration

work more directly with producers motivated by the democratization of local organic products in catering, promoting access to quality food for all, regular market insurance and fair remuneration for producers, and finally, raising young people's awareness of issues related to their daily food choices.

FH5 is a regional Food Hub for organic products created in 2005 by five producers who were looking for a robust offer for the institutional catering market. In the same year, the local government created a program to promote the local market structure for organic products to improve the quality of meals served to students from local colleges. Since 2015 FH5 has had non-profit status.

Comparison with the Literature Review

Based on a systematic literature review, an analysis and classification of FH according to their key attributes (e.g., stakeholder focus, structure and ownership, logistics functions and commercial services) was performed. To understand the role of these FH in the short supply chain, the classification was applied to them, following the snowball sampling method, as previously explained (see Tables 4, 5, and 6).

According to L. P. Palacios-Argüello and Gonzalez-Feliu (2016), the deployment of a Food Hub depends on the primary stakeholder of focus. The Food Hub can be focused on the producer, the retailer, the customer, or it can be a hybrid model, with a focus on several stakeholders. Table 4 shows results from the interview analysis relating to differences in stakeholder focus between each of the regional FH.

Table 4 Stakeholder focus of regional FH

Stakeholder focus	FH1	FH2	FH3	FH4	FH5
Focus on the retailer		x			
Focus on the producer	x		x	x	x
Farm - business (institutional model)	x	x	x	x	x
Signed commitment with producers	x		x	x	x
Led by the producer entrepreneur	x		x	x	x

Source: Authors' own elaboration

Table 5 Structure of regional FH

Structure	FH1	FH2	FH3	FH4	FH5
Cooperative owned by producers	x		x	x	x
Non-profit organizations	x		x	x	x
Privately owned		x			

Source: Authors' own elaboration

Table 6 Functions of regional FH

	Functions	FH1	FH2	FH3	FH4	FH5
Commercial	• Local producer markets who sales the producers' products		x			
	• Retail or diversified (wholesalers and retailers)		x			
	• Processing of convenience		x			x
	• Physical Services: physical aggregation, classification, packaging, sale and delivery of products	x	x	x	x	x
	• Intangible Services: specialists in coordinating, payment, marketing and product promotion	x	x	x	x	x
Logistical	• Producer cooperatives offer logistics services in order to increase processed volumes and reduce logistics costs	x	x	x	x	x
	• First mile consolidation: Works directly with producers to collect and store different products from various communities to centralized locations.	x	x	x	x	x
	• Last-mile Distribution: Distributes products to end customers.	x	x	x	x	x
	• Processing conservation: Processes food for product preservation including canning, pickling, and preserving in cold rooms, among others.	x	x	x	x	x

Source: Authors' own elaboration

Regarding the structure of FH, Palacios-Argüello and Gonzalez-Feliu (2016) affirm that FH can be structured as: non-profit organizations based on community initiatives; private organizations with a corporate structure; organizations owned by producers and/or customers making decisions in a decentralized manner and sharing risks; or public organizations led by the public sector. Table 5 shows the differences in structure among the five regional FH studied.

FH are collaborative systems between producers, distributors, and traders with the middlemen removed to shorten the food supply chain, and their main function is to strengthen the supply of agro-industrial products, so the nature and characteristics of FH can be commercial or logistical, which aims to add value to the final product.

Among their commercial functions, some FH: perform the marketing for local producers, adding value to the product by processing for convenience (i.e. food hub processes products to make them more convenient for the end customer. It includes washing, peeling, chopping, and/or bagging); are in charge of the producers' commercial training; offer physical services, such as aggregation, classification, packaging, sales and deliveries; and intangible services, such as payment coordination, marketing and product promotion. Among the logistics functions, there are FH that manage the first and/or the last mile distribution, those that perform the processing conservation (especially cold rooms), and those that offer logistics service to increase volumes and reduce logistics costs. Table 6 shows the differences in functions between each of the five regional FH.

Finally, the deployment of FH must consider a sustainable development logic, bringing positive economic, social, and environmental impacts to communities. Those impacts are listed below: economic impacts in Table 7; social impacts in Table 8; and environmental impacts in Table 9.

All of the FH work with a social integration enterprise, which promotes fair prices, consistent with the market and remunerative for producers, maintaining local employment. The organic farms that supply these platforms employ about 25% more manpower than regular farms. They work closely with the producers to ensure the right prices and stability of volume delivered every season.

All of the FH participate in the development of locally responsible agriculture. They develop the social bond by encouraging exchanges between agricultural actors and inhabitants. These FH give priority to farms on a human scale, farms are oriented or willing to move towards short circuits. They work with a social integration enterprise; exchanging best practices among the producers and training them.

Table 7 Economic impacts of regional FH

Economic impacts	FH1	FH2	FH3	FH4	FH5
• Improves rural/local economy through job creation and local production increase, strengthening agricultural communities and food systems in a sustainable way over time.	++	++	++	++	+++
• Increases farmers' income, profitability, and viability for producers by making them more attractive to buyers and having the advantages of larger-scale economies.	+++	++	+++	+++	+++
• Promotes good communication and information traceability, allowing producers to understand the operational costs of production, processing, transportation, and marketing. Promotes diverse product differentiation strategies to get better prices.	++	+	++	++	+++

Source: Authors' own elaboration
Note: "+" describes a slight impact, "++" describes a medium impact, and "+++" a large impact

Table 8 Social impacts of regional FH

Social impacts	FH1	FH2	FH3	FH4	FH5
• Provides training and professional development for farmers, promoting the creation of the next generation of farmers.	+		+	+	++
• Provides access to local food markets and promotes the availability of fresh and healthy food products to reduce health care costs.	+ +	+	+ +	++	+++
• Works closely with the producers, improving their living conditions.	++	+	++	+	++

Source: Authors' own elaboration

Table 9 Environmental impacts of regional FH

Environmental impacts	FH1	FH2	FH3	FH4	FH5
• Reduces environmental costs for food transportation that travels over long distances and/or through complex distribution networks	+	++	+	+	+
• Encourages farmers to improve their productive capacities to develop more reliable supplies of local and regional products grown sustainably	++	+	++	++	+++

Source: Authors' own elaboration

These FH encourage producers to go towards certified good agricultural practices, particularly organic farming. In so doing, FH are able to offer a set of healthy local-organic products with a high nutritional quality, obtained more naturally, without treatment or synthetic chemical additives, and avoiding pesticide residues. Such producers' practices help to preserve the producers' soil fertility, water quality, air quality and biodiversity, respecting animal welfare. Shorter circuits also encourage reduced packaging and fewer transport externalities, by optimizing routes of food collection and distribution.

FH are considered collaborative systems between producers, distributors, and customers that in fact shorten the food supply chain in geographical proximity by eliminating middlemen that do not add value to the final product. All of the FH studied perform various commercial and logistics functions that allow the producers to increase their benefits by adding value to the final product.

4.2 Design Assessment of a Food Hub-based Distribution System

General Considerations

This subsection explains the organic local producers considered, the FH's distribution scenarios considered, and the VRP's parameters configuration. The data was collected from Regional Organic Agriculture Observatory, and 153 producers offering organic products (e.g. vegetables and fruits) were considered for the construction of the Vehicle Routing Problem. Table 10 shows the locations of the producers considered for this study.

Table 11 shows the location of the FH, the department where the products come from (product origin) and the department where the majority of the FH' customers are located (product destination).

Nowadays the food distribution is performed according to each FH policy. The proposed distribution scenarios are: (1) Producer delivers the products to the Food Hub (Scenario 1), and (2) Coordinated collection of products from the Food Hub (Scenario 2). Figure 3 shows the distribution configuration analysed by scenario.

Table 10 Producers per department

	Number of producers			
Department	Fruits	Vegetables	Multi-product (F + V)	Total producers
Ain	2	2	–	4
Ardèche	11	17	2	30
Cantal	0	1	1	2
Drôme	21	28	7	56
Isère	1	4	–	5
Loire	3	14	1	18
Haute-Loire	2	10	–	12
Puy-de-Dôme	2	7	2	11
Rhône	3	3	2	8
Savoie	2	1	–	3
Haute-Savoie	1	3	–	4
			Total	*153*

Source: Authors' own elaboration

Table 11 FH locations, product origins and customer locations

	FH for organic local products														
	FH1			FH2			FH3			FH4			FH5		
Department	L	O	D	L	O	D	L	O	D	L	O	D	L	O	D
Ain											X	X			
Allier					X	X									
Ardèche		X	X								X				
Cantal					X	X									
Drôme	X	X	X								X				
Isère											X	X	X	X	X
Loire								X	X		X				
Haute-Loire					X	X									
Puy-de-Dôme				X	X	X									
Rhône							X	X	X		X				
Savoie										X	X	X			
Haute-Savoie											X	X			

Source: Authors' own elaboration
L FH location, *O* Origin, *D* Destination

The VRP was configured according to the following parameters:

- a fleet of vehicles serves a set of producers;
- the product collected from each customer per visit must be satisfied by a vehicle assigned on the route;
- each vehicle leaves and returns to the FH;

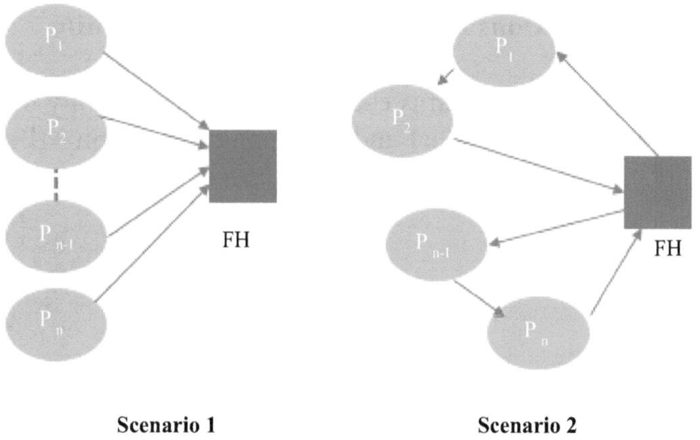

Fig. 3 FH distribution scenarios. (Source: Authors' own elaboration)

- the vehicle capacity should not exceed 900 kg (light vehicles of 3.5 tons).
- the working time constraints for the driver and vehicle on each route should not be violated, and cannot exceed 10 hours.

For the proposed test case, it is important to note that the time window restriction for the collections was not considered for this study.

Producers' Allocations According to the Administrative Subdivision

An interesting finding, which determines the design of the distribution system and thus its efficiency and sustainability, is the allocation of local producers to each Food Hub. This allocation is currently based on administrative subdivisions in France (i.e., Departments) and is not based on proximity or convenience.

The FH' and producers' locations, in relation to administrative subdivision, is shown in Fig. 4. This means that all of the FH deserve only producers located in their own department.

Mapping of the producers and FH was done using QGIS.

To compare the effects on vehicle-kilometres travelled of Scenario 1 and Scenario 2 based on the current allocation Table 12 shows the results in distance obtained with the VRP Spreadsheet Solver based on Excel.

Producers' Allocations Based on Geographic Proximity

The short food supply chain is characterized by a proximity dimension that considers both relational and geographical proximity (Bosona &

Fig. 4 Producers' allocations based on administrative subdivision. (Source: Authors' own elaboration)

Table 12 Distance results for producers' allocations based on administrative subdivision

FH	Nb producers	Distance Sc 1	Distance Sc 2
FH1	87	10,096	1384
FH2	26	3483	604
FH3	24	5308	1061
FH4	5	436	260
FH5	11	1715	631

Source: Authors' own elaboration

Gebresenbet, 2011). Geographical proximity is defined as "*the explicit spatial/geographical locality and distance/radius, within which food is produced, retailed, consumed, and distributed*". It refers to a specific area, community, place or geographical boundary that the food can come from, and it implies minimizing distance and reducing food kilometres.

To analyse the producers' allocations based on geographical proximity, it is necessary to calculate the distance between each producer and each Food Hub with the aim of minimizing the food transport distance.

The producers were first clustered in terms of their geographical proximity. In the clustering procedure, using QGIS software, five clusters of producers were formed, and allocated to the closest Food Hub. Figure 5 presents the five FH and the producers allocated to them, according to the Ellipsoidal distance, which iterates through each feature on the producers and the FH locations, and finds the closest distance.

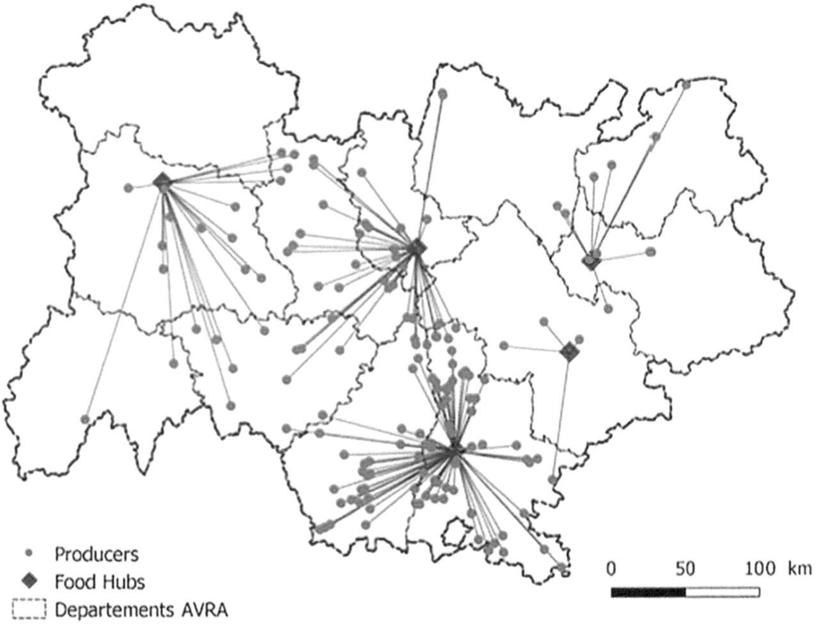

Fig. 5 Producers' allocations based on geographical proximity, using QGIS. (Source: Authors' own elaboration)

To compare the effects on vehicle-kilometres travelled of the geographical proximity-based allocation strategy, Table 13 shows the results in distance obtained with the VRP.

Figure 6 show the producers' allocations based on geographical proximity using optimised routes obtained with the VRP.

In summary, to increase the competitiveness of local food producers, it is necessary to improve the performance of their logistics by promoting the sustainability of local food systems and the development of FH. Indeed, the results show that after comparing the effects on vehicle-kilometres travelled of the current allocation and that of a geographical proximity-based allocation strategy, the latter could lead to reductions of about 8% of vehicle-kilometres travelled, using coordinated collection of products from the FH.

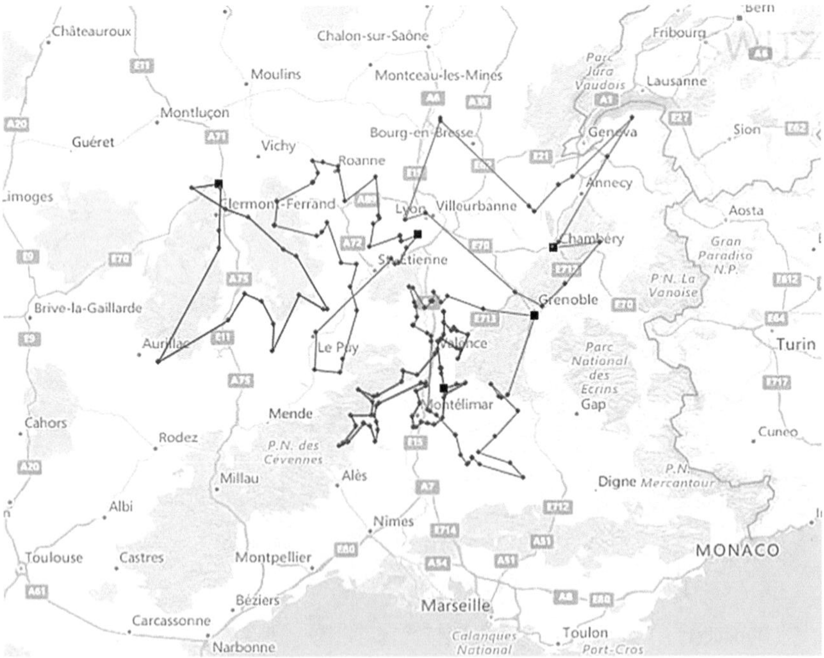

Fig. 6 Producers' allocations based on geographical proximity using VRP. (Source: Authors' own elaboration)

Table 13 Distance results for producers' allocations based on geographic proximity

FH	Nb producers	Distance Sc 1	Distance Sc 2
FH1	55	5992	721
FH2	17	3190	628
FH3	34	8536	747
FH4	16	2532	771
FH5	31	5265	763

Source: Authors' own elaboration

5 Discussion

Food Hubs (FH) are an important solution to respond to the demand for organic local products, because they allow producers to fill the demand of institutional catering's stakeholders regarding the limited number of intermediaries between the local producers and the consumer, and they strengthen the social connections between the organic local offer and the demand by offering a fresh product.

By working with a Food Hub, a producer could reduce the food distance travelled by consolidating freight, maintaining the high quality of the products and decreasing transportation costs; avoiding producers' trips that may be poorly conceived, with vehicles whose loading is not optimised.

Once a Food Hub is developed, it is important to analyse the allocation strategy. The notion of restricted territory is not clearly defined, but geographic or political limitations may be considered. When referring to local products, the distances between the producers and consumers are generally on the order of 100 km to 200 km, which corresponds to a departmental scale and refers to producers' allocation according to the administrative subdivision.

However, based on the VRP analysis, there are some distance issues to consider. Considering geographical proximity defined as the distance within which food is produced, retailed, consumed and distributed, another producer's allocation must be considered. The VRP shows that there is a distance gap between the allocation made by geographic proximity and administrative subdivision. Taking into account that the regional FH collect the producers' products, the average distance is: (1)

26 km per producer in an allocation by administrative subdivision; and (2) 24 km in an allocation by geographical proximity. This means that the best strategy is not always to allocate producers according to administrative subdivision in terms of reducing food distance travelled.

Policymakers could intervene to foster collaboration among the FH to have a more efficient solution for producers and consumers, by analysing the producers' allocations. These collaborations will help FH take advantage of their location, by conceiving optimized networks to minimize the producers' transport costs, and thus becoming more competitive than they may be within conventional distribution channels.

5.1 Perspectives of FH's Role for Organic Products Distribution

Within the next few years, the demand for organic foods in institutional catering is estimated to increase very quickly, influenced as it is by local policy. Will current FH be capable to supply this demand?

Based on the interviews conducted, it is clear that some FH currently have enough suppliers to supply the increasing demand for organic products. The suppliers see an increase in demand as a natural trend (ambitious but realistic) in the current market; they have already started to work to respond to the potential increase in demand for organic and local products.

In fact, today there is a French policy that is looking for a national requirement of 50% quality foods, including 20% organic products, in school canteen menus for 2022, and that this 20% represents the school canteen that is one-third of the institutional catering sector. Nevertheless, there are several points of vigilance that must be maintained to meet these goals:

- In terms of the volume of organic demand requirement, it will be necessary to start looking for organic product offerings in other regions in France or in Europe, which will decrease the local initiative and decrease the value or relevance of proximity. As a result, an overall balance must be struck between the organic and local demand.

- Age of producers: Currently, more than 50% of organic local producers are people over 60, which means that in ten years those producers will no longer be active producers for the institutional catering sector. Should demographic trends of young people going to the city to work, and not being drawn to or continuing with farm jobs, the aging of the organic agricultural work force will be a weakness.
- The number of organic farms must increase. To meet organic demand, a high percentage of local producers need to begin a process of organic conversion using the current lands.
- Local and organic demand fulfilment: Considering the organic and local products demand, it is necessary the FH work together to have a robust offer for all departments.
- Meal menu costs: local and organic products will increase the meal menu cost. One reason for this is increased logistic costs, which can be reduced by increasing food consolidation, economics of scale and route optimization as strategies of collaboration between the food supply chain stakeholders.

6 Conclusion

This research describes the role of FH in enabling short supply chains and allowing an increase in the share of local and organic products available for institutional catering. The research also provides a classification of different FH, and an assessment of producers' allocation strategies to FH.

This chapter identified and defined the roles and characteristics of five Food Hubs (FH) in the Auvergne-Rhône-Alpes region in France in relation to catering sector demands for local-organic products. The five FH were characterised according to the typology proposed by Palacios-Argüello et al. (2017). All are considered collaborative systems between producer, distributor, and customer, which shorten the food supply chain both in geographical proximity and by removing middlemen that do not add value to the final product.

Concerning the stakeholder focus of the FH, four out of five are focused on the producer, and are conceived as a farm business led by a collective of producer entrepreneurs. Regarding the FH structure, one out of five is managed privately, while the others are non-profit organizations or cooperatives owned by producers. All five of the FH perform various commercial and logistics functions that allow the producers to increase their benefits by adding value to the final product.

Several other positive sustainable impacts are generated after the FH deployment: there are some differences among the FH concerning the economic impacts based on structure; 4 out of 5 have the same social impacts (except the one that is privately owned); and the same environmental impacts were recalled during the interviews with the five FH managers.

Regarding the producers' allocations, two allocation strategies were discussed: producers allocation based on administrative subdivision, and producers allocation based on geographical proximity. In terms of distance efficiency, the second scenario provides better results; the proximity-based allocation represents about a 8% of gain compared to an allocation based on administrative subdivision, using the strategy of collection of products by the FH. This suggests that to strengthen the supply of agro-industrial products, one must consider the producers' allocation in terms of geographical proximity and accessibility of the FH, which influences the food distance travelled and the food collection efficiency in terms of distance.

In terms of research limitations, the assessment of the role of the FH and their benefits in this chapter is based on limited number of interviews with stakeholders involved in the distribution system, and does not measure real impacts on local producers. The second phase of this research is limited to the study of potential producer allocations to FH, assuming that the current allocation is the consequence of current legislation. However, it is possible that there are some connections between producers and FH that are not considered. It is also possible that a change in the allocation of producers could also have some effects on collaborations and trip consolidations for the first mile. This research does not consider the costs associated with the operation of FH, costs that could influence the decision to assign producers to a particular Food Hub.

The results of this research provide valuable information to distribution companies designing short supply chains with FH, as well as to policymakers aiming to increase their share of local and organic food in institutional catering. This research will help policy-makers to revise the legislation that allocates producers to FH based on administrative borders instead of geographical proximity. Finally, the results of this research could be extrapolated to other regions, and to a broader aim to increase organic and local foods, or short supply chains.

Acknowledgements The authors would like to acknowledge the support from Labex IMU via the project ELUD Project (AAP2017) "Efficacity of sustainable urban food logistics" and from the Volvo Research and Educational Foundations (VREF) that funded the Urban Freight Platform (UFP) through its Future Urban Transport research program.

Note

1. Ho.Re.Ca: Hotels, restaurants and cafes.

References

ADEME. (2016). Alimentation et environnement: Champs d'actions pour le professionnels.

Aubry, C., & Kebir, L. (2013). Shortening food supply chains: A means for maintaining agriculture close to urban areas? The case of the French metropolitan area of Paris. *Food Policy, 41*, 85–93. https://doi.org/10.1016/j.foodpol.2013.04.006

Barham, J., Tropp, D., Enterline, K., Farbman, J., Fisk, J., & Kiraly, S. (2012). *Regional food hub resource guide*. USDA Agricultural Marketing Service.

Behrends, S. (2016). Recent developments in urban logistics research – a review of the proceedings of the International Conference on City Logistics 2009. *Transportation Research Procedia, 12*(June 2015), 278–287. https://doi.org/10.1016/j.trpro.2016.02.065

Bosona, T. G., & Gebresenbet, G. (2011). Cluster building and logistics network integration of local food supply chain. *Biosystems Engineering, 108*(4), 293–302. https://doi.org/10.1016/j.biosystemseng.2011.01.001

Bryman, A., & Bell, E. (2015). Business research methods (Vol. 4th). Glasgow: Bell & Bain Ltd.

Cunha, A., & Elmes, M. (2015). From food bank to food hub: Challenges and opportunities.

De Fazio, M. (2016). Agriculture and sustainability of the welfare: The role of the short supply chain. *Agriculture and Agricultural Science Procedia, 8*, 461–466. https://doi.org/10.1016/j.aaspro.2016.02.044

Eisenhardt, K. M. (1989). Agency theory: An assessment and review. *Academy of Management Review, 14*(1), 57–74.

Erdoğan, G. (2017). An open source spreadsheet solver for vehicle routing problems. *Computers & Operations Research, 84*, 62–72.

Eriksson, M., Osowski, C. P., Malefors, C., Björkman, J., & Eriksson, E. (2017). Quantification of food waste in public catering services—A case study from a Swedish municipality. *Waste Management, 61*, 415–422.

Fredriksson, A., & Liljestrand, K. (2015). Capturing food logistics: A literature review and research agenda. *International Journal of Logistics Research and Applications, 18*(1), 16–34.

Holguín-Veras, J., Sánchez-Díaz, I., & Browne, M. (2015). *Freight demand management: Role in sustainable urban freight system.* 9th International Conference on City Logistics, Tenerife, Spain, pp. 43–57.

Lessirard, J., Patier, C., Perret, A., & Richard, M. (2017). Sociétés de restauration collective en gestion concédée, en restauration commerciale et approvisionnements de proximité.

Matson, J., & Thayer, J. (2013). The role of food hubs in food supply chains. *Journal of Agriculture, Food Systems, and Community Development, 3*(4), 1–6.

Maxwell, J. A. (2013). *Qualitative research design: An interactive approach* (218pp). SAGE.

Migliore, G., Schifani, G., & Cembalo, L. (2015). Opening the black box of food quality in the short supply chain: Effects of conventions of quality on consumer choice. *Food Quality and Preference, 39*, 141–146. https://doi.org/10.1016/j.foodqual.2014.07.006

Morganti, E. (2011). *Urban food planning, city logistics and sustainability: The role of the wholesale produce market. The cases of Parma and Bologna food hubs*. PhD.

Morganti, E. (2019). Food and urban logistics: A fast-changing sector with significant policy and business implications. In J. Browne, M. Behrends, S. Woxenius, J. Giuliano, & G. Holguin-Veras (Eds.), *Urban logistics. Management, policy and innovation in a rapidly changing environment* (pp. 196–209). London: Kogan Page.

Morganti, E., & Gonzalez-Feliu, J. (2015). *City logistics for perishable products. The case of the Parma's food hub*. Case Studies on Transport Policy.
Morley, A., Morgan, S., & Morgan, K. (2008). Food hubs: The "missing middle" of the local food infrastructure? *Context*, 1–25. Retrieved from http://www.brass.cf.ac.uk/uploads/Food_HubKM0908.pdf
Nicholson, C. F., Gómez, M. I., & Gao, O. H. (2011). The costs of increased localization for a multiple-product food supply chain: Dairy in the United States. *Food Policy, 36*(2), 300–310. https://doi.org/10.1016/j.foodpol.2010.11.028
Nordmark, I., Ljungberg, D., Gebresenbet, G., Bosona, T., & Jüriado, R. (2012). Integrated logistics network for the supply chain of locally produced food, part II: Assessment of E-trade, economic benefit and environmental impact. *Journal of Service Science and Management, 5*(3), 249–262. https://doi.org/10.4236/jssm.2012.53030
Palacios-Argüello, L., Morganti, E., & Gonzalez-Feliu, J. (2017). Food hub: Una alternativa para alimentar las ciudades de manera sostenible. *Revista Transporte Y Territorio, 17*, 10–33.
Palacios-Argüello, L. P., & Gonzalez-Feliu, J. (2016). *Food hub as an efficient alternative to sustainably feed the cities*. RIODD 2016. Saint-Etienne, France.
Rodriguez Reyes, M. (2012). *FAO: La función de los mercados mayoristas en los centros urbanos de Colombia*. Organización De Las Naciones Unidas Para La Agricultura Y La Alimentación Fao.
Sánchez-Díaz, I. (2018). *Potential of implementing urban freight strategies in the accommodation and food services sector*. Transportation Research Record: Journal of the Transportation Research Board.
Yin, R. K. (1994). Case study research: Design and methods (Applied social research methods, Vol. 5). Beverly Hills, CA: Sage Publications. *Rick Rantz Leading Urban Institutions of Higher Education in the New Millennium Leadership & Organization Development Journal, 23*(8), 2002.

Food Distribution in School Feeding Programmes in Brazil

João Roberto Maiellaro, João Gilberto Mendes dos Reis, Laura Palacios-Argüello, Fernando Juabre Muçouçah, and Oduvaldo Vendrametto

1 Introduction

The importance of food supply chains and school feeding programmes has been recognized in an increasing number of studies (Chauhan, 2015; Soares, Davó-Blanes, Martinelli, Melgarejo, & Cavalli, 2017; Sonnino, Lozano Torres, & Schneider, 2014; Torres & Benn, 2017). Moreover, in recent years, several studies have focused on food security (Bilska, Wrzosek, Kołożyn-Krajewska, & Krajewski, 2016; Kummu et al., 2012;

J. R. Maiellaro • J. G. M. dos Reis (✉) • O. Vendrametto
Postgraduate Programme in Production Engineering,
Universidade Paulista, São Paulo, Brazil
e-mail: joao.maiellaro@fatec.sp.gov.br; joao.reis@docente.unip.br

L. Palacios-Argüello
Institut Henri Fayol, École des Mines de Saint-Étienne, Saint-Étienne, France
e-mail: laura.palacios@emse.fr

F. J. Muçouçah
Fatec Ferraz de Vasconcelos, Ferraz de Vasconcelos, Brazil

E. Aktas, M. Bourlakis (eds.), *Food Supply Chains in Cities*,
https://doi.org/10.1007/978-3-030-34065-0_9

Lebersorger & Schneider, 2014) and school feeding of children (Belik & Fornazier, 2017; Kristjansson et al., 2016; López-Olmedo et al., 2018).

School feeding programmes are a persistent concern worldwide. They provide protection against food insecurity, preventing its negative impact on education attainment (Belik & Fornazier, 2017; Harvey, 2016; Kristjansson et al., 2016; Lebersorger & Schneider, 2014; López-Olmedo et al., 2018). Effective interventions to improve child nutrition can ensure healthy living and promote well-being. Decision makers need to identify the cost-effectiveness of these interventions (Kristjansson et al., 2016).

The Brazilian school feeding programme, being stimulated by law 11,947, has been prominent both nationally and internationally in recent years (Lopes & Diniz, 2018). Brazilian law guarantees that 30% of the transferred value by the Brazilian National School Feeding Programme (PNAE) must be invested in direct buying of familiar agriculture products (Brasil, 2009). This law stimulates economic development and sustainable communities (Fundo Nacional de Desenvolvimento da Educação [FNDE], 2017). Currently, 5570 municipalities in the country (Instituto Brasileiro de Geografia e Estatística [IBGE], 2017) receive federal investment for school feeding programmes development.

Here, the city of Mogi das Cruzes is selected as a case study because of its importance as a food producer city in Brazil. Using the resources of the PNAE, the city runs a food supply system that integrates 209 resident schools and 58 local food producers.

The new opportunities opened by the school meals programme procurement have revealed that the farming organizations are yet fragile regarding delivery logistics and quality patterns of food provided (Belik & Fornazier, 2017). The mayoralty purchases come directly from farming organizations, including those originating from land reform programmes and family farmers. One of the main challenges facing the school feeding programmes is transportation (Lopes & Diniz, 2018).

Logistics issues have been identified, and the transportation cost is very high. The producers are forced to transport the food from the current distribution point, which is located far from the schools, making food delivery to the furthest point very difficult. Due to the transhipments, quality controls are tough when the food finally arrives at the schools, and the reception and storage operations are long and tedious.

Considering that the food is bought with public resources, this study raises some questions:

- Is it possible to improve the operation by changing the location of the distribution point?
- In a logical before-after scenario assessment, what changes can be implemented to help schools regarding logistics issues?

We applied a method of spatial analysis using a geographic information system (GIS) to identify clusters of schools in Mogi das Cruzes city. This analysis enabled us to better understand the distribution system.

In this chapter, the school feeding programme is analysed by characterizing the food supply chain, which involves public schools and family food producers in Mogi das Cruzes, in order to propose a new distribution point that improves the current distribution system.

To attain this objective, this chapter is organized as follows: First, a literature review about school feeding programmes and studies that investigate distribution and logistics operations is presented. Then, the method used and the results obtained are discussed. Finally, as a conclusion, a new distribution point is assessed to improve the current logistics operation of the food supply chain for the school feeding system.

2 Literature Review

2.1 Mogi das Cruzes City and the Green Belt of São Paulo

Until the middle of the twentieth century, Brazilians were not aware of the wide variety of vegetables and characteristically fed on diet from the sea and plants before the arrival of Japanese immigrants. Many Japanese immigrants live in an area called Alto-Tietê region, and they have dedicated their entire lives to agriculture. The arrival of these immigrants was established through an immigration agreement between Japan and Brazil at the beginning of the twentieth century (Instituto Florestal, 2013).

The immigrants and their descendants introduced many agricultural techniques, such as the development of new species of vegetables and flowers, including orchids and various foreign dishes (da Silva, de Souza, & Moretti, 2018). de Queiroz (2013) demonstrated the importance of the green belt of São Paulo to preserve the environment, that is, the watershed protection areas, and produce water.

One of the reasons that prompted the declaration of the green belt of São Paulo as a biosphere reserve is that it surrounds one of the planet's largest cities that concentrates 10% of the Brazilian population, with very low rates of green area per inhabitant. The green belt supplies water and a variety of food to the metropolis of São Paulo (Instituto Florestal, 2013).

Mogi das Cruzes is considered as the main town in the region of São Paulo's green belt, and it has the largest centre of production of vegetables, fruits and flowers in Brazil, with a relevant share of the national market of mushrooms, persimmon, loquats, vegetables and flowers (Santos & Bello, 2014). The city is one of the leading mushroom producers in Brazil (da Silva et al., 2018) and is the principal persimmon producer. Its persimmon production totalled 28% of the Brazilian production in 2016 (IBGE, 2018). About 2000 farmers that supply 35% of the greengrocer market are in São Paulo, and 5% are in Rio de Janeiro city (Prefeitura de Mogi das Cruzes, 2018).

Brazil has 5570 cities, of which Mogi das Cruzes is the 900th largest by area, with 713 km^2 (IBGE, 2017). Over 200 schools spread in the urban and rural areas consume food from family agriculture.

2.2 Local Food Production

There is a growing demand for alternatives to what consumers perceive as industrialized food in the developed markets around the world (Zepeda & Li, 2006). Consumers prefer healthy diets, environmental sustainable diets, organic foods, natural foods and raw foods (Lang, Stanton, & Qu, 2014).

Local food attracts considerable policies and public interests; however, the local food sector features some emerging contours, even though the evidences are lacking (Buller & Morris, 2003).

In Brazil, there is a relevant inequality in accessing services and infrastructure that ensure urban well-being, especially regarding quality. The wealthier areas, where those with greater purchasing power concentrate, have at their disposal an abundant supply of goods and services, whereas the areas populated by the poor are supplied with less-quality goods and services (Machado Bógus & Pasternak, 2010).

It is difficult to draw conclusions about the local economic impact of local food systems, and it is challenging to make comparisons among operations, logistics and government support aspects (Migliore, Schifani, Guccione, & Cembalo, 2014; Nie & Zepeda, 2011).

Scaling food production is a global challenge. Encouraging the involvement of mid-sized farms seems to be a classic solution because it can expand the accessibility of local food while providing alternative revenue streams for troubled family farms. Limited understanding of the way in which scale developments affect the perception and legitimacy of local food systems can be a relevant matter (Mount, 2012).

Many families that defend the local food production might suggest that important pieces of added value within local food systems are generated by the reconnection of producer and consumer, the direct exchange through which this occurs and the shared goals and values that provide the basis for reconnection (Mount, 2012).

Acquiring the needed data to quantify the economic local impact is highly difficult, and researchers must agree on a standard method of accounting for the costs involved in producing and purchasing local foods or on a standard set of economic modelling conventions. Many questions surrounding the economic impact of local foods remain unanswered and could be raised by future researches; an example is the question of if local food systems help the city-rural economy or if the economic benefits of expanding local food systems might be unevenly distributed (United States Department of Agriculture, 2017).

Managing the distribution of perishable products, such as food, requires perpetual control of the environmental conditions of the storage and transport facilities to avoid those stresses that may affect the products' shelf life and quality (Accorsi, Gallo, & Manzini, 2017).

2.3 School Feeding Programmes

Schools have been considered as favourable environments to reverse the health risks that accompany excess weight in children through providing healthy meals to the child population (Soares, Martinelli, Melgarejo, Cavalli, & Davó-Blanes, 2017). Persistent malnutrition and the increase in overweight and obesity are sources of concern, and schools are considered as sites where children can generally learn about health, but with more emphasis on healthy eating (Torres & Simovska, 2017).

Several researchers have recognized the significance of the school feeding programmes (Chauhan, 2015; Soares, Davó-Blanes, et al., 2017; Torres & Benn, 2017), substantiating that school feeding is an important issue for food security through childhood.

School feeding programmes, which are present in different countries, promote healthy feeding through direct food purchase from local farmers and maintaining sustainable food systems. A relevant number of countries adopt school feeding programmes, such as United States, Ghana, Indonesia, Brazil, El Salvador, Honduras, Nicaragua, Paraguay, Niger, Senegal, Ethiopia, Mozambique, Malawi (Soares, Davó-Blanes, et al., 2017) and India (Gajpal, Roy, & Sahay, 2019).

Global estimations demonstrate that about 370 million children were school-fed in 2012 (Kristjansson et al., 2016; World Food Programme [WFP], 2013). School feeding programmes intend to alleviate short-term hunger, improve nutrition and cognition of children and transfer income to families (Jomaa, McDonnell, & Probart, 2011).

School feeding programmes can help countries to tackle crises periods. A survey of 77 countries show that 38 of them have scaled up their programmes in response to social shocks such as armed conflict, natural disasters and food and financial crises (WFP, 2013).

Incorporating family farms as suppliers of school feeding programmes seems to contribute to an increase in the presence of healthy foods in school meals. The public purchase of food stimulates an increase in agricultural production, in addition to allowing for more stable and structured food markets (Soares, Martinelli, et al., 2017). Linking programmes to the agriculture sector can benefit the entire community as well as the children (WFP, 2013).

In Brazil, the development of direct purchase policies began in 2003 as a part of the Zero Hunger Programme. Up until 2003, procurement was based on economic criteria that made participation of family farmers difficult. The purchase of family farm products for use in Brazilian public schools began with the implementation of the Food Procurement Programme (Soares, Martinelli, et al., 2017).

The federal Brazilian government encourages a proper nutrition in public schools and the purchase of locally produced foods in small rural properties. The PNAE provide food and nutritional education of all basic education stages. The PNAE offer free meals to over 40 million students in Brazil. When it began in the 1950s, the programme's goal was to address malnutrition. The current diet of most Brazilians is of low nutritional quality. The daily consumption of vegetables and fruits is little, and the consumption of unhealthy foods is high. The rise in overweight among children aged five to nine has become a new challenge of the programme (Soares, Martínez-Mián, Caballero, Vives-Cases, & Davó-Blanes, 2017).

Federal government transfers the financial resources in 10 monthly instalments (from February to November) to cover 200 school days, according to the number of students enrolled in each school network. Brazilian law 11,947 guarantees that 30% of the transferred value by the PNAE are invested in the direct buying of familiar agriculture products. This law stimulates economic development and sustainable communities (FNDE, 2018).

Many poor children that live in urban and rural areas depend on the school meals to prevent starving. The PNAE buy food from familiar local farmers located in the city of Mogi das Cruzes to feed students of public schools maintained by the municipal government.

In 1994, Brazilian law 8913 transferred to municipal governments the responsibilities of organizing daily menus, purchasing the ingredients, performing quality control and monitoring the use of resources through the operation of the Conselhos de Alimentação Escolar (CAEs—School

Nutrition Councils), which were designed to enhance civil society participation in school feeding policies (Sonnino et al., 2014).

2.4 Logistics Application in School Feeding Context

School feeding programmes in some cases need additional attention on infrastructure or logistics to ensure their efficiency (WFP, 2013). The logistics aspects of the school feeding of the municipality of São Paulo, Brazil, for instance, have been assessed in a study (Belik & Fornazier, 2017), whose methodology comprised bibliography, document review and consultation with public representatives.

A research conducted in Jamaica showed patterns by region of schools that participated in school feeding programmes in poor, distant rural areas, and determined whether there was a difference in attendance over a 10-year period between children (Jennings, 2016).

In case of perishable food products, distribution can be abstracted as a vehicle routing problem. Managing perishable food, such as the distribution of vegetables, milk, meat and flowers, has since been recognized as a difficult problem (X. Wang, Wang, Ruan, & Zhan, 2016). A study conducted in India considered the distribution system of a school feeding programme (mid-day meals), and the resultant problem was identified as a vehicle routing problem with a common due date (Gajpal et al., 2019).

2.5 Spatial Analysis

A hot spot analysis identifies considerable spatial groups or spatial clusters among characteristics or values through a data screening process that includes assessing the global data; structure, defining the scale of the analysis and analysing local association patterns. A hot spot is essentially a local concentration or cluster of significantly higher values of a certain indicator within a defined neighbourhood (De Valck et al., 2016).

Due their interdisciplinary nature, GIS methods are widely used in numerous management fields (Madi & Srour, 2019). However, GIS methods present challenges in analysing and processing spatiotemporal data (S. Wang, Zhong, & Wang, 2019).

Researchers have used GIS-based spatial analysis to elaborate hot spot maps in a wide range of applications such as hydrologic studies (Keum et al., 2018), marine debris and pollution accumulation (Martens & Huntington, 2012), identification of road accidents patterns (Aghajani, Dezfoulian, Arjroody, & Rezaei, 2017), evaluation of track condition and the dynamic performance of the rolling stock instrumented revenue vehicles (Lingamanaik et al., 2017) and prevention of youth dating violence and sexual harassment in middle school students (Taylor, Mumford, & Stein, 2015).

Quantum GIS (QGIS) is an open-source platform that is widely adopted in different analyses for software modelling (Meyer & Riechert, 2019), hydrodynamic systems evaluation (Nielsen, Bolding, Hu, & Trolle, 2017), quantification of potential shadow casts on lakes (Nielsen, Bolding, & Trolle, 2018), urban planning (Pelorosso, Gobattoni, Geri, & Leone, 2017) and also for teaching, learning, and using digital sharing (Eitzel et al., 2018).

3 Methods

As mentioned previously, the main purpose of this study is to analyse the distribution of food purchased by the local government. The financial resources are obtained from the PNAE in Mogi das Cruzes city. Moreover, we investigate how the adoption of a new distribution point can improve the overall logistics operation.

We conduct the research by following the steps below:

- A new distribution point was suggested by the local government official food distribution centre, and at this facility is established the headquarter of the Department of School Feeding. The team handles the food distribution related to the school feeding programme. The shed is in the urban area of the city. As the distribution point location is already defined, it was not possible to apply traditional localisation hub methods.

- We opted to perform a spatial analysis using a GIS. This new distribution point was suggested because the municipality owns the property. In addition, the team responsible for the quality of food can conduct an inspection before distribution, and there is the future possibility of processing the food within the facility so that it is packaged after fresh-cutting and made suitable for the storage capacity of the schools.
- We used the QGIS 2.18© software to perform a geospatial analysis that provided understanding of the current situation. We adopted the buffer method (de Oliveira et al., 2014; Li, Zhang, & Liang, 2010), aiming to verify the reachable radius of the new distribution point, and the hotspot mapping using Kernel density estimation (Kernohan, Millspaugh, Jenks, & Naugle, 1998; Okabe, Satoh, & Sugihara, 2009) to identify the concentration of the schools.

4 Findings

4.1 Characterization of the Logistics Issues

Mogi das Cruzes is the 900th largest city by area in Brazil, with 713 km^2 (IBGE, 2017), and over 200 schools distributed in the urban and rural

Fig. 1 Current distribution point. (Source: Authors)

areas consume food from family agriculture. An important logistics issue is the location of the current distribution point.

The current distribution point and storage shed is far away from most schools, and the quality of the roads is poor (Fig. 1). Many of them are not paved, bringing an extra challenge to the distribution operation.

Based on data collected and geographic coordinates we plotted a map with the current distribution point and distance to the schools using QGIS 2.18 (Fig. 2).

Note that the current distribution point was established by the producers in the territory as part of the cooperative system. They did not consider the distance to schools and logistics impacts. They probably did have not a previous plan or analysis of the impact of this solution.

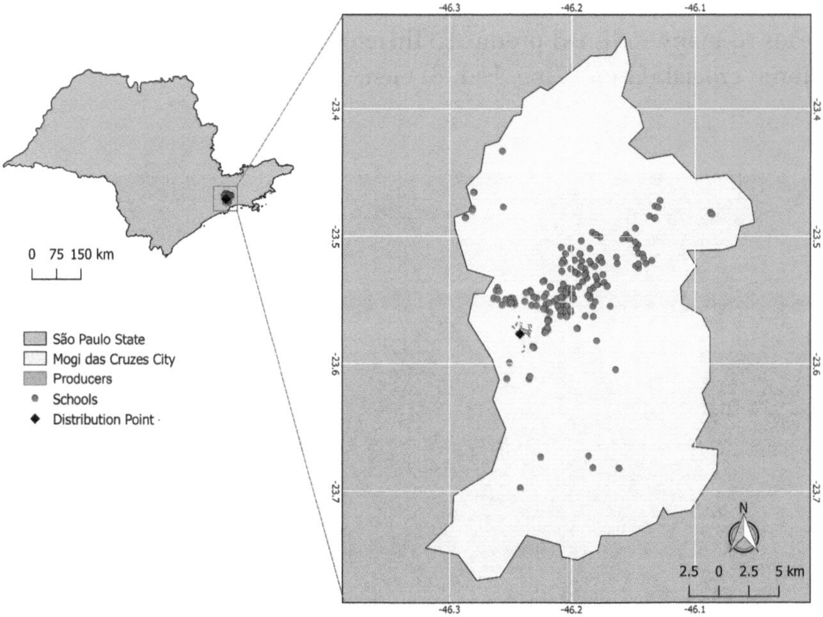

Fig. 2 Producers, current distribution point and schools. (Source: Adapted from QGIS 2.18 (2018))

4.2 Proposal of a New Distribution Point

The main idea is to locate a new transit point near the schools. Thus, vehicles with higher payload capacities can deliver the food from the current distribution point, and moreover, serving the supply network with new routes with higher-quality pavements can reduce food losses. The proposed new distribution point, which was suggested by the agricultural secretary, is located at the municipality's school feeding department.

Vehicles with higher payload capacity can send the food from the current distribution point to the new distribution point, and new routes will serve the supply network. At this facility, an industrial kitchen with approximately 300 square meters could perform the process of fresh-cutting the food received from the local producers (Fig. 3).

The fresh-cut is defined as fresh leafy greens whose leaves have been cut, shredded, sliced, chopped or torn. The term 'leafy greens' includes lettuce, escarole, cabbage and chard (Choi, Norwood, Seo, Sirsat, & Neal, 2016). The fresh goods may be easily spoilt during delivery, which leads to many returned products. Therefore, an efficient logistics operation is crucial (Guo, Wang, Fan, & Gen, 2017).

Fig. 3 New distribution centre at School Feeding Department. (Source: Authors)

The school feeding department performs a wide range of activities: they standardize menus for children's schools and elementary education; organize annual schedules for the delivery of foodstuffs (such as breads, cakes, meats and sausages, vegetables, grocery products and juice); create special menus for children with pathologies such as diabetes, phenylketonuria, lactose intolerance and malnutrition; elaborate and distribute the Handbook of Good Food Practices to all schools; research on the general food consumption per region; conduct monthly visits to supervise and guide the units served by the department and conduct quality control of food.

The team can verify the quality of the purchased food before sending to schools. This is relevant considering the current transportation operation whereby the purchased food from the local producers is sent directly to schools without any inspection. In the hierarchical organization, a managing director commands a group of procurement analysts and the inbound logistics employees (forklift operators and inbound logistics assistants). Figure 4 shows the location of the proposed distribution point and distance from school using QGIS 2.18.

We performed a spatial analysis, using a hot spot Kernel map analysis to determine the concentration of the served schools and the buffer method to determine the differences between both distribution points.

The new distribution point is in an area with a higher concentration of schools, compared with the current distribution point. In any case, a few schools served by the familiar food producers are still far from the new distribution point. Figure 5 shows the schools on the hot spot map.

4.3 Solution Assessment

We used a buffer mapping approach to highlight the smaller distances between schools and the proposed distribution point when compared to the current distribution point. The buffers show that only 26% of the

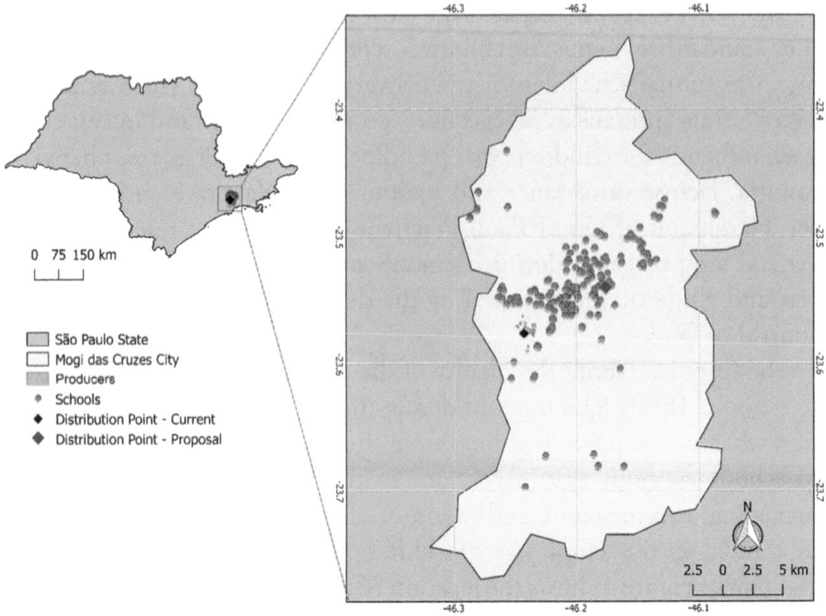

Fig. 4 The proposed distribution point. (Source: Adapted from QGIS 2.18 (2018))

schools are located at a distance up to 5 km from the current distribution point, while 74% of the schools are located within 5 km from the proposed distribution point (Table 1).

A significant number of schools are near the proposed distribution point, considering a distance of up to 2 km.

The current distribution point is surrounded by the food producers, and consequently, the total distance for inflows from food producers is short. Many producers are neighbours of the association, and they deliver the food using small vehicles. Eight urban delivery vehicles attend to 209 schools. The longest route, which measures about 70 km, serves the most distant schools in the south region of the city. The other shorter routes are most traversed in the urban area.

Fig. 5 Kernal hot spot map analysis. (Source: Adapted from QGIS 2.18 (2018))

Table 1 Schools distances from the distribution points

	Distribution point	
Distance	Current	Proposed
Up to 2 km	1.0%	10.0%
Up to 5 km	26.0%	74.0%
Up to 10 km	85.0%	93.0%
Up to 15 km	99.0%	99.5%

Source: Authors

The longest route in the north region measures about 40 km. Moving the distribution point, the longest route in the south of the city will not be improved. Figure 6 shows the buffer mapping analysis, comparing the current and proposed distribution points.

Note that there is a striking difference between the current and proposed distribution points. The proposed distribution point is more balanced and provides a better distribution system. Moreover, the

Fig. 6 Buffer method analysis. (Source: Adapted from QGIS 2.18 (2018))

proposed distribution point is more prepared to attend to the logistics needs, and foods can be semi-processed at this point before sent to schools. This can extend the expiry date of agricultural products. On the other hand, delivering fresh-cut food improves the vehicle's payload capacity.

5 Discussion and Conclusions

The findings of this research provide insights into improving the logistics of school food distribution in Mogi das Cruzes city, Brazil. Even with a high number of local producers, many logistics problems make the logistics operation expensive and inefficient. We can infer that cities in general have operations and logistics issues to solve in the context of school feeding programmes.

The present research should prove to be valuable to raise the need for new studies in this area in Brazil. New approaches could include operational research methods such as discrete event simulation and linear programming. As the scenario is complex, with diverse players, it is feasible to apply multi-criteria methods, such as analytic hierarchy process, to determine alternatives that can improve the short food supply chain, considering different points of view (Choi et al., 2016; Rabello Quadros & Nassi, 2015; Zhao et al., 2018).

The discussion of local food aspects can bring about advances such as strengthening the social and cultural sharing of values and providing direct incomes for families.

Reducing the distances between production and consumption contributes to enhancing horizontal relationships and ethical values in food production and distribution (Hinrichs, 2003). Moreover, it decreases fossil fuel-based transportation (Rothwell, Ridoutt, Page, & Bellotti, 2016) and considers the economic and environmental impact of city food logistics systems.

The PNAE plays an important role in the local economy. Between 2016 and 2018, more than BRL 9,000,000,000 was transferred by the Brazilian government to the state and municipal schools and day care centres. In 2017, about 41 million students benefited (Brasil, 2018). The programme has the huge challenge of improving the logistics difficulties due to long distances, bad quality of roads and the massive training needs of the players (Lopes & Diniz, 2018).

Suppliers of agricultural inputs, familiar producers, carriers and the served schools are ties of these short food supply chains. Operational and logistical improvements in this programme can motivate new familiar producers to provide food for the local government and motivate current food producers to continue providing food in larger quantities, providing large-scale gains for the system.

Our research demonstrated that spatial analysis can help to better understand the distribution issues, and future studies with quantitative approaches can improve operations and logistics efficiency. With the insights of our research, a simulation model, for instance, can analyse the impacts of the new distribution point in decreasing distances and time of the distribution operation.

For future research, operations research methods, such as linear programming, should be investigated, as they can help to improve the overall distribution system. An important limitation of this research is the lack of additional indicators such as system cost and total distance travelled.

This study shows that move delivering food from the current distribution point to the new centre facility, the food can be transported within a distance of 5 km to 74% of the served schools in Mogi das Cruzes city. Moreover, fresh-cut processing can be conducted in the new facility, which can decrease food losses and improve the vehicles' payload capacity. As result of this research Mogi das Cruzes city intends to transfer distribution centre for the new facility and start the fresh-cut processing.

Furthermore, this study may help other Brazilian cities with similar geographic characteristics and that buy food from family farmers to offer schools. The difficulties of managing logistics operations are similar. Reducing logistics costs and providing a better understanding of this system can result in public transparency of the PNAE. In this sense, the practical results of Mogi das Cruzes new distribution centre will be crucial to motivate other cities in the same direction. Therefore, in the continuation of this research will intend to measure the results obtained by Mogi das Cruzes with new distribution centre and fresh-cut processing.

References

Accorsi, R., Gallo, A., & Manzini, R. (2017). A climate driven decision-support model for the distribution of perishable products. *Journal of Cleaner Production, 165*, 917–929. https://doi.org/10.1016/j.jclepro.2017.07.170

Aghajani, M. A., Dezfoulian, R. S., Arjroody, A. R., & Rezaei, M. (2017). Applying GIS to identify the spatial and temporal patterns of road accidents using spatial statistics (case study: Ilam Province, Iran). *Transportation Research Procedia, 25*, 2126–2138. https://doi.org/10.1016/j.trpro.2017.05.409

Belik, W., & Fornazier, A. (2017). Chapter Three - Public policy and the construction of new markets to family farms: Analyzing the case of school meals in São Paulo, Brazil. In D. Barling (Org.), *Advances in food security and sustainability* (Vol. 2, pp. 69–86). Elsevier. Retrieved from https://doi.org/10.1016/bs.af2s.2017.09.001

Bilska, B., Wrzosek, M., Kołożyn-Krajewska, D., & Krajewski, K. (2016). Risk of food losses and potential of food recovery for social purposes. *Waste Management, 52*, 269–277. https://doi.org/10.1016/j.wasman.2016.03.035

Brasil. (2009). L11947. Retrieved from http://www.planalto.gov.br/ccivil_03/_Ato2007-2010/2009/Lei/L11947.htm

Brasil. (2018). Mais de R$ 9 bilhões foram investidos em merenda saudável para escolas. Retrieved from http://www.brasil.gov.br/editoria/educacao-e-ciencia/2018/07/mais-de-r-9-bilhoes-foram-investidos-em-merenda-saudavel-para-escolas

Buller, H., & Morris, C. (2003). The local food sector: A preliminary assessment of its form and impact in Gloucestershire. *British Food Journal, 105*(8), 559–566. https://doi.org/10.1108/00070700310497318

Chauhan, A. (2015). Plates for slates: The impact of a school feeding programme on community representations of schools. *International Journal of Educational Development, 41*(Suppl C), 292–300. https://doi.org/10.1016/j.ijedudev.2014.07.013

Choi, J., Norwood, H., Seo, S., Sirsat, S. A., & Neal, J. (2016). Evaluation of food safety related behaviors of retail and food service employees while handling fresh and fresh-cut leafy greens. *Food Control, 67*, 199–208. https://doi.org/10.1016/j.foodcont.2016.02.044

De Valck, J., Broekx, S., Liekens, I., De Nocker, L., Van Orshoven, J., & Vranken, L. (2016). Contrasting collective preferences for outdoor recreation and substitutability of nature areas using hot spot mapping. *Landscape and Urban Planning, 151*, 64–78. https://doi.org/10.1016/j.landurbplan.2016.03.008

Eitzel, M. V., Mhike Hove, E., Solera, J., Madzoro, S., Changarara, A., Ndlovu, D., et al. (2018). Sustainable development as successful technology transfer: Empowerment through teaching, learning, and using digital participatory mapping techniques in Mazvihwa, Zimbabwe. *Development Engineering, 3*, 196–208. https://doi.org/10.1016/j.deveng.2018.07.001

Fundo Nacional de Desenvolvimento da Educação. (2017). Sobre o Pnae - Portal do FNDE. Retrieved from http://www.fnde.gov.br/programas/pnae/pnae-sobre-o-programa/pnae-sobre-o-pnae

Fundo Nacional de Desenvolvimento da Educação. (2018). FNDElegis - Sistema de Legislação do FNDE. Retrieved from https://www.fnde.gov.br/fndelegis/action/UrlPublicasAction.php?acao=abrirAtoPublico&sgl_tipo=LEI&num_ato=00011947&seq_ato=000&vlr_ano=2009&sgl

Gajpal, Y., Roy, V., & Sahay, B. S. (2019). Vehicle routing for a mid-day meal delivery distribution system. *Heliyon, 5*(1), e01158. https://doi.org/10.1016/j.heliyon.2019.e01158

Guo, J., Wang, X., Fan, S., & Gen, M. (2017). Forward and reverse logistics network and route planning under the environment of low-carbon emissions: A case study of Shanghai fresh food E-commerce enterprises. *Computers & Industrial Engineering, 106*, 351–360. https://doi.org/10.1016/j.cie.2017.02.002

Harvey, K. (2016). "When I go to bed hungry and sleep, I'm not hungry": Children and parents' experiences of food insecurity. *Appetite, 99*, 235–244. https://doi.org/10.1016/j.appet.2016.01.004

Hinrichs, C. C. (2003). The practice and politics of food system localization. *Journal of Rural Studies, 19*(1), 33–45. https://doi.org/10.1016/S0743-0167(02)00040-2

Instituto Brasileiro de Geografia e Estatística. (2017). Panorama. Retrieved from https://cidades.ibge.gov.br/brasil/sp/sao-paulo/panorama

Instituto Brasileiro de Geografia e Estatística. (2018). Culturas temporárias e permanentes. Retrieved from https://www.ibge.gov.br/estatisticas-novoportal/economicas/agricultura-e-pecuaria/9117-producao-agricola-municipal-culturas-temporarias-e-permanentes.html?=&t=downloads

Instituto Florestal. (2013). O Cinturão Verde. Retrieved from http://iflorestal.sp.gov.br/o-instituto/rbcv/o-cinturao-verde/

Jennings, Z. (2016). Impact of the provision of school lunch on attendance in remote rural Jamaican primary schools. *International Journal of Educational Development, 46*(Suppl C), 74–81. https://doi.org/10.1016/j.ijedudev.2015.09.006

Jomaa, L. H., McDonnell, E., & Probart, C. (2011). School feeding programs in developing countries: Impacts on children's health and educational outcomes. *Nutrition Reviews, 69*(2), 83–98. https://doi.org/10.1111/j.1753-4887.2010.00369.x

Kernohan, B. J., Millspaugh, J. J., Jenks, J. A., & Naugle, D. E. (1998). Use of an adaptive kernel home-range estimator in a GIS environment to calculate habitat use. *Journal of Environmental Management, 53*(1), 83–89. https://doi.org/10.1006/jema.1998.0198

Keum, J., Coulibaly, P., Razavi, T., Tapsoba, D., Gobena, A., Weber, F., et al. (2018). Application of SNODAS and hydrologic models to enhance entropy-based snow monitoring network design. *Journal of Hydrology, 561*, 688–701. https://doi.org/10.1016/j.jhydrol.2018.04.037

Kristjansson, E. A., Gelli, A., Welch, V., Greenhalgh, T., Liberato, S., Francis, D., et al. (2016). Costs, and cost-outcome of school feeding programmes and feeding programmes for young children. Evidence and recommendations. *International Journal of Educational Development, 48*(Suppl C), 79–83. https://doi.org/10.1016/j.ijedudev.2015.11.011

Kummu, M., de Moel, H., Porkka, M., Siebert, S., Varis, O., & Ward, P. J. (2012). Lost food, wasted resources: Global food supply chain losses and their impacts on freshwater, cropland, and fertiliser use. *Science of the Total Environment, 438*, 477–489. https://doi.org/10.1016/j.scitotenv.2012.08.092

Lang, M., Stanton, J., & Qu, Y. (2014). Consumers' evolving definition and expectations for local foods. *British Food Journal, 116*(11), 1808–1820. https://doi.org/10.1108/BFJ-03-2014-0117

Lebersorger, S., & Schneider, F. (2014). Food loss rates at the food retail, influencing factors and reasons as a basis for waste prevention measures. *Waste Management, 34*(11), 1911–1919. https://doi.org/10.1016/j.wasman.2014.06.013

Li, X., Zhang, L., & Liang, C. (2010). A GIS-based buffer gradient analysis on spatiotemporal dynamics of urban expansion in Shanghai and its major satellite cities. *Procedia Environmental Sciences, 2*, 1139–1156. https://doi.org/10.1016/j.proenv.2010.10.123

Lingamanaik, S. N., Thompson, C., Nadarajah, N., Ravitharan, R., Widyastuti, H., & Chiu, W. K. (2017). Using instrumented revenue vehicles to inspect track integrity and rolling stock performance in a passenger network during peak times. *Procedia Engineering, 188*, 424–431. https://doi.org/10.1016/j.proeng.2017.04.504

Lopes, S. R. S., & Diniz, P. R. (2018). *Boas práticas de agricultura familiar para a alimentação escolar / Programa Nacional de Alimentação Escolar*. Brasilia: Ministério da Educação/Fundo Nacional para o Desenvolvimento Escolar.

López-Olmedo, N., Jiménez-Aguilar, A., Morales-Ruan, M. d. C., Hernández-Ávila, M., Shamah-Levy, T., & Rivera-Dommarco, J. A. (2018). Consumption of foods and beverages in elementary schools: Results of the implementation of the general guidelines for foods and beverages sales in elementary schools in Mexico, stages II and III. *Evaluation and Program Planning, 66*(Suppl C), 1–6. https://doi.org/10.1016/j.evalprogplan.2017.08.009

Machado Bógus, L. M., & Pasternak, S. (2010). Changing urbanization patterns in the Brazilian metropolis. In *Suburbanization in global society* (Vol. 10, pp. 231–251). Emerald Group Publishing Limited. Retrieved from https://doi.org/10.1108/S1047-0042(2010)0000010012

Madi, N., & Srour, I. (2019). Managing emergency construction and demolition waste in Syria using GIS. *Resources, Conservation and Recycling, 141*, 163–175. https://doi.org/10.1016/j.resconrec.2018.10.018

Martens, J., & Huntington, B. E. (2012). Creating a GIS-based model of marine debris "hot spots" to improve efficiency of a lobster trap debris removal program. *Marine Pollution Bulletin, 64*(5), 949–955. https://doi.org/10.1016/j.marpolbul.2012.02.017

Meyer, D., & Riechert, M. (2019). Open source QGIS toolkit for the advanced research WRF modelling system. *Environmental Modelling & Software, 112*, 166–178. https://doi.org/10.1016/j.envsoft.2018.10.018

Migliore, G., Schifani, G., Guccione, G. D., & Cembalo, L. (2014). Food community networks as leverage for social embeddedness. *Journal of Agricultural and Environmental Ethics, 27*(4), 549–567. https://doi.org/10.1007/s10806-013-9476-5

Mount, P. (2012). Growing local food: Scale and local food systems governance. *Agriculture and Human Values, 29*(1), 107–121. https://doi.org/10.1007/s10460-011-9331-0

Nie, C., & Zepeda, L. (2011). Lifestyle segmentation of US food shoppers to examine organic and local food consumption. *Appetite, 57*(1), 28–37. https://doi.org/10.1016/j.appet.2011.03.012

Nielsen, A., Bolding, K., Hu, F., & Trolle, D. (2017). An open source QGIS-based workflow for model application and experimentation with aquatic ecosystems. *Environmental Modelling & Software, 95*, 358–364. https://doi.org/10.1016/j.envsoft.2017.06.032

Nielsen, A., Bolding, K., & Trolle, D. (2018). A GIS-based framework for quantifying potential shadow casts on lakes applied to a Danish lake experimental facility. *International Journal of Applied Earth Observation and Geoinformation, 73*, 746–751. https://doi.org/10.1016/j.jag.2018.08.022

Okabe, A., Satoh, T., & Sugihara, K. (2009). A kernel density estimation method for networks, its computational method and a GIS-based tool. *International Journal of Geographical Information Science, 23*(1), 7–32. https://doi.org/10.1080/13658810802475491

de Oliveira, F. B., de Oliveira, C. H. R., Lima, J. S. d. S., Filho, R. B. R., Miranda, M. R., Neves, L. Z., et al. (2014). Application of geoprocessing and fuzzy logic for the establishment of zoning in Parque Estadual da Cachoeira. *Revista Brasileira de Cartografia, 1*(66/3).

Pelorosso, R., Gobattoni, F., Geri, F., & Leone, A. (2017). PANDORA 3.0 plugin: A new biodiversity ecosystem service assessment tool for urban green infrastructure connectivity planning. *Ecosystem Services, 26*, 476–482. https://doi.org/10.1016/j.ecoser.2017.05.016

Prefeitura de Mogi das Cruzes. (2018). Agronegócios. Retrieved from http://www.agricultura.pmmc.com.br/

de Queiroz, G. C. M. (2013). *Fortalecimento e fragmentação do cinturão verde do Alto Tietê: perspectivas sobre a atividade agrícola da região*. Master's thesis, USCS, São Caetano do Sul, Brazil.

Rabello Quadros, S. G., & Nassi, C. D. (2015). An evaluation on the criteria to prioritize transportation infrastructure investments in Brazil. *Transport Policy, 40*, 8–16. https://doi.org/10.1016/j.tranpol.2015.02.002

Rothwell, A., Ridoutt, B., Page, G., & Bellotti, W. (2016). Environmental performance of local food: Trade-offs and implications for climate resilience in a developed city. *Journal of Cleaner Production, 114*, 420–430. https://doi.org/10.1016/j.jclepro.2015.04.096

Santos, V. N., & Bello, E. M. (2014). Aspectos gerais da cultura alimentar do município de Mogi das Cruzes - SP. *Revista Científica Linkania Master, 1*(9). Retrieved from http://linkania.org/master/article/view/207

da Silva, T. T., de Souza, O. S., Jr, L. T. K., & Moretti, T. T. d. S. (2018). Mapeamento da cadeia produtiva do cogumelo no Alto Tietê. *South American Development Society Journal, 4*(11), 121. https://doi.org/10.24325/issn.2446-5763.v4i11p121-145

Soares, P., Davó-Blanes, M. C., Martinelli, S. S., Melgarejo, L., & Cavalli, S. B. (2017). The effect of new purchase criteria on food procurement for the Brazilian school feeding program. *Appetite, 108*(Suppl C), 288–294. https://doi.org/10.1016/j.appet.2016.10.016

Soares, P., Martinelli, S. S., Melgarejo, L., Cavalli, S. B., & Davó-Blanes, M. C. (2017). Using local family farm products for school feeding programmes: Effect on school menus. *British Food Journal, 119*(6), 1289–1300. https://doi.org/10.1108/BFJ-08-2016-0377

Soares, P., Martínez-Mián, M. A., Caballero, P., Vives-Cases, C., & Davó-Blanes, M. C. (2017). Alimentos de producción local en los comedores escolares de España. *Gaceta Sanitaria, 31*(6), 466–471. https://doi.org/10.1016/j.gaceta.2016.10.015

Sonnino, R., Lozano Torres, C., & Schneider, S. (2014). Reflexive governance for food security: The example of school feeding in Brazil. *Journal of Rural Studies, 36*(Suppl C), 1–12. https://doi.org/10.1016/j.jrurstud.2014.06.003

Taylor, B. G., Mumford, E. A., & Stein, N. D. (2015). Effectiveness of "shifting boundaries" teen dating violence prevention program for subgroups of middle school students. *Journal of Adolescent Health, 56*(2, Suppl 2), S20–S26. https://doi.org/10.1016/j.jadohealth.2014.07.004

Torres, I., & Benn, J. (2017). The rural school meal as a site for learning about food. *Appetite, 117*(Suppl C), 29–39. https://doi.org/10.1016/j.appet.2017.05.055

Torres, I., & Simovska, V. (2017). Community participation in rural Ecuador's school feeding programme: A health promoting school perspective. *Health Education, 117*(2), 176–192. https://doi.org/10.1108/HE-02-2016-0009

United States Department of Agriculture. (2017). *USDA ERS - Trends in U.S. local and regional food systems*. A Report to Congress. Retrieved from https://www.ers.usda.gov/publications/pub-details/?pubid=42807

Wang, S., Zhong, Y., & Wang, E. (2019). An integrated GIS platform architecture for spatiotemporal big data. *Future Generation Computer Systems, 94*, 160–172. https://doi.org/10.1016/j.future.2018.10.034

Wang, X., Wang, M., Ruan, J., & Zhan, H. (2016). The multi-objective optimization for perishable food distribution route considering temporal-spatial distance. *Procedia Computer Science, 96*, 1211–1220. https://doi.org/10.1016/j.procs.2016.08.165

World Food Programme. (2013). 2013 - State of school feeding worldwide. Retrieved from https://www1.wfp.org/publications/state-school-feeding-worldwide-2013

Zepeda, L., & Li, J. (2006). Who buys local food? *Journal of Food Distribution Research, 37*(3), 1–11.

Zhao, L., Li, H., Li, M., Sun, Y., Hu, Q., Mao, S., et al. (2018). Location selection of intra-city distribution hubs in the metro-integrated logistics system. *Tunnelling and Underground Space Technology, 80*, 246–256. https://doi.org/10.1016/j.tust.2018.06.024

A Descriptive Analysis of Food Retailing in Lebanon: Evidence from a Cross-Sectional Survey of Food Retailers

Rachel A. Bahn and Gumataw K. Abebe

1 Introduction

The topic of food retail transformation is important because of its potential impacts on the wider agri-food system, including the efficiency and structure of agri-food supply chains and the participation of value chain actors. The term *supermarketisation* describes the expansion of modern food retailers—chains of large-format, self-service stores, selling a variety of food and household goods—in competition with traditional food retailers such as specialised retailers (greengrocers, bakers, butchers) and independent, 'mom and pop' grocery stores (Dries, Reardon, & Swinnen, 2004; Reardon, Barrett, Berdegué, & Swinnen, 2009). Modern food retailers include supermarkets, hypermarkets, convenience stores, and discount stores.

R. A. Bahn (✉) • G. K. Abebe
Faculty of Agricultural and Food Sciences, American University of Beirut, Beirut, Lebanon
e-mail: rb89@aub.edu.lb; ga81@aub.edu.lb

E. Aktas, M. Bourlakis (eds.), *Food Supply Chains in Cities*,
https://doi.org/10.1007/978-3-030-34065-0_10

Cross-country evidence suggests that consumer food expenditure patterns have shifted significantly in recent decades, from traditional to modern food retail formats. Modern food retailing has generally spread from upper-income to middle-income and poorer consumer markets, and from large and intermediate cities to smaller towns (Reardon, Henson, & Berdegué, 2007). This pattern is predicted to enhance spatial access to modern food retailers, improve local food system development by linking small farmers to modern food retail outlets, and bring about dietary changes by availing industrial commodities to consumers residing in less urban environments (Rischke, Kimenju, Klasen, & Qaim, 2015; Timmer, 2009).

However, limited evidence about the extent and process of supermarketisation or its effects within the agri-food chain has been reported in the Middle East and North Africa (MENA) region, specifically in urban areas. Accordingly, this case study of Lebanon seeks to explore the extent of food retail transformation and the effect this may have on traditional food retailing, by analysing the extent and process of supermarketisation using the following dimensions—in terms of availability, product assortments, and quality as well as supply chain integration. Modern food retail formats are hypothesised to have larger product assortments and more coordinated supplier relationships. To test these hypotheses, the study analyses data from a survey of 49 Lebanese food retailers that have been classified as traditional or modern, and whose market orientation (urban/less urban environments) is identifiable.

This study aims to address three research questions: (1) What are the characteristics of the current Lebanese food retail landscape? (2) What, if any, systematic differences exist within the food retail sector in terms of operations, marketing, and supply chain structures across modern and traditional retailers? (3) Is there any significant difference in these food retailing structures across urban and rural areas in Lebanon?

This chapter first provides a brief literature review summarising the available evidence about food retail spatial patterns and strategies within the geographic landscape of MENA and specifically Lebanon. Methods and materials are briefly addressed. Findings are presented, followed by a discussion. Conclusions and further research suggestions are presented at the end.

2 Literature Review

2.1 Definitions: Food Retail Formats

The food environment may include food retail stores, restaurants, and school and work-site food environments (McKinnon, Reedy, Morrissette, Lytle, & Yaroch, 2009). This chapter considers food retailers, including both traditional and modern retail stores.

The academic literature does not apply uniform definitions of modern and traditional food retail formats, because definitions vary across time and context. For example, the CCRRCA (2007) has distinguished modern from traditional formats based on three dimensions—service (scale and range of products), degree of independence, and ownership structure—defining modern retailers as "self-service offerings that typically are part of a chain of stores and belong to an organisation that has a corporate structure" and traditional retailers as "family-owned, small over-the-counter stores that are independent in nature" (p. 15). Others define food retail formats in terms of "visible" and "hidden" features (Alexander, 2008; Goldman, 2001), with the former focusing on retail offerings—location, product assortment, price, and marketing and promotion characteristics—and the latter on internal aspects including cultural, technological, and organisational characteristics. Euromonitor International (2018) provides one of the most widely used definitions: Modern food retail formats refer to the aggregation of hypermarkets (e.g., Carrefour, Tesco Extra), supermarkets (e.g., Champion, Tesco), discounters (e.g., Aldi, Lidl), forecourt retailers (e.g., BP Connect, Shell Select), and convenience stores (e.g., 7-Eleven, Spar); while traditional food retail formats include only the non-chained and family- or individually-owned retail stores. Following the existing literature, modern food retail formats are generally distinguished based on five key dimensions: product assortment, technological requirements (e.g., number of cash registers), supply systems (e.g., number of retail outlets), organisational factors (e.g., number of full-time employees), and operational factors (e.g., volume of sales).

2.2 Retail Expansion Patterns and Strategies

The topic of food retailing attracted significant attention from economic geographers and development planners following the "supermarket revolution" that began in the 1990s (Reardon & Berdegué, 2002). This period was marked by a rising level of concentration in the retail sector, with global food retailers gaining the advantage over domestic food retailers in terms of cost savings (economies of scale), range of products (economies of scope), management support systems (innovation), and international experience (economies of replication) (Lowe & Crewe, 1996; Reardon, Barrett, et al., 2009; Vias, 2004). The spatial distribution of the food environment has gained attention because of its implications for, *inter alia*, local food system development (Eckert & Shetty, 2011; Gatrell, Reid, & Ross, 2011; McEntee & Agyeman, 2010). To the best of our knowledge, however, there are few studies exploring the spatial distribution of food retail formats in the Middle Eastern context (Omer & Goldblatt, 2016; Rotem-Mindali, 2012) or Lebanon specifically (Bahn & Abebe, 2017; Seyfert, Chaaban, & Ghattas, 2014).

The geographic expansion of modern food retail is generally considered to have diffused from developed to developing countries or from larger to smaller cities (Reardon, Henson, et al., 2007; Reardon, Timmer, & Minten, 2012). Existing studies have identified three common food retail expansion patterns, according to geography, socioeconomic characteristics, and product category (Goldman, Ramaswami, & Krider, 2002; Reardon, Henson, et al., 2007). Geographic expansion occurs when penetration of new geographic segments leads to an increase in market share: Modern food retailers are hypothesised to establish their businesses in larger urban centres and gradually extend to regional centres and rural areas. Socioeconomic expansion occurs as modern food retailers focus first on upper-income consumers, then middle-income consumers, and finally the poor. Product category-based expansion occurs when modern food retailers penetrate the market by selling processed products, then semi-processed products, and eventually fresh products. In the study context, a recent study by Bahn and Abebe (2017) identifies two food retail expansion patterns in Lebanon—geographic expansion (with increasing presence in peri-urban and rural areas) and across socioeconomic groups (with increasing presence in relatively poorer areas).

Multiple methodologies have been applied in the literature to analyse the spatial distribution and expansion of food environments, including geographic information systems (GIS) and other spatial resource mapping, demographic analysis, market basket analysis, historical analysis, observations, interviews, surveys, and focus groups (Charreire et al., 2010; McKinnon et al., 2009; Pothukuchi, 2004; Short, Guthman, & Raskin, 2007). Studies show a strong link between the type and the concentration of food environment, the socioeconomic status of the population (Apparicio, Cloutier, & Shearmur, 2007; Richardson, Boone-Heinonen, Popkin, & Gordon-Larsen, 2012; Russell & Heidkamp, 2011), and the level of local food system development (Eckert & Shetty, 2011; Gatrell et al., 2011).

Globalisation of food retailing and increasing distances between food production and consumption on the one hand, rising food quality and food safety concerns on the other have led food supply chains toward concentration and integration. Increasing concentration and integration may allow modern food retailers to gain competitive advantages over traditional food retailers in terms of responding to the laws, policies, and standards pertaining to food quality and food safety and in terms of efficiency gains (economies of scale) and product offerings (range of products) (Lawrence & Dixon, 2015; Neven, Odera, Reardon, & Wang, 2009). Looking upstream, supermarketisation has led to consolidation of supply chains in the agri-food sector. In some cases such as that of growers/shippers of tomatoes in Mexico and of bananas and mangoes in Guatemala, the supermarket supply chain has excluded small farmers, food producers, and processors, thereby preventing them from accessing larger, more profitable market segments (Berdegué, Balsevich, Flores, & Reardon, 2005; Reardon, Barrett, et al., 2009; Reardon, Henson, et al., 2007). Exclusion of smaller actors from within the agri-food value chain has negative implications for inclusive growth and sustainable development as well as on the availability and affordability of healthy and nutritious food to urban consumers. A key for such a transformation is the availability of modern and sustainable food value chains. According to Gómez and Ricketts (2013), modern value chains[1] drive the expansion of modern food retailers and thereby the availability of a wide assortment of food products year-round at affordable prices due to economies of scale for urban consumers.

Outstanding questions related to supermarketisation include the extent of food retail penetration (Humphrey, 2007; McEntee & Agyeman, 2010); the pattern and the extent of food retail spatial concentration within a specific geographic area (Gustafson, Hankins, & Jilcott, 2012); the effect of modern food retail expansion on traditional food establishments (replacement or co-existence) (Guarín, 2013; Minten, 2008; Schipmann & Qaim, 2010); and the relative nutritional access to larger product assortments and/or healthier diets and dietary intakes of consumers relying on modern and traditional food retail formats (Moore & Diez Roux, 2006). Because spatial distribution of food retailing is dynamic (Joseph & Kuby, 2013), additional evidence is needed to guide development planners as they tailor context-specific interventions. For example, in a food desert, policy may promote the establishment or expansion of modern food retail chains so as to increase consumer access to affordable and nutritious food (Walker, Keane, & Burke, 2010).

2.3 Lebanon's Geography, Demography, and Economy

Lebanon is small Middle Eastern country, located along the eastern Mediterranean Sea. Lebanon was home to approximately four million Lebanese residents in 2016.[2] The country has experienced significant changes in its rural-urban[3] population dynamic since the 1960s, notably a sharp increase in the share of population living in smaller and intermediary cities (Fig. 1). The share of the total population living in the largest city, Beirut, has declined particularly since the 1980s (during and after the Lebanese civil war (1975–1990)), falling from more than 60% in the early 1980s to 34% in 2016. The rural population also declined in relative terms during this period. However, the share of population living in smaller and intermediary cities more than quadrupled from 1980 (11%) to 2016 (54%).[4]

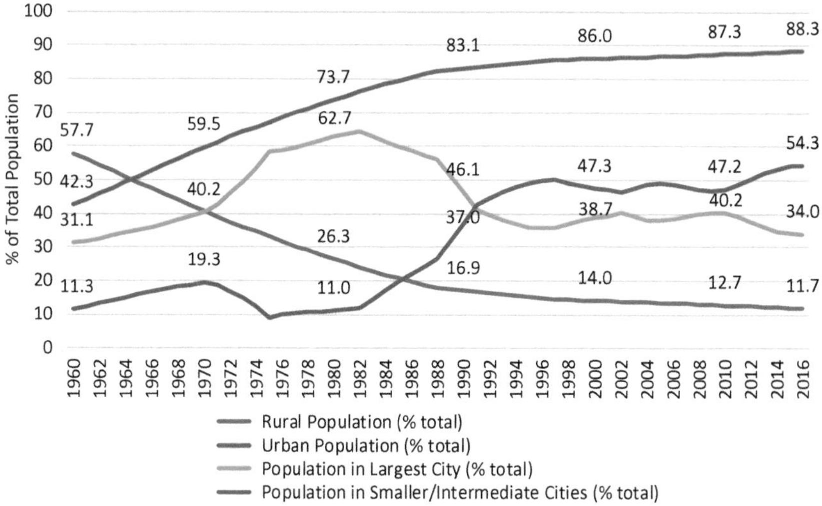

Fig. 1 Lebanese population living in urban, peri-urban, and rural areas. (Source: World Bank (2019) and authors' calculations based on World Bank data)

The Lebanese economy (nominal GDP) exceeded $49 billion in 2016. It grew an average of 1.8% annually over the period 2011–2016, a deceleration from more robust growth in the previous decade (5.9% annually over the period 2001–2010). Lebanon's economy is highly dependent upon the services sector, which generates approximately 79% of GDP; industry and agriculture contribute 17% and 4%, respectively (World Bank, 2018).

Lebanon is an upper middle-income country with an average per capita income of approximately $13,268 (PPP, 2011 international dollars) in 2016 (Fig. 5). Per capita GDP has increased significantly since 1990, with both positive and negative fluctuations over time (World Bank, 2018). Average income figures are misleading due to high income inequality: Lebanon ranks 129 among 141 countries in terms of income inequality, as measured by the Gini coefficient (Ministry of Finance and UNDP, 2017). In 2011, approximately 27% of the Lebanese population lived in poverty (CAS and World Bank, 2015).

Studies of Lebanese consumer shopping patterns—particularly in urban areas—are scant. Vignal (2007) attributes the emergence of a consumer society in Middle Eastern cities including Beirut since the 1990s to economic liberalisation, globalisation, and regional investment and trade, facilitated by the 2005 establishment of the Greater Arab Free Trade Area. Seyfert et al. (2014) find that Lebanese consumers continue to rely on traditional food retailers, making it harder for large retail formats to expand into several territories and socioeconomic groups; the authors attribute this to the 'institutional landscape'. Alternatively, Cleveland, Laroche, and Hallab (2013) highlight the importance of cultural factors like religion in influencing the shopping behaviour of Lebanese consumers. More specifically, their research examined the relationships between *inter alia*, culture, religion, consumption-related values, and consumption behaviours among coexisting religious groups (Christians and Muslims) in Lebanon, and identified the importance of religion in shaping consumption behaviours as well as the differential impact of globalization on subcultures within countries. In a recent study of Lebanon's food retail patterns, Bahn and Abebe (2017) observe absolute growth in both modern and traditional food retail sales but find that modern-format food retailers have not definitively replaced traditional food retail formats in recent decades, as had been reported in other developing countries such as Argentina, Brazil, and Thailand (Hawkes, 2008). Rather, modern food retail formats were most prevalent in large cities such as Beirut and for upper- and middle-income socioeconomic groups, while traditional retailers were more equally presented across urban, peri-urban, and rural settings and in areas marked by higher prevalence of poverty.

3 Methods

To address the first research question (i.e., provide an overview of the Lebanese food retail landscape), a descriptive analysis of total and per capita value of retail sales by retail category is provided. In addition, based on a primary survey of Lebanese food retailers, the study presents residence location of primary customers and retailers' market orientation by retail category. The second research question rests on analysis of the operational and marketing activities of Lebanese retailers. For this, a description of retailers' operational characteristics, including operational efficiency, market orientation, marketing programs, demand characteristics, spatial distribution, and perceptions of customer behaviour is reported. The third research question relates to the structure of food supply chains involving Lebanese food retailers. Accordingly, the study analyses the upstream relationships of Lebanese food retailers.

A secondary data set from Euromonitor International (2018) is used to answer the first research question (i.e., conduct an initial, macro-level assessment of the Lebanese food retail landscape). For the second and third questions our analysis relies on primary data collected from 49 retailers in Lebanon. Primary data was collected through a survey questionnaire of food retailers operating in Lebanon. The questionnaire was developed from similar questionnaires used in previous studies of food retail and wholesaling operations in MENA (Seyfert et al., 2014), but modified to reflect the research questions of interest and solicit information on retailers' operations, marketing, and customers.

To reach the survey population, a list of food retailers and their contact information was obtained from three chambers of commerce operating in Lebanon: Chamber of Commerce of Beirut and Mount Lebanon (CCIABML); Chamber of Commerce of Saida and South (CCIAS); and Chamber of Commerce of Zahle and Bekaa (CCIAZ). The lists indicated the operating size of the food retailer (from largest to smallest: supermarket, minimarket, or grocery store).[5] The total population of the combined lists was 1820; of these, a minority were classified as supermarkets (4%), while the far larger share classified as minimarkets (49%) or grocery stores (47%). Food retailer owners, managers, or personnel with visibility over the retailers' operations were contacted and completed the survey by phone or through an internet-based platform between April and May 2016. The survey questionnaire did not collect unique, identifying information so as to encourage truthful and candid responses. Respondents were not required to indicate the chamber with which they are registered and/or their retail classification, as this might have violated anonymity given small population sizes for selected chambers.

The research team obtained 49 complete surveys, corresponding to 2.7% of the population of registered retailers, within the limited time period for survey administration. Accordingly, the results reported below are representative only of those retailers responding. Caution should be applied in interpreting these findings as representative of all food retailers in Lebanon. Table 14 displays the number of retailers registered by chamber and classification, versus the number of complete responses received.

Survey responses were coded into Excel. Numeric coding corresponded to the survey questionnaire, including both continuous data and categorical data (e.g., Likert scales). A rapid review using visual inspection and summary statistics was performed to verify quality of responses. Quantitative analysis was performed using Stata/SE 15.1 statistical software. Analysis focused particularly on a comparison of operations, marketing and customer profile, and supply chain relationships (research questions 2 and 3). Self-reported differences across retailers permitted stratified analysis and comparative means testing across retailer classifications and across geographic profiles. The analysis applied the classifications used by the Lebanese chambers of commerce when data was available: Responding retailers were classified as modern-format (supermarkets) and traditional-format (minimarkets, grocery stores) per the definitions applied by Bahn and Abebe (2017). The remaining retailers were categorized as 'unspecified' and were retained for analytical purposes. The ANOVA test with the Bonferroni option was used to test differences in means of continuous variables, while the Chi-square test was used to test differences in means of categorical variables.

4 Findings

4.1 Food Retail Landscape in Lebanon

According to the latest Euromonitor International (2018) dataset, Lebanon's food retail market was estimated at approximately $6.3 billion in 2017, corresponding to approximately 10% of GDP. By comparison, Lebanon's food retail market was about the same size as that of Iraq, Oman, and Jordan; but represented only 3% of the value of that of the United Kingdom (Fig. 2) (Euromonitor International, 2018). Lebanon's

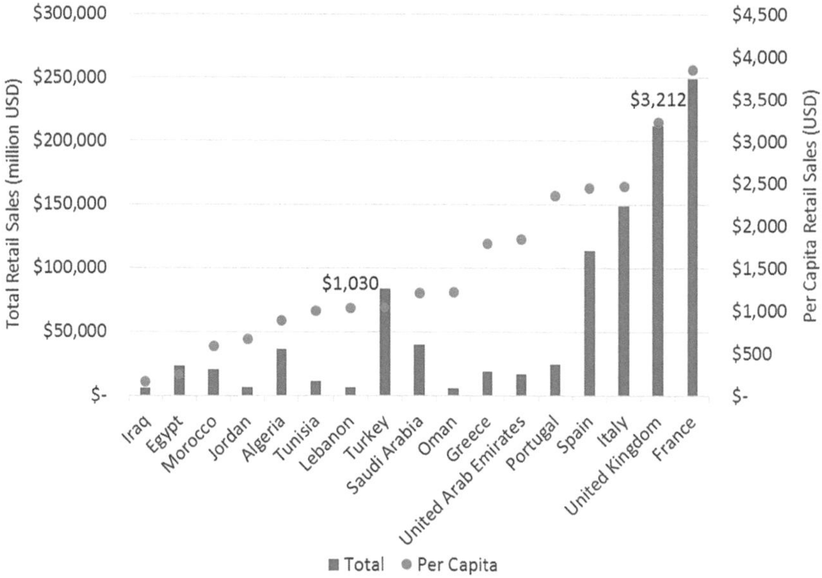

Fig. 2 Total and per capital value of retail sales by grocery retailers in selected countries, 2017. (Source: Euromonitor International, 2018)

food retail sales per capita were approximately $1030, roughly the same as reported for Tunisia, Turkey, and Saudi Arabia.

In 2017, the traditional retail segment[6] accounted for 67% of Lebanon's total retail value ($4.2 billion) and the modern food retail segment the remaining 33% ($2.1 billion) (Fig. 3). Modern grocery retailers' sales recorded a nominal increase of 252% between 2003 and 2017, demonstrating continuous growth. The corresponding increase for traditional grocery retailers was approximately 65% over the same period, although growth appeared to level-off after 2012.

Survey respondents are principally classified as minimarkets (n = 18 or 37%), grocery stores (n = 8 or 16%), or supermarkets (n = 6 or 12%); and accordingly, as traditional- (n = 26 or 53%) or modern-format food retailers (n = 6 or 12%). Remaining respondents (n = 17) are unspecified and unclassified, as classification was not possible based on retailers' survey responses. Respondents principally serve urban and suburban customers: Approximately 32% of retailers reported that the majority of their customers live in urban communities, with another 22% reporting that the majority of customers live in suburban communities (Table 1). This result is at a level lower than expected given that Lebanon's population is highly urbanised (Fig. 1). The discrepancy might be explained by

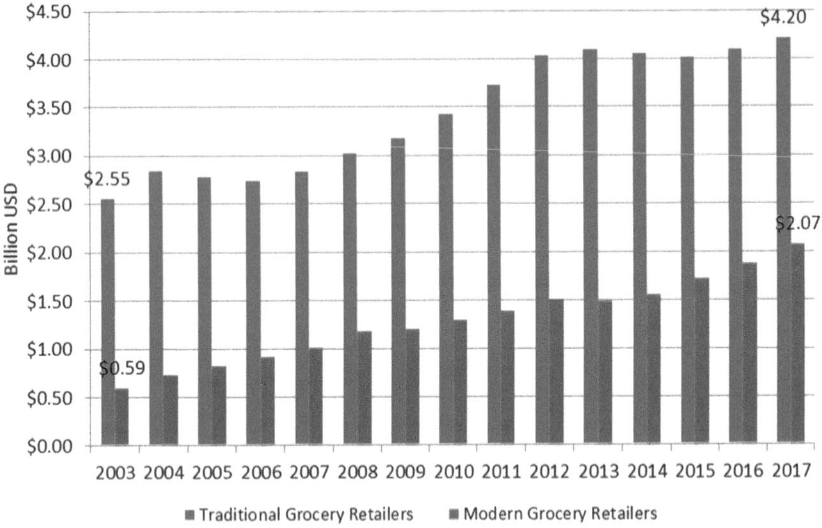

Fig. 3 Retail value of traditional and modern grocery retailers' sales, 2003–2017. (Source: Euromonitor International, 2018)

Table 1 Retailer categories and market orientation: primary location of customers' residence

Mean geographic profile of customers	All retailers	Supermarket	Minimarket	Grocery store	Unspecified
Urban area	12 (32.4%)	2 (40.0%)	2 (13.3%)	4 (66.7%)	4 (36.4%)
Suburban area	8 (21.6%)	1 (20.0%)	7 (46.7%)	–	–
Rural area	17 (45.9%)	2 (40.0%)	6 (40.0%)	2 (33.3%)	7 (63.4%)
No response	12 –	1 –	3 –	2 –	6 –
Total	49	6	18	8	17

Table 1 indicates the number of total responses; and the share of responses among those who had a substantive response per retailer category in parentheses
Source: Authors

Table 2 Retailer classifications and market orientation: *p*-value resulting from means testing (Chi-square)

Variable tested	Retailer categories (supermarket, minimarket, grocery store)	Retailer classifications (modern, traditional)	Retailer orientation (urban/ suburban, rural)
Store categories (supermarket, minimarket, grocery store)			0.958
Store classifications (modern, traditional)			0.937
Mean geographic profile of customers	0.114	0.815	

Source: Authors

the fact that most retailers in the sample are relatively small, and smaller-size formats are often used to expand to less urban environments. However, modern retailers do not exclusively serve urban and suburban communities, and traditional retailers do not solely serve rural communities: We find no statistical difference in the perceived geographic profile of customers across retailer categories or classifications when a Chi-square test of the primary location of customers' residence is performed (Table 2).[7]

4.2 Operational and Marketing Activities of Modern vs. Traditional Food Retail Formats

Nearly all retailers surveyed indicated local (Lebanese) ownership. The majority of retailers were self-owned (55%) or family-owned (37%) businesses. The average year of establishment was 2001, or 15 years of operation at the time surveyed. All respondents engage in food retail, with a minority also engaged in upstream (16% also engage in food production) or downstream stages of the food supply chain (33% carry out their own food packaging; 27% engage in food wholesaling). These results are consistent across the modern-traditional classifications and geographic orientation: We find no statistically significant differences in the mean results across either retail classifications or geographic orientation when ANOVA or Chi-square tests are performed (Table 15).

Retailers most commonly sell non-perishable food, and less frequently frozen or fresh foods. The food product categories most commonly sold are canned goods (96%), dry goods including cereals and pulses (92%), dairy products (86%), baked goods (83%), snack foods (75%), vegetables (67%), frozen goods (65%), and fruits (61%). Fewer food retailers sell meat (31%) or fish (9%). Retailers serving urban/suburban customers are more likely to sell fresh meat than retailers serving rural customers (Table 3), but we find no such differences across the modern-traditional classifications. Retailers sell on average 6.5 food product categories (among those categories listed above). There is no statistically significant difference in the mean number of product categories sold across retailer classification or geographic orientation when an ANOVA test is conducted (Table 15).

Food retailers surveyed reported an average of two locations (including stores, warehouses, and outlets), ranging from a minimum of 1 to a maximum of 11. The business premises are typically owned directly (38% total ownership, 14% primary ownership) or rented (35% total rental, 11% primary rental). We find no systematic difference in the mean number of locations or the form of ownership of the business premises across retail classification or geographic orientation (Table 3). The former result is unexpected, as we might have expected modern retailers to be chained with more than one location.

Across their locations, retailers reported an average total shelf space of approximately 300 square metres; average shelf space per location (total shelf space divided by the number of business locations) is 222 square metres. The average storage or warehouse space reported is 219 square metres; average storage space per location is 126 square metres. As a subcategory of storage space, retailers report an average total cold storage space of 70 square metres, and average cold storage space per location of 52 square metres. There is no systematic difference in the total or average shelf space, the total or average storage space, or the total or average cold storage space across retailer classifications (Table 3), which is unexpected

Table 3 Retailers' operations: mean values and *p*-values resulting from means testing

Variable tested	Test type	Retailer classifications (modern, traditional)			Retailer orientation (urban/suburban, rural)		
		Mean		*p*-value	Mean		*p*-value
		Modern	Traditional		Urb/sub	Rural	
Likelihood to sell fresh meat	Chi-square	0.333	0.346	0.952	0.526	0.176	0.029**
Mean number of locations	ANOVA	3.800	2.478	0.3704	2.500	1.824	0.4345
Mean ownership type of business premises	Chi-square	2.200	2.318	0.526	2.579	2.250	0.814
Mean shelf space—total	ANOVA	198.667	237.367	0.8116	322.731	290.546	0.8265
Mean shelf space per location	ANOVA	153.212	118.695	0.6686	211.379	250.876	0.7693
Mean storage space—total	ANOVA	130.000	294.382	0.6524	373.967	172.000	0.2420
Mean storage space per location	ANOVA	31.212	122.615	0.6000	206.242	143.409	0.5504
Mean cold storage space—total	ANOVA	41.750	46.711	0.9168	159.636	22.153	0.0827*
Mean cold storage space per location	ANOVA	28.962	11.428	0.2060	115.621	19.201	0.2170
Likelihood to use automated billing system	Chi-square	1.000	1.478	0.047**	1.300	1.412	0.478
Likelihood to use electronic scale	Chi-square	1.200	1.087	0.459	1.100	1.059	0.647
Likelihood to use electronic inventory	Chi-square	1.400	1.652	0.295	1.400	1.647	0.134
Mean number of owners	ANOVA	2.000	3.705	0.4612	3.675	2.25	0.2729
Mean number of employees—managers	ANOVA	1.600	1.636	0.9644	1.900	1.250	0.1758
Mean number of employees—full-time non-managerial	ANOVA	5.250	15.273	0.3800	16.455	2.600	0.0460**
Mean number of employees—casual daily laborers	ANOVA	1.667	4.143	0.5725	8.750	2.286	0.1646

(*continued*)

Table 3 (continued)

Variable tested	Test type	Retailer classifications (modern, traditional)			Retailer orientation (urban/suburban, rural)		
		Mean		*p*-value	Mean		*p*-value
		Modern	Traditional		Urb/sub	Rural	
Mean number of employees—total	ANOVA	8.900	14.295	0.6279	18.125	5.882	0.0590*
Mean expenditure—permanent employees	ANOVA	0	1,902,679	0.4928	2,673,438	1,362,917	0.2476
Mean expenditure—seasonal employees	ANOVA	208,300	251,667	0.9451	412,500	126,277	0.3950
Mean expenditure—electricity/gas	ANOVA	1,441,667	1,041,500	0.7668	1,523,250	1,137,680	0.6573
Mean expenditure—water	ANOVA	150,000	108,508	0.7971	97,757	115,281	0.8427
Mean expenditure—transport of goods	ANOVA	125,000	466,818	0.2822	262,143	394,642	0.5142
Mean expenditure—rent	ANOVA	3,750,000	739,706	0.0600*	1,609,091	363,462	0.1200
Mean expenditure—interest on money borrowed/mortgage	ANOVA	0	300,000	0.6568	450,000	25,000	0.3119

Source: Authors

Significance levels are reported as follows: * 10% level, ** 5% level, and *** 1% level

given that physical area (specifically shelf space) is one criterion for store classifications such as hypermarkets and supermarkets (e.g., Codron, Bouhsina, Fort, Coudel, & Puech, 2004; Reardon & Berdegué, 2002), albeit not the classification criteria of the Lebanese chambers of commerce. Similarly, we find no difference in average shelf or storage space across geographic orientation, though we might have expected to find smaller physical spaces in more urban environments where the price of land is higher, and the space is at a premium. However, retailers primarily serving urban/suburban customers report a significantly higher average total cold storage space than retailers serving rural customers (160 square metres versus 22 square metres) (Table 3).

The retailers surveyed use a range of technologies to facilitate operations, including an electronic scale (92% of retailers responding), an automated billing system (64%), and an electronic inventory (49%). Modern retailers are more likely to use an automated billing system than traditional retailers; however, there is no systematic difference in the use of electronic scales or electronic inventories across retail classifications (Table 3). Interestingly, systematic differences are found only for the supplier-oriented technology, but not for technologies oriented to internal operations management. We find no systematic difference in retailers' use of any of the three technologies across geographic orientation.

In terms of employment, the retailers surveyed were small employers with an average of three owners working directly in the company in addition to 1.6 managers, 6.7 full-time, non-managerial employees, and 2.5 casual, daily labourers.[8] An average of 14 employees corresponds to a minimarket classification according to Lebanese chambers of commerce. We find no difference in the mean number of employees, whether per type of employee or total number of employees, across retailer classifications (Table 3). This result is unexpected and suggests that classifications of the chambers of commerce are not principally based on the number of registered employees within the company. However, retailers serving urban/suburban customers report a higher average number of full-time, non-managerial employees (16.5) than rural-oriented retailers (2.6); and a correspondingly higher number of total employees (18.1 versus 5.9). This difference may reflect a greater availability of semi-skilled labour in Lebanon's more densely populated areas.

Table 4 Monthly business expenditures by line item per store

Line item[a]	Expenditure per store (USD)[b]			
	Average	Min	Max	Standard deviation
Permanent employees	1251	0	4700	1611
Electricity and gas	857	0	5802	1370
Rent	620	0	4973	1290
Transport of goods	232	0	995	282
Seasonal employees	156	0	2188	482
Interest on money borrowed/mortgage	145	0	2984	636
Water	72	0	663	143

Source: Authors

[a]Average among those food retailers reporting a value (including zero) for the line item expense

[b]Figures were reported in Lebanese lira (LL) and converted to USD at the official exchange rate of 1508 LL: 1 USD

In terms of business expenditures, the greatest expenses reported are for permanent employee salaries/wages; electricity and gas; and rent (Table 4). These results do not generally differ across retail classifications or geographic orientation, with one exception: Modern retailers report higher average expenditure on rent than traditional retailers (Table 3), which suggests that modern retailers may pay a premium for better space, for example in terms of location. Urban/suburban-oriented retailers do not report higher average expenditure on rent than rural-oriented retailers, which is surprising given that rural property values would be expected to be lower on a per-square-metre basis. To control for variation in store size, we calculate total monthly business expenditures per store, per square metre of shelf area, and per employee and compare across retailer classification and geographic orientation (Table 5). While modern retailers initially appear to spend more per store, per shelf area, and per employee than traditional retailers, these differences are not statistically significant. Similarly, urban/suburban-oriented retailers appear to spend less per square metre of shelf space than rural-oriented retailers, but more per store and more per employee; however, these differences are not statistically significant (Table 15).

Table 5 Average total monthly business expenditures per store, area, and employee

		Classification		Orientation	
	Total[a]	Modern	Traditional	Urban/ suburban	Rural
Expenditure per store	1733	2166	2026	2875	1847
	n = 43	*n* = 4	*n* = 23	*n* = 15	*n* = 17
	(2524)	(3425)	(2.707)	(3284)	(1996)
Expenditure per shelf area (m^2)	35	40	29	26	46
	n = 21	*n* = 2	*n* = 13	*n* = 9	*n* = 11
	(6)	(53)	(27)	(29)	(74)
Expenditure per employee	604	1409	636	714	543
	n = 33	*n* = 3	*n* = 20	*n* = 15	*n* = 17
	(1049)	(1434)	(1190)	(1377)	(709)

Source: Authors

[a]The summed number of responses across retailer classifications and orientations do not correspond to total number of respondents, as these are unspecified for some respondents

In terms of marketing and pricing strategies, food retailers typically set their selling prices based on a combination of strategies: More than half of retailers surveyed (58%) reported setting prices based on two or more criteria. The most common pricing strategies were a mark-up strategy (adding a pre-set profit margin or percentage to the cost of merchandise) (80% of respondents); and setting prices below the competition (43%). In contrast, fewer retailers reported setting prices according to the manufacturer's stated retail price (20%); psychological pricing techniques such as pricing goods at 1495LL instead of 1500LL (14%); pricing above the competition (2%); or a multiple pricing strategy such as offering two-for-one incentives (2%). We find two statistically significant differences in the likelihood of using different pricing strategies across retailer classifications or geographic orientation: Traditional retailers are more likely than modern retailers to set prices below the competition, while urban/suburban-oriented retailers are more likely than rural-oriented retailers to apply a psychological pricing strategy (Table 6).

Retailers report that the prices upon which they base their final retail prices (whether MSRP, competitors' prices, or other) are generally accurate, neither too low nor too high (45% of respondents). Another 29% of

Table 6 Retailers' marketing and customer demand/profile: *p*-value resulting from means testing

Variable tested	Test type	Retailer classifications (modern, traditional)			Retailer orientation (urban/suburban, rural)		
		Mean		*p*-value	Mean		*p*-value
		Modern	Traditional		Urb/sub	Rural	
Pricing strategy—price below competition	Chi-square	1.833	1.423	0.070*	1.400	1.529	0.431
Pricing strategy—psychological	Chi-square	1.833	1.846	0.938	1.650	2.000	0.007***
Mean perception of prices	Chi-square	2.800	3.000	0.078*	2.800	2.588	0.929
Mean perceived importance to customers—range of products available	Chi-square	4.200	4.300	0.693	4.700	3.625	0.059*
Mean perceived importance to customers—store reputation	Chi-square	4.500	4.750	0.044**	4.833	4.625	0.541

Source: Authors

Significance levels are reported as follows: * 10% level, ** 5% level, and *** 1% level

retailers report that prices are inaccurate but not systematically too high or too low. Modern retailers are more likely to report that prices are either too low or accurate than traditional retailers (Table 6). We find no systematic difference in the perception of pricing across retailers' geographic orientation.

Survey results also reveal retailers' perceptions of customer demand and profile. For example, retailers were asked about any change in total demand across food products over the past five years, as a means to assess possible changes in local food consumption habits; results are presented in Table 7. Varied answers were reported for every food group, but the average response was equivalent to 'no change' across most food groups (fresh fruits, fresh vegetables, dairy products, fresh meat, baked goods, canned goods, and dry goods). For example, retailers selling fruit are evenly split among those who report the volume of demand for fruit has increased, decreased, or remained steady. Retailers reported an increase in demand for fresh fish,[9] but this result is interpreted with caution given the small number of observations subject to analysis. We find no evidence of systematic differences in perceived demand changes across retailer classifications or geographic orientation (Table 15).

In terms of customer profile, retailers uniformly report that their customers are primarily drawn from the geographic area (administrative district) where the retail operation is located (results not shown). This finding is consistent with the observation that most retailers surveyed are smaller businesses, as food retail tends to be a local business with the exception of larger-format stores.

Retailers primarily serve middle- and low-income customers (self-defined by respondents): Most retailers surveyed (65%) reported that the majority of their customers are middle-income, while another 27% reportedly that the majority of their customers are low-income (Table 8). Despite apparent differences, the perceived customer income profile is statistically consistent across retailer classifications and geographic orientation as determined by a Chi-square test (Table 15).

Retailers also reported their perceptions of the factors of importance to customers in choosing where to shop for food (Table 9). Factors were, in descending order of importance, price (92% reported to be fairly or very important to customers), store reputation (89%), quality (86%), range of

Table 7 Retailers' perceived changes in demand across food groups

	Increased	No change	Decreased	Do not know	Do not sell product/ No response
Fruits	9	8	9	1	22
	(34.6%)	(30.8%)	(34.6%)	–	–
Vegetables	9	10	9	1	20
	(32.1%)	(35.7%)	(32.1%)	–	–
Dairy products	13	14	10	–	12
	(35.1%)	(37.8%)	(27.0%)		–
Meat	3	6	4	–	36
	(27.1%)	(46.2%)	(30.8%)		–
Fish	2	2	–	–	45
	(50.0%)	(50.0%)			–
Baked goods	8	19	7	1	14
	(23.5%)	(55.9%)	(20.6%)	–	–
Frozen goods	5	11	10	1	22
	(19.2%)	(42.3%)	(38.5%)	–	–
Canned goods	13	11	16	1	8
	(32.5%)	(27.5%)	(40.0%)	–	–
Dry goods	14	16	10	–	9
	(35.0%)	(40.0%)	(25.0%)		–
Snacks	10	14	9	1	15
	(30.3%)	(42.4%)	(27.3%)	–	–

Table 7 indicates the number of total responses and the share of responses among those who had a substantive response per retailer category in parentheses
Source: Authors

products (81%), food safety standards (75%), and store location (73%). Less than half of retailers considered either a customer loyalty program or the availability of specialty products to be important factors in guiding customer decisions as to where to shop for food.

The relative ranking of these perceived factors appears to differ somewhat across retailer classifications (Table 9). For example, 100% of modern retailers report that customers consider food quality very important or fairly important in determining where to shop, as compared to 81% of traditional retailers. Modern retailers are more likely to report that their customers consider price very or fairly important in determining where they shop (100% versus 86% of traditional retailers). Modern retailers are less likely to report that customers place importance on store

Table 8 Retailers' perception of customer income category

Mean income profile of customers	Total	Modern	Traditional	Urban/ suburban	Rural
High-income	2 (5.4%)	1 (20.0%)	1 (4.8%)	2 (10.0%)	–
Middle-income	24 (64.9%)	3 (60.0%)	13 (61.9%)	11 (55.0%)	13 (76.5%)
Low-income	10 (27.0%)	1 (20.0%)	6 (28.6%)	6 (30.0%)	4 (23.5%)
Do not know	1 (2.7%)	–	1 (4.8%)	1 (5.0%)	–
No response	12 –	1 –	5 –	–	–
	49	6	26	20	17

Table 8 indicates the number of total responses; and the share of responses among those who had a substantive response per retailer category in parentheses. The summed number of responses across retailer classifications and orientations may not correspond to total number of respondents, as these are unspecified for some respondents

Source: Authors

location (50%), as compared to traditional retailers (67%). However, these differences are not statistically significant across retail classifications, with the exception of modern and traditional retailers' perception of store reputation in influencing customer selection of food retailer (Table 6). Across geographic orientation, the only statistically significant difference concerns the range of products available: Rural-oriented retailers believe this factor is less important to customers than urban/suburban-oriented retailers.

Only 27% of retailers surveyed offer a customer loyalty program (Table 10). We find no statistically significant difference in response across retailer classification or geographic orientation (Table 15). Those retailers offering a loyalty program report a broadly even distribution of customers enrolled: Two retailers indicated participation rates below 25%, two retailers indicated participation rates of 25–50%, three retailers of 50–75%, and three retailers above 75%. We find no statistically significant difference in customer enrolment across either retailer classification or geographic orientation (Table 15).

Table 9 Retailers' perceptions of determinants of customer food outlet selection

	Total		Modern		Traditional		Urban/ suburban		Rural	
Factor	%	Rank	%	Rank	%	Rank	%	Rank	%	Rank
Price	92%	1	100%	1	86%	2	90%	2	94%	1
Store reputation	89%	2	80%	3	90%	1	90%	2	88%	2
Quality	86%	3	100%	1	81%	4	90%	2	82%	3
Range of products	81%	4	80%	3	81%	4	95%	1	65%	5
Food safety standards	75%	5	80%	3	86%	2	85%	5	63%	6
Store location	73%	6	50%	6	67%	6	75%	6	71%	4
Customer loyalty program	44%	7	40%	7	38%	7	50%	7	38%	7
Availability of specialty products	33%	8	40%	7	29%	8	40%	8	25%	8

Table 9 presents the percentage of respondents who reported the factor to be "fairly important" or "very important" to customers in selecting a food outlet (remaining choices were "indifferent," "not very important," or "not important at all"). Rank figures are the relative ranking of factors by order of most to least important

Source: Authors

Table 10 Retailers offering additional customer services

	Total	Modern	Traditional	Urban/ suburban	Rural
Customer loyalty program					
Yes	10 (27%)	2 (40%)	4 (19%)	7 (35%)	3 (18%)
No	27 (73%)	3 (60%)	17 (81%)	13 (65%)	14 (82%)
Delivery service					
Yes	21 (58%)	4 (80%)	13 (65%)	13 (68%)	8 (47%)
No	15 (42%)	1 (20%)	7 (35%)	6 (32%)	9 (53%)

Table 10 presents the number of total substantive responses, with the share of responses per retailer classification displayed in parentheses

Source: Authors

Approximately 58% of food retailers surveyed offer a delivery service (delivery of items selected by the customer in-store, by phone, or online) (Table 10). We find no statistical difference across retailer classification (Table 15). Those retailers offering delivery service report that generally few customers take advantage of the service: 85% indicated that fewer than half of customers take advantage of the service. Again, we find no systematic difference in the reported use of delivery service across retailer classifications. We find no significant difference in the likelihood of retailers to offer a delivery service, or its use by customers, across geographic orientation (Table 15), although we might have expected urban/suburban retailers more likely than rural retailers to offer delivery services given shorter distances.

4.3 Supply Chain Structure of Modern vs. Traditional Food Retail Formats

We next consider retailers' supply chain arrangements and structures. Retailers' principal suppliers vary by food product. Wholesale companies and traders are the most commonly reported suppliers of canned goods, dry goods, frozen goods, and snacks; the wholesale market is the most commonly reported supplier of fresh vegetables and fresh fruit; and direct-from-producer purchases supply most fish, dairy, baked goods, and meat. Results are consistent across retail classifications for perishable products but differ somewhat for non-perishable products: Modern retailers selling canned goods are more likely to report that they source them from a wholesale company while traditional retailers are more likely to source from other traders (Table 11). Modern retailers selling dry goods are more likely to report that they source them from a wholesale company while traditional retailers are more likely to source from other traders. Modern retailers selling snack foods are more likely to report that they source from a wholesale company than traditional retailers, which are more likely to source from other traders. In terms of geographic orientation, we find systematic differences across both perishable and non-perishable food categories: Urban/suburban-oriented retailers source fresh meat directly from the producer or a wholesaler, while rural-oriented

Table 11 Retailers' supply chain structures: mean value and *p*-value resulting from means testing

Variable tested	Test type	Retailer classifications (modern, traditional)			Retailer orientation (urban/suburban, rural)		
		Mean		*p*-value	Mean		*p*-value
		Modern	Traditional		Urb/ Sub	Rural	
Supplier of fresh meat (source)	Chi-square	1.000	1.857	0.693	1.111	6.000	0.007***
Supplier of baked goods (source)	Chi-square	1.000	2.000	0.555	1.111	2.583	0.054*
Supplier of canned goods (source)	Chi-square	2.000	3.720	0.047**	2.895	3.600	0.339
Supplier of dry goods (source)	Chi-square	2.333	3.500	0.065*	2.526	4.200	0.008***
Supplier of snack foods (source)	Chi-square	2.000	3.840	0.031**	2.944	3.250	0.666
Mean number of suppliers—snack foods	ANOVA	10.667	9.000	0.6907	13.353	5.909	0.0311**
Average shipping time to retailer—fresh fruit	ANOVA	0.363	1.188	0.1647	1.433	0.698	0.0859*
Average shipping time to retailer—canned goods	ANOVA	9.688	0.998	0.0321**	4.155	1.068	0.3455
Average shipping time to retailer—dry goods	ANOVA	0.730	0.975	0.4900	0.690	1.104	0.0953*
Average shipping time to retailer—snack foods	ANOVA	9.750	0.963	0.0255**	3.858	1.075	0.3982
Importance in selecting supplier—geography of supplier	Chi-square	2.400	2.520	0.095*	2.450	2.875	0.413
Importance in selecting supplier—food safety certification	Chi-square	5.000	4.750	0.706	4.842	4.412	0.063*
Requirement of food safety standards for—fresh fruit	Chi-square	1.000	1.400	0.188	1.400	1.000	0.056*

Mean response—long-term contractual agreement with suppliers	Chi-square	1.000	1.458	0.055*	1.250	1.706	0.006***
Mean duration of long-term contractual agreement with suppliers	Chi-square	3.600	3.833	0.966	4.154	2.200	0.076*
Likelihood of contract terms—time of delivery	Chi-square	1.600	1.333	0.309	1.462	1.000	0.063*
Likelihood of contract terms—grade of product	Chi-square	1.600	1.333	0.301	1.692	1.200	0.060*

Source: Authors

Significance levels are reported as follows: * 10% level, ** 5% level, and *** 1% level

retailers (only one respondent) source from own production. Urban/suburban-oriented retailers source baked goods directly from the producer or a wholesaler, while rural-oriented retailers source from a wider range of suppliers. Urban/suburban-oriented retailers source dry goods principally from wholesalers, other traders, or the producer directly; rural-oriented retailers source mostly from other traders and less frequently from wholesalers.

Retailers report a greater number of suppliers for non-perishable food items than fresh food items. The average number of suppliers reported by retailers varies across product categories, from a maximum of 14 suppliers of canned goods to a minimum of one supplier of fresh fish (Table 12). These results imply that suppliers of fish, meat, baked goods, and dry goods are fewer and may therefore exert a higher degree of market power

Table 12 Average number of suppliers, by food product, across retailer types

	Total	Modern	Traditional	Urban/ suburban	Rural
Fresh fruit	7.96	5.00	10.75	8.73	8.55
	(*n* = 26)	(*n* = 3)	(*n* = 12)	(*n* = 11)	(*n* = 11)
Fresh vegetables	7.70	5.00	9.40	8.23	9.30
	(*n* = 27)	(*n* = 3)	(*n* = 15)	(*n* = 13)	(*n* = 10)
Dairy	4.47	4.00	4.30	4.88	4.15
	(*n* = 36)	(*n* = 6)	(*n* = 20)	(*n* = 17)	(*n* = 13)
Fresh meat	1.42	1.00	1.50	1.39	1.00
	(*n* = 13)	(*n* = 2)	(*n* = 7)	(*n* = 9)	(*n* = 2)
Fresh fish	1.00	1.00	1.00	1.00	–
	(*n* = 4)	(*n* = 1)	(*n* = 1)	(*n* = 3)	
Baked goods	2.92	3.17	2.52	2.82	3.38
	(*n* = 36)	(*n* = 6)	(*n* = 23)	(*n* = 17)	(*n* = 13)
Frozen goods	4.26	4.00	3.59	5.08	3.36
	(*n* = 27)	(*n* = 2)	(*n* = 17)	(*n* = 13)	(*n* = 11)
Canned goods	14.63	9.33	12.40	19.05	9.29
	(*n* = 41)	(*n* = 6)	(*n* = 25)	(*n* = 19)	(*n* = 14)
Dry goods	3.55	2.50	3.08	3.12	3.43
	(*n* = 40)	(*n* = 6)	(*n* = 24)	(*n* = 19)	(*n* = 14)
Snacks	9.09	10.67	9.00	13.35	5.91
	(*n* = 35)	(*n* = 6)	(*n* = 24)	(*n* = 17)	(*n* = 11)

Source: Authors

Among retailers selling the food product. The summed number of responses across retailer classifications and orientations do not correspond to total number of respondents, as these are unspecified for some respondents vis-à-vis retailers; while retailers may have significantly more market power over suppliers of canned goods, snack foods, vegetables, and fruits, where alternative suppliers may be available. We find no statistically significant differences in the mean number of suppliers across retailer classifications (Table 15). In terms of geographic orientation, urban/suburban-oriented retailers report a statistically significantly higher average number of suppliers of snacks (13) than rural-oriented retailers (6) (Table 11).

Suppliers are nearly always Lebanese, with only a minor share of non-perishable goods sourced from foreign suppliers. Indeed, retailers reported only Lebanese suppliers of fruit, vegetables, dairy, meat, and fish, which may reflect a substantial domestic agricultural sector producing these products for sale through the wholesale market (fruit, vegetables) or for direct sale to market (dairy, meat, fish). A minor share of retailers (less than 7% for each food product) sourced the following non-perishable goods from foreign suppliers: dry goods, baked goods, snacks, and canned goods from Europe, and frozen goods from non-GCC Arab countries. This finding may indicate limited competition within the non-perishable food sub-sector from non-Lebanese suppliers, though it is unclear if Lebanese suppliers are truly more competitive or if there are barriers to the entry of non-Lebanese firms. We find no evidence of systematic differences in the location of suppliers across retailer classifications or geographic orientation (results not shown); this may be attributable to the limited number of responses and the lack of classification information for some retailers.

Retailers report relatively modest shipping times to receive food products from suppliers, which seems consistent with the reported reliance on Lebanese suppliers and the generally small size of the country. Fruits reach food retailers in an average of 1.1 truck hours (as opposed to other forms of transport); vegetables, 1.0 truck hours; dairy products, 2.7 truck hours; meat products, 0.76 truck hours; fish, 2.8 truck hours; baked goods, 0.65 truck hours; frozen goods, 0.96 truck hours; canned goods, 4.0 truck hours; dry goods, 1.8 truck hours; and snack foods, 2.3 truck hours.

Longer delivery times for non-perishable items are expected. The slightly longer delivery times for dairy products and fresh fish, particularly as compared to the delivery time for frozen goods, are less expected as these products must be kept cool during delivery. We find few statistically significant differences in delivery times across retailer classifications (Table 11): Average delivery times are significantly shorter for traditional retailers than modern retailers for canned goods (1.0 hours versus 9.7 hours) and for snack foods (1.0 hours versus 9.8 hours). Longer delivery times for modern retailers might be the result of sophisticated logistics systems or long supply chains. In terms of geographic orientation, delivery times for fresh fruit to urban/suburban-oriented retailers are longer than for rural-oriented retailers (1.4 hours versus 0.7 hours); however, delivery times for dry goods to urban/suburban-oriented retailers are shorter than for rural-oriented retailers (0.7 hours versus 1.1 hours).

Retailers generally are able to source food products from alternative suppliers (85% responding affirmatively). There is no statistically significant difference in availability of alternative suppliers across retailer classifications or geographic orientation (Table 15).

While retailers report a greater number of suppliers for non-perishable food items than fresh food items (Table 12), they change suppliers less frequently for non-perishable than fresh food items (Fig. 4). Notably, retailers report a higher frequency of change in suppliers of fresh fruits and vegetables, changing suppliers of fruits daily (27%) or weekly (9%) and changing suppliers of vegetables daily (29%) or weekly (8%). In the case of fruits and vegetables, this likely reflects rotating purchases among existing suppliers, for example suppliers selling at the wholesale market, on the basis of availability and differences in price and quality. For all other food products, a majority of retailers report that they do not change suppliers more frequently than once every five years. We find no statistically significant difference in the frequency of change in supply across retailer classifications or geographic orientation (Table 15).

Retailers rate quality, food safety, and price as the most important factors when selecting a supplier (Table 13). Indeed, nearly all food retailers report that they care most about the quality (98% of those responding rank as very important or fairly important), food safety certification (95%) and price (91%). Other features are less important in selecting suppliers, including acquaintance with business partner (50%); organic certification (46%); the size of the delivery (43%); supplier location (38%); and price predictability (36%). We find no statistically significant

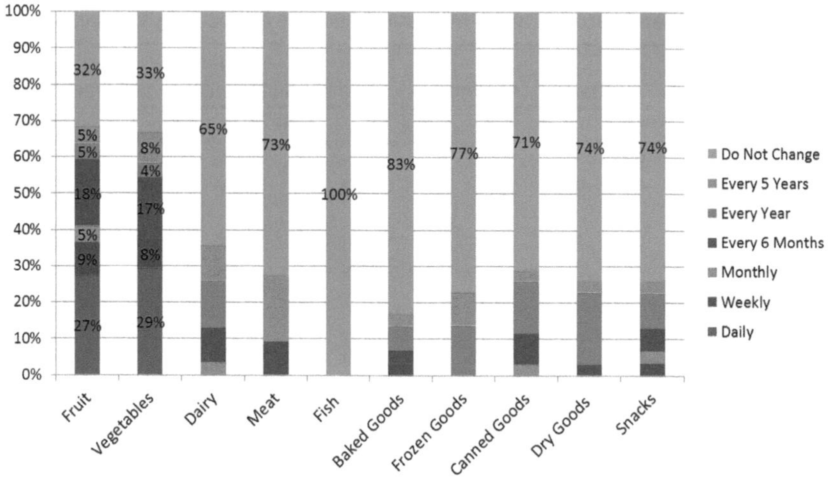

Fig. 4 Frequency of change of suppliers, by food category (Among retailers selling the food product, excluding "do not know" responses. The numbers of responses per food product category are as follows: fruits (n = 22), vegetables (n = 24), dairy (n = 31), fresh meat (n = 11), fresh fish (n = 2), baked goods (n = 30), frozen goods (n = 22), canned goods (n = 35), dry goods (n = 35), and snacks (n = 31).). (Source: Authors)

difference in the selection criteria for suppliers across retailer classifications or geographic orientation with two exceptions: Traditional retailers rank the location of the supplier as more important on average than modern retailers, and urban/suburban-oriented consumers rank food safety certification to be more important on average than rural-oriented retailers (Table 11).

Approximately 79% of retailers surveyed require suppliers to meet food safety standards,[10] principally for dairy products, fish, and non-perishable foods but less frequently for fresh fruits, fresh vegetables, fresh meat, and frozen goods. We find no statistically significant difference in the likelihood of requiring food safety standards, generally or by food category, across retailer classifications (Table 15). Across geographic orientation, we find only one statistically significant difference: Rural-oriented retailers uniformly require suppliers of fresh fruit to meet food safety standards, while relatively fewer urban/suburban-oriented retailers impose this requirement (Table 11).

Table 13 Retailers' perceptions of important features in selecting suppliers

	All retailers		Modern		Traditional		Urban/ suburban		Rural	
Factor	%	Rank	%	Rank	%	Rank	%	Rank	%	Rank
Quality	98%	1	100%	1	96%	1	95%	1	100%	1
Food safety certification	95%	2	100%	1	96%	1	95%	1	94%	2
Price	91%	3	80%	3	88%	3	95%	1	82%	3
Acquaintance with business partner	50%	4	40%	6	68%	4	55%	4	69%	4
Organic certification	46%	5	60%	4	36%	6	39%	6	47%	6
Size of delivery	43%	6	40%	6	38%	5	35%	7	38%	8
Supplier location	38%	7	20%	8	36%	6	30%	8	50%	5
Price predictability	36%	8	60%	4	32%	8	42%	5	47%	6

Table 13 reports the percentage of respondents who consider the factor to be "fairly important" or "very important" in selecting a supplier. Rank figures are the relative ranking of factors by order of most to least important
Source: Authors

Most retailers (57%) report having contractual agreements with a duration of at least two weeks with their suppliers. Modern retailers are more likely to have a contractual agreement with suppliers than traditional retailers; similarly, urban/suburban-oriented retailers are more likely than rural-oriented retailers to have a contractual agreement with suppliers (Table 11). The majority of retailers with such a contractual arrangement have written contracts (67%), with another 29% using verbal agreements. Contractual agreements most frequently cover a period of at least six months (64% of responses). While there is no difference in contract duration across retailer classifications, urban/suburban-oriented retailers are more likely to have a contract that exceeds six months than rural-oriented retailers, who tend to report shorter contracts (Table 11). Contracts stipulate a range of provisions, most commonly the time of delivery (67%) and the quantity (52%), and less frequently the selling price (41%), grade of product (41%), and packaging (40%). We find no difference in the content of the contract terms across retailer classifications (Table 15). In terms of geographic orientation, however, rural-oriented retailers are more likely than urban/suburban-oriented retailers to specify the time of delivery and the grade of the product within the contract (Table 11).

The majority of retailers do not pay suppliers in advance (93%), rather upon or after receiving the product. This result may indicate that retailers exert some degree of market power over the food producers and suppliers. However, food retailers reported that payment is made reasonably promptly: Approximately 32% of respondents indicated that payment is made upon receiving the product, another 10% within one week of receipt, and a further 27% within one month of receipt. The remaining respondents indicated that payment is made within three months of receiving the product (27%), or within three to six months of receipt (5%). We find no evidence of systematic differences in average payment period across retailer classifications or geographic orientation (Table 15), though we might have expected modern retailers to exert more power over suppliers than smaller, un-chained, traditional retailers.

In terms of payment mechanism, retailers most commonly report paying suppliers by cash on delivery (40% of retailers responding to the question); by signing off at delivery and paying later through bank transfer or check (28%); or by signing off at delivery and paying by instalments (15%). We find no systematic differences in payment mechanism across retailer classifications or geographic orientation (Table 15). Nearly all food retailers (98% of those responding to the question) reported that the supplier provided an invoice or bill for transactions conducted in the previous month.

5 Discussion

This section discusses key findings related to the current Lebanese food retail landscape, as well as differences in the operations and marketing activities and supply chain structures of modern versus traditional retailers and retailers in urban and rural areas in Lebanon.

In terms of food retail landscape, macro-level evidence suggests that modern-format retail continues to grow in both absolute and relative terms, which should indicate some displacement of traditional retailers. In terms of expansion, our survey results indicate no systematic alignment of modern-format retailers with an urban/suburban orientation, and therefore no or weak evidence of a geographic expansion pattern of supermarketisation in Lebanon, although the small sample size should be kept in mind when concluding these observations. Similarly, we find no systematic alignment of modern-format retailers with a high- or middle-income customer profile. Accordingly, our micro-level survey does not provide a strong indication of a socio-economic expansion pattern of supermarkets in Lebanon. Perhaps most surprisingly, we find no systematic differences in the food products sold across retail classifications, and thus no robust evidence for a food product expansion pattern among modern retailers. Nevertheless, only six modern retailers were included in the sample and surveying more modern retailers may lead to a different conclusion.

In terms of operational profile, micro-level evidence reveals limited differences between modern and traditional retailers operating in Lebanon. What is not reported may be more revealing: The modern retailers surveyed are not generally chained or larger in terms of either physical space or employment, suggesting that the retailer classifications applied by Lebanon's chambers of commerce may not align with the wider supermarketisation literature. In terms of geographic orientation, the finding that urban/suburban retailers are systematically larger employers may reflect local labour market availability.

In terms of product offerings, modern and traditional retailers carry the same number of product categories (6.5 on average). This suggests that traditional retailers may be diversifying their product offerings to compete with modern retailers; and provides weak evidence that modern retailers offer a significantly more diverse food basket to their customers, as was reported in other countries (Rischke et al., 2015). Indeed, retailers generally perceived little change in demand across food products over the past five years, which may further explain why we find no significant differences in the number of product categories sold across retailer classifications.

In terms of customer profile, retailers' customers are primarily drawn from the geographic area in which the retail operation is located. This is indeed expected as most of the retailers surveyed are smaller businesses. The surveyed retailers considered price, store reputation, quality, range of products, food safety standards, and store location as important attributes for their customers, in that order. These perceptions do not meaningfully vary across geographic orientation except for range of products available, which rural-oriented retailers believe is less important to customers than urban/suburban-oriented retailers.

In terms of supply chain arrangements, retailers' principal suppliers vary by type of food. Wholesale companies and traders are the main suppliers of canned goods, dry goods, frozen goods, and snacks, and the wholesale market for fresh fruits and vegetables. Fish, dairy, baked goods, and meat are often directly sourced from producers. This generally holds across retail classifications for perishable products but differs somewhat for non-perishable products. Suppliers also systematically vary by geographic orientation across both perishable and non-perishable food categories. Furthermore, retailers source non-perishable items from more suppliers compared to fresh food items. This is expected as the production and supply of highly perishable food products may require coordinated action and fewer, preferred supplier relationships (Berdegué et al., 2005). Also, food items represented by a few supply sources may lead to a higher degree of market power (Hingley, 2005). This may be particularly true in rural areas, for example, where retailers report a statistically lower average number of suppliers of snacks.

Despite the availability of alternative suppliers for non-perishable food items, retailers change their suppliers for non-perishable food items infrequently, which may suggest a trend toward long-term supply relationships. Conversely, contractual relationships involving the supply of fresh food items appeared to be short-term, perhaps due to the seasonality of supply for such produce. In general, modern-format retailers are more likely to have a contractual agreement with suppliers than traditional retailers. Contracts may involve a range of provisions including delivery time, quantity, price, product grade, and packaging, in that order. The most important factors that retailers consider when selecting a supplier are quality, food safety, and price, which is unsurprising given major quality and safety concerns affecting the modern food industry (Aung & Chang, 2014), including in Lebanon (Abebe et al., 2017).

We unexpectedly find no evidence of systematic differences in average payment period across retailer classifications. Our results suggest that suppliers can expect reasonably prompt payment for their goods in the Lebanese food retail sector, regardless of retailer type, and therefore we find no evidence of eroding supplier livelihoods at the hands of modern retailers. This is unexpected, because larger retailers have been reported to exert market power by paying their suppliers at a longer delay, and the issue of delayed payment has been highlighted as one factor that may exclude smaller producers or farmers from entering the supermarket supply chain in some countries (Maglaras, Bourlakis, & Fotopoulos, 2015).

6 Conclusions

This chapter provides insights into the food retail landscape of Lebanon, a country generally overlooked by the literature on food retail transformation. A commercial link between the Middle East and Europe, Lebanon holds several bilateral and regional trade agreements with the European Union including the Euro-Mediterranean partnership. This commercial relationship is growing in value: Lebanese-EU trade has increased by an average of 7.6% annually since 2006 (European Commission, 2017). In 2016, Lebanon imported approximately $441 million in goods from the United Kingdom and exported $37.5 million to the UK. Beverages, food products, and household goods comprised a significant portion of these trade flows (International Trade Centre, 2018). Accordingly, international and regional actors can gain from this study a better understanding of Lebanon's food retailing landscape for the purposes of guiding investment in its retail sector, or tracking changes in the domestic market with regard to food safety or other standards that may affect import supply quality.

This study has sought to analyse the extent of food retail transformation in Lebanon by exploring (1) the current Lebanese food retail landscape; (2) the operational and marketing activities and supply chain structures of selected modern and traditional retail formats; and (3) the operational and marketing activities and supply chain structures of retailers across geographies within the country. Both the macro- and micro-level datasets show early evidence of food retail transformation in Lebanon: a transformation from traditional to modern, and from short-term or transaction-based relations to long-term, contractually-based exchanges. This implies that the food retail segment in Lebanon if guided by appropriate policies could support the expansion of sustainable food supply chains, which are critical to enhance the availability and accessibility of a year-round, wide assortment of nutritious food products to urban consumers (Gómez & Ricketts, 2013). In terms of private sector supply chains, this study suggests that non-Lebanese retailers sourcing directly from suppliers rather than the local wholesale market may have an advantage vis-à-vis more traditional retailers still dependent upon wholesale market transactions: They might raise the standards required of suppliers to deliver higher quality (e.g., food safety, convenience packaging) particularly for items such as fresh fruits and vegetables in order to gain market share, as was done in more advanced markets (Gereffi, Humphrey, & Sturgeon, 2005). In terms of limitations, this study is based on a narrow survey of food retailers. The geographic orientation is derived from self-reporting of retailers and was not possible to independently verify given anonymity requirements when implementing the survey. Most participating retailers report an operational profile that would be classified as traditional retail according to definitions applied within the food retail literature. Accordingly, this study provides a clearer view of the current food retail landscape in Lebanon, specifically the operations and supply chain structures of traditional retailers. Future research might expand the research sample to a larger number of retailers and gather more detailed information on the location of both the retailer and their respective supplier and customer bases.

This study may potentially contribute to policies guiding sustainable food system development, particularly with regard to monitoring the consolidation of value chains, the participation of preferred or selected actors within the upstream value chain, and the availability of diverse and nutritious foods for urban and non-urban consumers; and to future research exploring the impact of modern food retailing on value chain relationships, evolving supply chain dynamics, and shifts within the food system in Lebanon and the MENA region.

Acknowledgements This study was funded by the University Research Board (URB project #22687) at the American University of Beirut. There is no financial interest to report. The authors acknowledge the contribution of Ms. Lory Boutchakdjian for her support in the collection of data that underpins this chapter. The authors thank Dr. Ali Chalak for his input on conduct of appropriate statistical tests. The authors thank participants in a research seminar held at the American University of Beirut, Faculty of Agricultural and Food Sciences in February 2017 for valuable comments on initial research findings.

Appendix 1

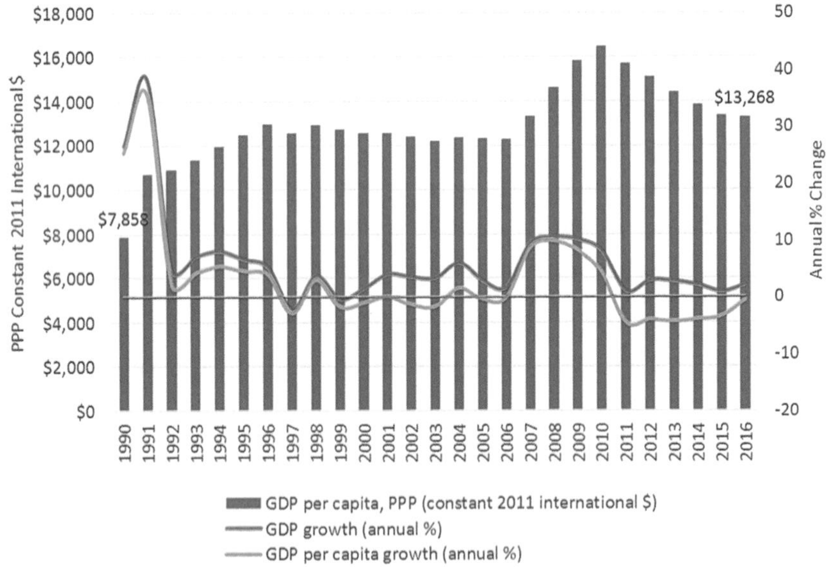

Fig. 5 Selected economic statistics for Lebanon. (Source: World Bank, 2018)

Appendix 2

Table 14 Population and survey responses across registered retailers

	CCIABML					CCIAS				CCIAZ				Unspecified				Total				
	Super-market	Mini-market	Grocery Store	Unspecified	Total	Super-market	Mini-market	Grocery Store	Total	Super-market	Mini-market	Grocery Store	Total	Super-market	Mini-market	Grocery Store	Total	Super-market	Mini-market	Grocery Store	Unspecified	Total
Number registered	40	110	832		982	4	2	24	30	26	782	–	808	*N/A*				70	894	856		1820
Responses received	1	9	7	1	18	–	–	1	1	–	9	5	14	–			16	6	18	8	17	49

Source: Authors

Appendix 3

Table 15 Complete statistical analysis results: *p*-values resulting from means testing

Variable tested	test type	Retailer classifications (modern, traditional)			Retailer orientation (urban/suburban, rural)		
		Mean		*p*-value	Mean		*p*-value
		Modern	Traditional		Urb/Sub	Rural	
Store categories (supermarket, minimarket, grocery store)	Chi-square	3.000	1.692	0.000	1.938	2.000	0.958
Store classifications (modern, traditional)	Chi-square	1.000	2.000	0.000	1.813	1.800	0.937
Likelihood to engage in food production	Chi-square	1.833	1.769	0.732	1.800	1.765	0.795
Likelihood to engage in food wholesaling	Chi-square	1.833	1.692	0.489	1.650	1.824	0.236
Mean ownership structure	Chi-square	1.333	1.423	0.524	1.550	1.471	0.826
Year of establishment	ANOVA	2007	2000	0.2767	2000	1999	0.7336
Mean number of product categories sold	ANOVA	6.667	6.654	0.9907	7.150	6.647	0.4728
Likelihood to sell fresh fruit	Chi-square	0.500	0.500	1.000	0.600	0.706	0.501
Likelihood to sell fresh vegetables	Chi-square	0.500	0.615	0.604	0.700	0.706	0.969
Likelihood to sell dairy products	Chi-square	1.000	0.769	0.192	0.850	0.941	0.373

Likelihood to sell fresh meat	Chi-square	0.333	0.346	0.952	0.526	0.176	0.029**
Likelihood to sell fresh fish	Chi-square	0.167	0.038	0.242	0.158	0.000	0.107
Likelihood to sell baked goods	Chi-square	0.833	0.885	0.732	0.947	0.824	0.238
Likelihood to sell frozen food	Chi-square	0.667	0.654	0.952	0.650	0.750	0.517
Likelihood to sell canned goods	Chi-square	0.833	0.962	0.242	0.950	1.000	0.364
Likelihood to sell dry goods	Chi-square	0.833	0.923	0.497	0.950	1.000	0.350
Likelihood to sell snacks	Chi-square	1.000	0.962	0.625	0.900	0.688	0.109
Supplier of fresh fruit	Chi-square	3.000	3.231	0.837	2.750	2.818	0.284
Supplier of fresh vegetables	Chi-square	3.000	3.125	0.716	2.714	2.818	0.360
Supplier of dairy products	Chi-square	1.167	1.950	0.613	1.471	2.154	0282
Supplier of fresh meat	Chi-square	1.000	1.857	0.693	1.111	6.000	0.007***
Supplier of fresh fish	Chi-square	1.000	2.000	0.157	1.333	–	–
Supplier of baked goods	Chi-square	1.000	2.000	0.555	1.111	2.583	0.054*
Supplier of frozen goods	Chi-square	2.000	2.412	0.555	2.083	2.833	0.301
Supplier of canned goods	Chi-square	2.000	3.720	0.047**	2.895	3.600	0.339
Supplier of dry goods	Chi-square	2.333	3.500	0.065*	2.526	4.200	0.008***
Supplier of snack foods	Chi-square	2.000	3.840	0.031**	2.944	3.250	0.666
Mean number of suppliers—fruits	ANOVA	5.000	10.750	0.3018	8.727	8.545	0.9557
Mean number of suppliers—vegetables	ANOVA	5.000	9.400	0.4044	8.231	9.300	0.7342

(*continued*)

Table 15 (continued)

Variable tested	test type	Retailer classifications (modern, traditional) Mean			Retailer orientation (urban/suburban, rural) Mean		
		Modern	Traditional	*p*-value	Urb/Sub	Rural	*p*-value
Mean number of suppliers—dairy products	ANOVA	4.000	4.300	0.8333	4.882	4.154	0.5275
Mean number of suppliers—meat	ANOVA	1.000	1.500	0.4622	1.389	1.000	0.5167
Mean number of suppliers—fish	ANOVA	1.000	1.000	–	1.000	–	–
Mean number of suppliers—baked goods	ANOVA	3.167	2.522	0.3984	2.824	3.385	0.4214
Mean number of suppliers—frozen goods	ANOVA	4.000	3.588	0.8124	5.077	3.364	0.1767
Mean number of suppliers—canned goods	ANOVA	9.333	12.400	0.7180	19.053	9.286	0.1817
Mean number of suppliers—dry goods	ANOVA	1.517	2.244	0.5538	3.316	3.429	0.8843
Mean number of suppliers—snack foods	ANOVA	10.667	9.000	0.6907	13.353	5.909	0.0311**
Average shipping time to retailer—fresh fruit	ANOVA	0.363	1.188	0.1647	1.433	0.698	0.0859*
Average shipping time to retailer—fresh vegetables	ANOVA	0.363	1.001	0.2545	1.204	0.698	0.2218

Average shipping time to retailer—dairy products	ANOVA	0.630	1.039	0.4127	0.760	3.030	0.3209
Average shipping time to retailer—fresh meat	ANOVA	0.210	0.540	0.4548	0.430	1.000	0.2900
Average shipping time to retailer—fresh fish	ANOVA	–	1.000	–	1.000	–	–
Average shipping time to retailer—baked goods	ANOVA	0.533	0.651	0.7022	0.428	0.683	0.1911
Average shipping time to retailer—frozen goods	ANOVA	0.723	0.873	0.6508	0.895	0.850	0.8780
Average shipping time to retailer—canned goods	ANOVA	9.688	0.998	0.0321**	4.155	1.068	0.3455
Average shipping time to retailer—dry goods	ANOVA	0.730	0.975	0.4900	0.690	1.104	0.0953*
Average shipping time to retailer—snack foods	ANOVA	9.750	0.963	0.0255**	3.858	1.075	0.3982
Availability of alternative suppliers	Chi-square	1.200	1.167	0.858	1.222	1.133	0.510
Frequency of change of suppliers—fruits	Chi-square	5.000	3.750	0.868	3.818	4.000	0.366
Frequency of change of suppliers—vegetables	Chi-square	5.000	4.000	0.836	4.077	4.000	0.409
Frequency of change of suppliers—dairy products	Chi-square	6.000	6.389	0.290	6.313	6.091	0.561
Frequency of change of suppliers—meat	Chi-square	7.000	6.714	0.391	6.875	6.500	0.236

(continued)

Table 15 (continued)

Variable tested	test type	Retailer classifications (modern, traditional)			Retailer orientation (urban/suburban, rural)		
		Mean		*p*-value	Mean		*p*-value
		Modern	Traditional		Urb/Sub	Rural	
Frequency of change of suppliers—fish	Chi-square	7.000	7.000	–	7.000	–	–
Frequency of change of suppliers—baked goods	Chi-square	7.000	6.400	0.755	6.688	6.273	0.124
Frequency of change of suppliers—frozen goods	Chi-square	6.500	6.875	0.102	6.417	6.875	0.307
Frequency of change of suppliers—canned goods	Chi-square	7.000	6.478	0.857	6.722	5.917	0.116
Frequency of change of suppliers—dry goods	Chi-square	7.000	6.682	0.586	6.556	0.856	0.388
Frequency of change of suppliers—snack foods	Chi-square	6.000	6.435	0.338	6.235	6.500	0.327
Importance in selecting supplier—quality	Chi-square	5.000	4.800	0.630	4.750	4.824	0.639
Importance in selecting supplier—price	Chi-square	4.200	4.480	0.707	4.750	4.294	0.108
Importance in selecting supplier—size of delivery	Chi-square	3.200	2.958	0.835	2.900	2.938	0.121
Importance in selecting supplier—price predictability	Chi-square	3.000	2.720	0.339	3.053	1.588	0.289

Importance in selecting supplier—acquaintance with business partner	Chi-square	2.800	3.720	0.453	3.350	2.941	0.903
Importance in selecting supplier—geography of supplier	Chi-square	2.400	2.520	0.095*	2.450	2.875	0.413
Importance in selecting supplier—food safety certification	Chi-square	5.000	4.750	0.706	4.842	4.412	0.063*
Importance in selecting supplier—organic certification	Chi-square	3.400	2.560	0.130	2.778	3.000	0.238
Method of payment to supplier	Chi-square	3.500	3.130	0.886	2.833	3.313	0.395
Timing of payment to supplier	Chi-square	4.200	3.435	0.328	3.850	3.250	0.446
Mean response—long-term contractual agreement with suppliers	Chi-square	1.000	1.458	0.055*	1.250	1.706	0.006***
Mean form of contractual agreement with suppliers	Chi-square	1.200	1.308	0.648	1.333	1.600	0.226
Mean duration of long-term contractual agreement with suppliers	Chi-square	3.600	3.833	0.966	4.154	2.200	0.076*
Likelihood of contract terms—selling price	Chi-square	1.800	1.667	0.582	1.692	1.400	0.255

(*continued*)

Table 15 (continued)

Variable tested	test type	Retailer classifications (modern, traditional)			Retailer orientation (urban/suburban, rural)		
		Mean		*p*-value	Mean		*p*-value
		Modern	Traditional		Urb/Sub	Rural	
Likelihood of contract terms—quantity to delivery	Chi-square	1.600	1.583	0.949	1.615	1.200	0.114
Likelihood of contract terms—time of delivery	Chi-square	1.600	1.333	0.309	1.462	1.000	0.063*
Likelihood of contract terms—grade of product	Chi-square	1.600	1.333	0.301	1.692	1.200	0.060*
Likelihood of contract terms—production methods	Chi-square	2.000	2.000	–	1.923	1.800	0.457
Likelihood of contract terms—packaging	Chi-square	1.800	1.636	0.513	1.692	1.400	0.255
Likelihood of contract terms—external product specifications	Chi-square	1.800	1.909	0.541	1.769	1.800	0.888
Likelihood of requiring suppliers to meet food safety standards	Chi-square	1.000	1.174	0.314	1.105	1.286	0.184
Requirement of food safety standards for—fresh fruit	Chi-square	1.000	1.400	0.188	1.400	1.000	0.056*
Requirement of food safety standards for—fresh vegetables	Chi-square	1.000	1.250	0.333	1.250	1.000	0.149

Requirement of food safety standards for—dairy products	Chi-square	1.000	1.000	–	1.000	1.000	–
Requirement of food safety standards for—fresh meat	Chi-square	1.000	1.143	0.571	1.111	1.000	0.621
Requirement of food safety standards for—fresh fish	Chi-square	1.000	1.000	–	1.000	–	–
Requirement of food safety standards for—baked goods	Chi-square	1.000	1.000	–	1.000	1.000	–
Requirement of food safety standards for—frozen goods	Chi-square	1.250	1.067	0.288	1.167	1.000	0.176
Requirement of food safety standards for—canned goods	Chi-square	1.200	1.053	0.289	1.118	1.000	0.284
Requirement of food safety standards for—dry goods	Chi-square	1.200	1.056	0.311	1.118	1.000	0.284
Requirement of food safety standards for—snack foods	Chi-square	1.200	1.050	0.269	1.118	1.000	0.312
Pricing strategy—mark-up on cost	Chi-square	1.333	1.115	0.185	1.050	1.000	0.350
Pricing strategy—MSRP	Chi-square	2.000	1.769	0.192	1.850	1.647	0.152

(continued)

Table 15 (continued)

Variable tested	test type	Retailer classifications (modern, traditional)			Retailer orientation (urban/suburban, rural)		
		Mean		*p*-value	Mean		*p*-value
		Modern	Traditional		Urb/Sub	Rural	
Pricing strategy—price below competition	Chi-square	1.833	1.423	0.070*	1.400	1.529	0.431
Pricing strategy—price above competition	Chi-square	2.000	1.962	0.625	1.950	2.000	0.350
Pricing strategy—psychological	Chi-square	1.833	1.846	0.938	1.650	2.000	0.007***
Pricing strategy—multiple pricing	Chi-square	2.000	1.962	0.625	1.950	2.000	0.350
Mean perception of prices	Chi-square	2.800	3.000	0.078*	2.800	2.588	0.929
Mean number of locations	ANOVA	3.800	2.478	0.3704	2.500	1.824	0.4345
Mean ownership type of business premises	Chi-square	2.200	2.318	0.526	2.579	2.250	0.814
Mean shelf space—total	ANOVA	198.667	237.367	0.8116	322.731	290.546	0.8265
Mean shelf space per location	ANOVA	153.212	118.695	0.6686	211.379	250.876	0.7693
Mean storage space—total	ANOVA	130.000	294.382	0.6524	373.967	172.000	0.2420
Mean storage space per location	ANOVA	31.212	122.615	0.6000	206.242	143.409	0.5504
Mean cold storage space—total	ANOVA	41.750	46.711	0.9168	159.636	22.153	0.0827*
Mean cold storage space per location	ANOVA	28.962	11.428	0.2060	115.621	19.201	0.2170

Likelihood to use automated billing system	Chi-square	1.000	1.478	0.047**	1.300	1.412	0.478
Likelihood to use electronic scale	Chi-square	1.200	1.087	0.459	1.100	1.059	0.647
Likelihood to use electronic inventory	Chi-square	1.400	1.652	0.295	1.400	1.647	0.134
Likelihood of carrying out own packaging	Chi-square	1.600	1.652	0.825	1.600	1.688	0.587
Mean number of owners	ANOVA	2.000	3.705	0.4612	3.675	2.25	0.2729
Mean number of employees—managers	ANOVA	1.600	1.636	0.9644	1.900	1.250	0.1758
Mean number of employees—full-time non-managerial	ANOVA	5.250	15.273	0.3800	16.455	2.600	0.0460**
Mean number of employees—casual daily labourers	ANOVA	1.667	4.143	0.5725	8.750	2.286	0.1646
Mean number of employees—total	ANOVA	8.900	14.295	0.6279	18.125	5.882	0.0590*
Mean expenditure—permanent employees	ANOVA	0	1,902,679	0.4928	2,673,438	1,362,917	0.2476
Mean expenditure—seasonal employees	ANOVA	208,300	251,667	0.9451	412,500	126,277	0.3950
Mean expenditure—electricity/gas	ANOVA	1,441,667	1,041,500	0.7668	1,523,250	1,137,680	0.6573
Mean expenditure—water	ANOVA	150,000	108,508	0.7971	97,757	115,281	0.8427

(continued)

Table 15 (continued)

Variable tested	test type	Retailer classifications (modern, traditional)			Retailer orientation (urban/suburban, rural)		
		Mean		*p*-value	Mean		*p*-value
		Modern	Traditional		Urb/Sub	Rural	
Mean expenditure—transport of goods	ANOVA	125,000	466,818	0.2822	262,143	394,642	0.5142
Mean expenditure—rent	ANOVA	3,750,000	739,706	0.0600*	1,609,091	363,462	0.1200
Mean expenditure—interest on money borrowed/mortgage	ANOVA	0	300,000	0.6568	450,000	25,000	0.3119
Mean expenditure per store	ANOVA	3,266,650	3,055,071	0.9273	4,335,355	2,784,766	0.2866
Mean expenditure per shelf area (m^2)	ANOVA	59,671	44,337	0.6600	39,843	69,097	0.4711
Mean expenditure per employee	ANOVA	369,214	501,696	0.8006	523,577	458,633	0.8112
Mean geographic profile of customers	Chi-square	2.000	2.095	0.815	1.400	3.000	–
Mean income profile of customers	Chi-square	2.000	1.750	0.526	1.789	1.764	0.292
Mean perceived importance to customers—quality	Chi-square	5.000	4.238	0.513	4.550	4.118	0.260
Mean perceived importance to customers—price	Chi-square	4.800	4.619	0.575	4.650	4.647	0.752

Mean perceived importance to customers—range of products available	Chi-square	4.200	4.300	0.693	4.700	3.625	0.059*
Mean perceived importance to customers—availability of specialty products	Chi-square	3.500	2.368	0.297	2.833	2.600	0.413
Mean perceived importance to customers—food safety	Chi-square	4.000	4.400	0.822	4.263	3.500	0.363
Mean perceived importance to customers—store location	Chi-square	4.250	4.050	0.884	4.556	3.765	0.101
Mean perceived importance to customers—customer loyalty program	Chi-square	3.667	2.833	0.346	3.733	2.688	0.264
Mean perceived importance to customers—store reputation	Chi-square	4.500	4.750	0.044**	4.833	4.625	0.541
Likelihood of offering a customer loyalty card or program	Chi-square	1.600	1.810	0.318	1.650	1.824	0.236

(continued)

Table 15 (continued)

Variable tested	test type	Retailer classifications (modern, traditional)			Retailer orientation (urban/suburban, rural)		
		Mean		*p*-value	Mean		*p*-value
		Modern	Traditional		Urb/Sub	Rural	
Mean customer participation in retailer's loyalty card or program	Chi-square	3.500	3.000	0.290	3.571	4.000	0.736
Likelihood of offering a delivery service	Chi-square	1.200	1.350	0.520	1.316	1.529	0.194
Mean customer use of store delivery service	Chi-square	2.250	2.500	0.852	2.571	2.500	0.102

Source: Authors

Significance levels are reported as follows: * 10% level, ** 5% level, and *** 1% level

Notes

1. In terms of structure, supply chains are traditionally associated with upstream relationships (optimization of costs) while the notion of value chains emphasizes demand-driven, customer-focused chains. This distinction is becoming blurred in today's environment, and these terms are at times used interchangeably.
2. Lebanon is also home to approximately 320,000 Palestinian refugees and 1.5 million Syrian refugees (CAS, 2015).
3. Lebanon does not strictly define urban, semi-urban, or rural at an administrative or legal level. However, in most developing countries, urban areas are identified by a combination of factors including land-use, population density, and economic activity (a large majority of the population is not engaged in agriculture or fishing) (UN-HABITAT, 2011). This study follows the conceptualization provided by Thapa and Murayama (2008) and Lerner, Eakin, and Sweeney (2013), which define a peri-urban setting as a special case of urban-rural interaction and a transitional area between urban and rural zones. In a peri-urban setting, residents share some characteristics with rural areas by engaging in agricultural activities while increasingly relying on non-agricultural employment.
4. The population living in smaller and intermediary cities was calculated by deducting the number of people living in large cities and rural areas from the reported total population.
5. The classification is in theory based on objective measures including number of employees registered with the National Social Security Fund and the enterprise's capital, value added tax, and profit levels.
6. Euromonitor International applies a categorization of food retailers (refer to Sect. 2.1) that differs from but appears to be broadly consistent with the classifications of the Lebanese chambers of commerce.
7. Significant or remarkable results of statistical means tests across retailer classifications and geographic orientation are indicated within the chapter text (Tables 2, 3, 7, and 11). Complete results are reported in full in Table 15.
8. Corresponding standard deviations are as follows: owners (3.9), managers (1.4), full-time, non-managerial employees 13.6), and casual, daily labourers (6.4).

9. A one-way test of the mean perception of change in demand for fish as >0 (increasing) yields a *p*-value of 0.0908.
10. The ability of retailers to verify these standards is beyond the scope of the current study.

References

Abebe, G. K., Chalak, A., & Abiad, M. G. (2017). The effect of governance mechanisms on food safety in the supply chain: Evidence from the Lebanese dairy sector. *Journal of the Science of Food and Agriculture, 97*, 2908–2918.

Alexander, A. (2008). Format development and retail change: Supermarket retailing and the London co-operative society. *Business History, 50*(4), 489–508.

Apparicio, P., Cloutier, M.-S., & Shearmur, R. (2007). The case of Montreal's missing food deserts: Evaluation of accessibility to food supermarkets. *International Journal of Health Geographics, 6*(1), 4.

Aung, M. M., & Chang, Y. S. (2014). Traceability in a food supply chain: Safety and quality perspectives. *Food Control, 39*, 172–184.

Bahn, R. A., & Abebe, G. K. (2017). Analysis of food retail patterns in urban, peri-urban and rural settings: A case study from Lebanon. *Applied Geography, 87*, 28–44.

Berdegué, J. A., Balsevich, F., Flores, L., & Reardon, T. (2005). Central American supermarkets' private standards of quality and safety in procurement of fresh fruits and vegetables. *Food Policy, 30*(3), 254–269.

Central Administration of Statistics (CAS). (2015). Planning figures for LCRP 2016 – total Lebanese. Retrieved January 5, 2016, from https://data.hdx.rwlabs.org/dataset/lebanon-estimated-figures-for-the-resident-population-in-lebanon

Central Administration of Statistics (CAS) and World Bank. (2015). *Measuring poverty in Lebanon using 2011 HBS*. Lebanon: World Bank. Retrieved June 12, 2018, from http://www.cas.gov.lb/images/Excel/Poverty/Measuring%20poverty%20in%20Lebanon%20using%202011%20HBS_technical%20report.pdf

Charreire, H., Casey, R., Salze, P., Simon, C., Chaix, B., Banos, A., et al. (2010). Measuring the food environment using geographical information systems: A methodological review. *Public Health Nutrition, 13*(11), 1773–1785.

Cleveland, M., Laroche, M., & Hallab, R. (2013). Globalization, culture, religion, and values: Comparing consumption patterns of Lebanese Muslims and Christians. *Journal of Business Research, 66*(8), 958–967.

Coca-Cola Retailing Research Council Asia (CCRRCA). (2007). *Food retail formats in Asia – understanding format success*. A study conducted for the Coca-Cola Retailing Research Council Asia by IBM.

Codron, J.-M., Bouhsina, Z., Fort, F., Coudel, E., & Puech, A. (2004). Supermarkets in low-income Mediterranean countries: Impacts on horticulture systems. *Development Policy Review, 22*(5), 587–602.

Dries, L., Reardon, T., & Swinnen, J. F. (2004). The rapid rise of supermarkets in Central and Eastern Europe: Implications for the agrifood sector and rural development. *Development Policy Review, 22*(5), 525–556.

Eckert, J., & Shetty, S. (2011). Food systems, planning and quantifying access: Using GIS to plan for food retail. *Applied Geography, 31*(4), 1216–1223.

Euromonitor International. (2018). Market sizes – grocery retailers – retail value RSP excl. sales tax. Retrieved June 12, 2018, from http://www.portal.euromonitor.com/portal/statisticsevolution/index

European Commission (2017, November 29). *Countries and regions: Lebanon*. Retrieved from http://ec.europa.eu/trade/policy/countries-and-regions/countries/lebanon/

Gatrell, J. D., Reid, N., & Ross, P. (2011). Local food systems, deserts, and maps: The spatial dynamics and policy implications of food geography. *Applied Geography, 31*(4), 1195–1196.

Gereffi, G., Humphrey, J., & Sturgeon, T. (2005). The governance of global value chains. *Review of International Political Economy, 12*(1), 78–104.

Goldman, A. (2001). The transfer of retail formats into developing economies: The example of China. *Journal of Retailing, 77*(2), 221–242.

Goldman, A., Ramaswami, S., & Krider, R. E. (2002). Barriers to the advancement of modern food retail formats: Theory and measurement. *Journal of Retailing, 78*(4), 281–295.

Gómez, M. I., & Ricketts, K. D. (2013). Food value chain transformations in developing countries: Selected hypotheses on nutritional implications. *Food Policy, 42*, 139–150.

Guarín, A. (2013). The value of domestic supply chains: Producers, wholesalers, and urban consumers in Colombia. *Development Policy Review, 31*(5), 511–530.

Gustafson, A., Hankins, S., & Jilcott, S. (2012). Measures of the consumer food store environment: A systematic review of the evidence 2000–2011. *Journal of Community Health, 37*(4), 897–911.

Hawkes, C. (2008). Dietary implications of supermarket development: A global perspective. *Development Policy Review, 26*(6), 657–692.

Hingley, M. K. (2005). Power imbalanced relationships: Cases from UK fresh food supply. *International Journal of Retail & Distribution Management, 33*(8), 551–569.

Humphrey, J. (2007). The supermarket revolution in developing countries: Tidal wave or tough competitive struggle? *Journal of Economic Geography, 7*(4), 433–450.

International Trade Centre. (2018). Bilateral trade between Lebanon and United Kingdom. *TradeMap*. Retrieved from https://trademap.org/Bilateral_TS.aspx?nvpm=1|422||826||TOTAL|||2|1|1|1|2|1|1|1|1

Joseph, L., & Kuby, M. (2013). Regionalism in US retailing. *Applied Geography, 37*, 150–159.

Lawrence, G., & Dixon, J. (2015). The political economy of Agri-food: Supermarkets. In A. Bonanno & L. Busch (Eds.), *Handbook of the international political economy of agriculture and food* (pp. 213–231). Cheltenham: Edward Elgar Publishing.

Lerner, A. M., Eakin, H., & Sweeney, S. (2013). Understanding peri-urban maize production through an examination of household livelihoods in the Toluca Metropolitan Area, Mexico. *Journal of Rural Studies, 30*, 52–63.

Lowe, M., & Crewe, L. (1996). Shop work: Image, customer care and the restructuring of retail employment. In N. Wrigley & M. Lowe (Eds.), *Retailing, consumption, and capital: Towards the new retail geography* (pp. 3–30). London: Longman.

Maglaras, G., Bourlakis, M., & Fotopoulos, C. (2015). Power-imbalanced relationships in the dyadic food chain: An empirical investigation of retailers' commercial practices with suppliers. *Industrial Marketing Management, 48*, 187–201.

McEntee, J., & Agyeman, J. (2010). Towards the development of a GIS method for identifying rural food deserts: Geographic access in Vermont, USA. *Applied Geography, 30*(1), 165–176.

McKinnon, R. A., Reedy, J., Morrissette, M. A., Lytle, L. A., & Yaroch, A. L. (2009). Measures of the food environment. *American Journal of Preventive Medicine, 36*(4), S124–S133.

Ministry of Finance and United Nations Development Programme (UNDP). (2017). Assessing labor income inequality in Lebanon's private sector. Retrieved June 12, 2018, from http://www.lb.undp.org/content/dam/lebanon/docs/Governance/Publications/Assessing%20Labor%20Income%20Inequality%20in%20Lebanon%E2%80%99s%20Private%20Sector.pdf?download

Minten, B. (2008). The food retail revolution in poor countries: Is it coming or is it over? *Economic Development and Cultural Change, 56*(4), 767–789.
Moore, L. V., & Diez Roux, A. V. (2006). Associations of neighborhood characteristics with the location and type of food stores. *American Journal of Public Health, 96*(2), 325–331.
Neven, D., Odera, M. M., Reardon, T., & Wang, H. (2009). Kenyan supermarkets, emerging middle-class horticultural farmers, and employment impacts on the rural poor. *World Development, 37*(11), 1802–1811.
Omer, I., & Goldblatt, R. (2016). Spatial patterns of retail activity and street network structure in new and traditional Israeli cities. *Urban Geography, 37*(4), 629–649.
Pothukuchi, K. (2004). Community food assessment a first step in planning for community food security. *Journal of Planning Education and Research, 23*(4), 356–377.
Reardon, T., Barrett, C. B., Berdegué, J. A., & Swinnen, J. F. (2009). Agrifood industry transformation and small farmers in developing countries. *World Development, 37*(11), 1717–1727.
Reardon, T., & Berdegué, J. A. (2002). The rapid rise of supermarkets in Latin America: Challenges and opportunities for development. *Development Policy Review, 20*(4), 371–388.
Reardon, T., Henson, S., & Berdegué, J. (2007). 'Proactive fast-tracking' diffusion of supermarkets in developing countries: Implications for market institutions and trade. *Journal of Economic Geography, 7*(4), 399–431.
Reardon, T., Timmer, C. P., & Minten, B. (2012). Supermarket revolution in Asia and emerging development strategies to include small farmers. *Proceedings of the National Academy of Sciences, 109*(31), 12332–12337.
Richardson, A. S., Boone-Heinonen, J., Popkin, B. M., & Gordon-Larsen, P. (2012). Are neighbourhood food resources distributed inequitably by income and race in the USA? Epidemiological findings across the urban spectrum. *BMJ Open, 2*(2), e000698.
Rischke, R., Kimenju, S. C., Klasen, S., & Qaim, M. (2015). Supermarkets and food consumption patterns: The case of small towns in Kenya. *Food Policy, 52*, 9–21.
Rotem-Mindali, O. (2012). Retail fragmentation vs. urban livability: Applying ecological methods in urban geography research. *Applied Geography, 35*(1), 292–299.
Russell, S. E., & Heidkamp, C. P. (2011). 'Food desertification': The loss of a major supermarket in New Haven, Connecticut. *Applied Geography, 31*(4), 1197–1209.

Schipmann, C., & Qaim, M. (2010). Spillovers from modern supply chains to traditional markets: Product innovation and adoption by smallholders. *Agricultural Economics, 41*(3–4), 361–371.

Seyfert, K., Chaaban, J., & Ghattas, H. (2014). Food security and the supermarket transition in the Middle East, two case studies. In Z. Babar & S. Mirgani (Eds.), *Food security in the Middle East.* C. Hurst & Co Publishers Ltd, Oxford Scholarship Online.

Short, A., Guthman, J., & Raskin, S. (2007). Food deserts, oases, or mirages? Small markets and community food security in the San Francisco Bay Area. *Journal of Planning Education and Research, 26*(3), 352–364.

Thapa, R. B., & Murayama, Y. (2008). Land evaluation for peri-urban agriculture using analytical hierarchical process and geographic information system techniques: A case study of Hanoi. *Land Use Policy, 25*(2), 225–239.

Timmer, C. P. (2009). Do supermarkets change the food policy agenda? *World Development, 37*(11), 1812–1819.

UN-HABITAT. (2011). *Lebanon urban profile: A desk review report.* Beirut, Lebanon: UN-HABITAT. Retrieved February 2, 2019, from https://unhabitat.org/lebanon-urban-profile-a-desk-review-report-october-2011/

Vias, A. C. (2004). Bigger stores, more stores, or no stores: Paths of retail restructuring in rural America. *Journal of Rural Studies, 20*(3), 303–318.

Vignal, L. (2007). The emergence of a consumer society in the Middle East: Evidence from Cairo, Damascus, and Beirut. In B. Drieskens, F. Mermier, & H. Wimmen (Eds.), *Cities of the south, citizenship and exclusion in the 21st century* (pp. 68–81). Beirut, Lebanon: Saqi Books.

Walker, R. E., Keane, C. R., & Burke, J. G. (2010). Disparities and access to healthy food in the United States: A review of food deserts literature. *Health & Place, 16*(5), 876–884.

World Bank. (2018). *World development indicators.* Washington, DC: World Bank. Retrieved May 31, 2018, from http://databank.worldbank.org/data/reports.aspx?source=World-Development-Indicators

World Bank. (2019). *World development indicators.* Washington, DC: World Bank. Retrieved January 29, 2020, from http://databank.worldbank.org/data/reports.aspx?source=World-Development-Indicators

Index[1]

A

Administrative, 41, 43, 48, 237, 243, 253–254, 257, 258, 260, 261, 309, 341n3
Agriculture, 33, 43, 171, 172, 174, 244, 249, 266–268, 271, 275, 295, 341n3
Agri-food, 154, 205, 233, 289, 290, 293
Algorithm, ix, 79, 82, 103, 104, 125–127, 130, 131, 133, 134, 243
Allocation, x, 14, 243, 253–260
Alternative, ix, 13, 36, 50, 52, 54, 74, 76, 79, 124, 125, 150–152, 178, 203, 204, 209, 218, 226, 243, 268, 269, 280, 317, 318, 323
Alternative food network (AFN), ix, xi, 145–164, 173, 174, 178, 187
Animal, 151–156, 159, 161–163, 251
ANOVA, 298, 302
Applied, 46, 71, 81, 211, 212, 243, 247, 267, 293, 298, 322, 325
Approximate Dynamic Programming (ADP), ix, 101–134
Assessment, 19, 35, 37, 45, 53, 72, 83, 84, 103, 104, 108, 114–122, 251–256, 259, 260, 267, 277–280, 297
Association, 146, 154, 155, 162, 184, 211, 244, 272, 278
Assumption, 22, 36, 38, 42, 43, 45, 46, 51, 53, 55, 85, 87–89, 92, 104, 133

[1] Note: Page numbers followed by 'n' refer to notes.

E. Aktas, M. Bourlakis (eds.), *Food Supply Chains in Cities*,
https://doi.org/10.1007/978-3-030-34065-0

Athens, viii, 72, 85, 87, 91, 92
Attribute, x, 9, 14, 178, 247, 295, 296, 323
Authorities, 12, 24, 73, 74, 83, 151, 155, 161, 233, 234, 236, 237, 244
Auvergne-Rhône-Alpes, x, 240, 241, 243, 244, 259
Availability, xi, 1, 2, 8, 9, 14, 20, 25, 27, 92, 172, 211, 212, 290, 293, 294, 305, 310, 318, 322, 323, 325
Average, 35, 42, 43, 45, 46, 49, 51, 75, 89, 92, 124, 128, 130, 237, 257, 295, 301, 302, 305–307, 309, 316–319, 321–324
Awareness, x, 102, 158, 159, 161, 216, 218, 225, 226, 247

B

Baked goods, 302, 309, 313, 316, 317, 319, 323
Barriers, ix, 15, 78, 146, 147, 161, 173, 181, 182, 206, 317
Behaviour, viii, 24, 81, 161, 202, 205, 210, 212, 226, 296, 297
Benefit, 23, 34, 52, 69–92, 103, 122, 123, 154, 156, 162–164, 173, 225, 234, 238, 251, 260, 269, 271
Beverage, 102, 103, 108, 145, 324
Bottles, 102, 103, 122, 123, 127, 128, 131, 133
Bottling, 120–122
Brazil, ix, x, 148–163, 203, 205, 211, 265–282, 296
Buyers, 15, 52, 53, 146, 205, 206, 209, 210, 214, 224

C

Canned goods, 302, 309, 313, 316–319, 323
Capacity, 41, 76, 78–80, 84, 87, 88, 124, 202, 253, 274, 276, 280, 282
Capital, 15, 146, 147, 150–152, 157, 159, 161, 181, 299, 341n5
Case study, ix, 72, 85, 89, 104, 122, 135, 148, 156, 162, 174, 181–184, 193, 195, 211, 212, 239, 240, 266, 290
Category, 5, 115–119, 292, 296, 300, 310, 311, 319
Catering, x, 234–239, 241, 243, 244, 247, 257–259, 261
Central, 9, 37, 74, 123, 124, 127, 171, 172, 186, 188, 205, 212, 221, 237
Centre, 8, 9, 37, 74–76, 82–85, 89, 102, 149, 151, 155, 157, 191, 203, 211, 214, 217, 237, 243, 268, 273, 276, 281, 282, 292
Certification, 172, 173, 187, 191–193, 318, 319
Challenge, vii, 1, 8, 9, 14, 15, 18, 21, 25, 37, 54, 70, 71, 77, 80, 81, 103, 149, 191, 203, 205, 233, 234, 266, 269, 271, 273, 275, 281
Channel, 78, 114, 150, 214, 216, 219–221, 224, 258
Characteristics, 13–15, 40, 55, 73, 80, 81, 103, 150, 173, 174, 194, 234, 235, 239, 245–246, 249, 259, 272, 282, 290–292, 297, 341n3
Chicago, 38–40, 42, 44–49, 53

Children, 149, 159, 187, 266, 270–272, 277
China, ix, 171–173, 176, 179, 181, 182, 187, 192–195
Chinese, ix, 171–195
Chi-square, 298, 301, 302, 309
City, 2, 9, 10, 14, 17, 22–24, 34, 74–76, 82–85, 89–92, 121, 127, 130, 145, 146, 148, 149, 152, 153, 156, 157, 159, 163, 164, 172, 174, 179, 211, 214, 259, 266–269, 271, 273, 274, 278–282, 294
City logistics, 72, 73, 76, 82, 83, 91
Classification, 79, 115, 238, 247, 249, 259, 297–299, 301, 302, 305–307, 309–313, 316–324, 341n5, 341n6, 341n7
Closed distribution network, 102
Collaboration, vi, 76–80, 82, 83, 85, 87, 89, 91, 181, 203, 204, 206, 210–213, 217, 224, 225, 227, 258–260
Collaborative, viii, 69–92, 188, 206, 213, 219–221, 224, 225, 239, 249, 251, 259
Collect, 151, 152, 154, 157–161, 191, 257, 297
Collection, 2, 108, 145–149, 151, 154, 155, 157–159, 161, 162, 206, 211–213, 237, 241, 251, 253, 260
Combination, 3, 4, 20, 75, 179, 202, 203, 307, 341n3
Combined transport, 81
Commercial, x, 33, 36, 37, 42, 43, 147–149, 151, 153–155, 157, 159, 161, 162, 179, 235, 238, 239, 244, 247, 249, 251, 260, 324
Communication, 35, 162, 163, 188, 194, 204–206, 219, 224
Community/communities, 8, 10, 13, 17, 18, 21, 27, 33, 36, 148, 150, 155, 163, 164, 178, 180, 187, 244, 248, 249, 255, 266, 271, 300, 301
Community supported agriculture (CSA), 178–180, 182, 186, 187, 192, 194
Company/companies, 13, 42, 72, 74, 76–80, 84, 85, 88, 89, 92, 103, 104, 108, 114, 115, 121–123, 127, 128, 149, 154, 157, 159, 189, 201, 204, 206–208, 212, 216–219, 236, 244, 260, 305, 313, 323
Compare, 34, 35, 38, 45, 83, 84, 92, 253, 256, 306
Compared, 34–37, 49, 52, 71, 84, 85, 90, 91, 150, 155, 187, 208, 225, 260, 277, 310, 311, 317, 323
Comparison, viii, ix, 33–54, 70, 79, 84, 184, 241, 247–251, 269, 298
Competition, x, 54, 101, 289, 307, 317
Competitive, 76, 79, 101, 121, 161, 234, 258, 293, 317
Complex, 1, 10, 15, 20, 36, 81, 121, 146, 280
Comprehensive, 4, 14, 27, 45, 53, 83, 134, 154, 182, 183
Concentration, 70, 74, 176, 216, 272, 274, 277, 292–294

Concept, 1, 14, 17, 19, 22, 23, 26, 71, 76, 79, 81, 82, 102, 114, 121, 172, 174, 202, 219, 238
Concerns, vii, 2, 9, 13, 102, 114, 115, 150, 156, 159, 176, 178, 180, 190, 194, 233, 266, 270, 293, 311, 323
Conditions, x, 10, 52, 53, 83, 103, 115, 120, 121, 145, 149, 151, 153, 162, 164, 172, 178, 202, 225, 269, 273
Configuration, 35–38, 43, 46, 50, 53, 54, 84, 204, 251
Congestion, viii, 70, 72–74, 82, 89, 103, 115, 120, 121, 123, 130
Consolidation, xi, 74, 75, 82, 237, 238, 259, 260, 293, 325
Consumer, v, ix–xi, 15, 16, 35, 52, 53, 66n2, 150, 160, 172–174, 176–179, 181, 187, 190, 193, 194, 202, 203, 208–210, 214, 216, 218–220, 224–226, 233–236, 243, 244, 257, 258, 268, 269, 290, 292–296, 319, 325
Consumption, viii, x, 11, 24, 36, 102, 115, 124, 125, 128, 129, 148, 150, 173, 178, 181, 194, 202, 210, 216, 221, 225, 244, 271, 277, 281, 293, 296, 309
Content analysis, 3–5, 7, 212
Contract, 147, 152, 205, 211, 224, 235, 320, 321, 323
Contractual, 320, 323
Control, 10, 15, 41, 42, 54, 74, 78, 81, 120, 146, 152, 205, 224, 225, 267, 269, 271, 277, 306
Controlled environment agriculture (CEA), viii, 33–54
Convenience, 173, 177, 194, 249, 253, 289, 291, 325
Conventional, 33–54, 76, 178, 180, 183, 238, 258
Cooking, 235
Cooking oil, ix, 156–161, 163
Cooperation, 78, 79, 82, 183, 187, 192, 219, 224
Coordination, ix, 115, 202–207, 209–217, 221, 224–227, 249
Corporate, 77, 78, 108, 178, 236, 248, 291
Cost, vi, viii, x, 34–39, 45, 46, 50–54, 57, 60–62, 73, 76–80, 82, 83, 86, 91, 92, 114, 115, 120, 122, 124, 125, 130, 131, 146, 150, 155, 160, 161, 173, 177, 179, 183, 186, 189, 193, 203, 206, 208, 214, 234, 238, 249, 257–260, 266, 269, 281, 282, 292, 307, 341n1
Country, vi, viii, x, 11, 14, 16, 17, 23, 26, 74, 76, 103, 148–161, 194, 202–204, 226, 266, 270, 290, 292, 294–296, 299, 317, 322, 324, 325, 341n3
Cross-docking, 85, 123, 124, 127, 128, 130, 131, 135, 136
Cultural, 10, 19, 21, 23, 174, 176, 181, 281, 291, 296
Customer, 52, 53, 74, 80, 83–85, 87–92, 101, 102, 104, 114, 122, 123, 172–174, 177, 179–182, 189, 191, 193–195, 218–221, 225, 235–238, 247–249, 251, 252, 259, 296–302, 305, 308–313, 322, 323, 325, 341n1
Cycle, 39, 76, 129, 130, 163, 181, 209

D

Daily, 11, 17, 71, 92, 102, 122, 148, 149, 154, 191, 213, 247, 271, 305, 318, 341n8
Data, vi, viii, x, 5, 11, 18, 20, 24, 25, 27, 36, 37, 42, 45, 49, 72, 77, 78, 80, 82, 84–85, 87, 89, 90, 92, 104, 114, 123, 127–131, 162, 182–184, 187, 191, 192, 195, 206, 211–213, 226, 239–243, 251, 269, 272, 273, 275, 290, 297, 298
Dealers, 122–124, 127–129, 131, 136, 146, 151
Decision, viii, 13, 24, 34, 35, 42, 52, 78, 80, 92, 103, 115, 120–122, 125, 127, 130, 131, 133, 134, 190, 248, 260, 266, 310
Decision maker, 71, 103, 120, 122, 125, 131, 133, 134, 266
Define/definition, 1, 8–10, 12, 19, 39, 72, 114, 146, 202, 208, 234, 236–239, 291, 298, 325, 341n3
Degree of market power, 316, 321, 323
Degree of standardisation, 224
Deliver/delivery, v, vi, viii, 4, 34–36, 38, 41, 42, 47, 48, 71–76, 79, 80, 82–85, 88, 89, 92, 101–134, 155, 164, 176–179, 181, 186, 187, 189, 191, 193, 194, 216, 219–221, 224, 235, 238, 249, 251, 266, 276–278, 313, 317, 318, 320, 321, 323, 325
Demand, x, 34, 37, 48, 79, 80, 92, 102, 114, 115, 121, 122, 124, 128, 129, 148, 162, 179, 182, 190, 191, 204, 207–210, 216, 217, 220, 224–226, 235, 237, 238, 257–259, 268, 297, 308–310, 322, 342n9
Department, 183, 188, 189, 237, 243, 251–253, 259, 276, 277
Depot, 74, 82, 89, 123, 124, 126, 127, 146
Design, x, 14, 26, 77, 81, 182–184, 192, 234, 237, 239–243, 251–256, 260
Diesel, 40, 44, 76
Difference, 18, 36–38, 41, 42, 46, 48, 53, 54, 66n2, 126, 149, 180, 208, 212, 219, 247, 249, 260, 272, 277, 279, 290, 298, 301, 302, 305–307, 309, 311, 313, 317–322, 324
Digital, ix, x, 201–227, 273
Digital business platform, ix, 201–227
Direct, 11, 39, 41, 53, 76, 103, 122, 152, 154, 161, 173, 179, 193, 214, 219, 234, 236, 266, 269–271, 281, 317
Direct sales, 174, 177, 178, 180, 187, 234, 317
Discarded, x, 149, 202, 203, 214, 216, 220, 224, 225
Discrete, 72, 147, 280
Distance, ix, x, 42, 48, 49, 52, 54, 82, 83, 87–90, 92, 102, 124, 127, 128, 130, 131, 135, 136, 172, 174, 176, 179, 203, 206, 213, 224, 234, 243, 253–258, 260, 275, 277–279, 281, 282, 293, 313

Distribution, v–viii, x, xi, 10, 20, 23, 35, 37, 46, 69–92, 102–104, 115, 120, 121, 123, 124, 127, 128, 133, 150, 173, 178, 181, 184, 187, 202, 203, 211, 214, 221, 225, 234, 235, 237, 238, 240, 242–244, 249, 251–256, 258–260, 265–282, 292–294, 297, 311

E

Ecological, 34, 102, 171–174, 178, 181, 182, 187, 194
E-commerce, 72–74, 77, 80, 179, 224, 234
Economic/economies/economy, vi, viii, ix, 1, 2, 8–13, 15, 19–22, 33–54, 70, 72, 73, 78, 82, 83, 89, 92, 102, 103, 121–123, 146, 149, 154, 156, 161–164, 171, 173, 203–205, 234, 244, 249, 250, 259, 260, 266, 269, 271, 281, 292, 294–296, 341n3
Education, 147, 159, 182, 183, 187, 205, 218, 266, 271, 277
Effect, 9, 11, 13, 15, 20, 24, 25, 27, 49, 53, 54, 71, 81, 114, 122, 127, 148, 153, 157, 216, 242, 253, 256, 260, 290, 294
Effective, 73, 77, 80, 151, 181, 209, 241, 266
Efficiency, 12, 40, 78, 80, 163, 183, 188, 191, 193, 253, 260, 272, 281, 289, 293, 297
Efficient, 3, 73, 80, 157, 206, 234, 241, 258, 276
Electricity, 42, 44–46, 53, 157, 306
Electronic, 145, 209, 305
Emerging, x, 15, 21, 23, 26, 121, 163, 172, 182, 194, 195, 226, 269
Emission, v, viii, ix, 34, 36, 37, 39, 44, 48, 65, 70, 71, 74, 76, 78, 81–83, 89, 92, 108, 115, 120, 124, 125, 128–131, 133
Empirical, 2, 23, 26, 27, 34, 36
Employees, 160, 182, 193, 277, 291, 305–307, 341n5, 341n8
Empty, 103, 122, 123, 131
Energy, 34–37, 39, 41–46, 48, 49, 52–55, 61–64, 72, 120, 177
Environment, vii, viii, 8–10, 18, 20–22, 24–26, 33–54, 70, 72, 73, 75, 80, 81, 83, 91, 101, 120, 146, 159, 164, 172, 176, 178, 182, 204, 268, 270, 290–293, 301, 305, 341n1
Environmental, viii, 2, 8, 9, 11, 13, 15, 20–22, 33–54, 70–75, 77, 82–84, 89, 92, 101–104, 108, 114, 115, 122, 123, 146, 148–150, 155–157, 159, 161–164, 173, 201, 203, 204, 209, 225, 234, 249, 250, 260, 268, 269, 281
Equipment, 36, 40, 41, 43–45, 49, 52, 53, 57, 60, 81
Estimate, 39, 41, 42, 45, 53, 66n1, 92, 125, 164, 190, 216, 218
Estimated, 8, 44, 45, 48, 128, 129, 157, 159, 163, 258, 298
Evaluate, viii, 43, 71, 77, 81, 83, 86, 89, 91
Evaluated, 36, 37, 71, 80, 86, 88, 89, 92
Event, 72, 218, 280

Evidence, x, 12, 46, 52, 54, 154, 179, 181, 182, 190, 269, 289–326
Examined, 4, 35, 37, 79, 84, 85, 296
Example, vi, 9, 10, 12, 13, 18, 19, 22, 23, 34, 35, 71, 72, 74, 77, 78, 91, 121, 122, 128, 133, 150, 152, 154, 162, 163, 179, 182, 189–192, 195, 207, 221, 225, 226, 269, 291, 294, 306, 309, 310, 318, 323
Exchange, 77, 78, 80, 149, 159, 173, 204, 208, 209, 217, 225, 249, 269, 306, 325
Exclusively, 148, 155, 193, 221, 237, 240, 301
Expansion, xi, 16, 186, 289, 292–294, 322, 325
Expected, v, xi, 69, 74, 85, 124, 188, 204, 209, 210, 300, 302, 305, 306, 313, 317, 321, 323
Expenditure, 17, 290, 306, 307

F

Facilitate, 34, 42, 74, 77, 208, 209, 217, 224, 225, 244, 305
Facility/facilities, 11, 43, 74, 75, 127, 135, 147–149, 202, 237, 270, 273, 274, 276, 282
Fact, 36, 49, 77, 90, 104, 114, 133, 161, 204, 206, 209, 221, 224, 237, 301
Factor, 2, 8, 10, 14–16, 18, 20–22, 24, 26, 44, 53, 73, 80, 83, 86–90, 114, 115, 122, 125, 146, 176, 187, 203, 205, 224, 291, 296, 309–312, 318, 320, 323, 324, 341n3
Family, 172, 179, 217, 221, 225, 266–271, 275, 281, 282, 291
Farmers, ix, 152, 154, 155, 162, 172, 174, 176, 177, 179, 180, 183, 186–191, 193, 194, 203, 209, 214, 216–218, 220, 225, 234, 244, 266, 268, 270, 271, 282, 290, 293, 324
Farm/farms, x, 34, 39, 44, 66n1, 163, 172, 174–183, 186–189, 191–195, 217, 220, 221, 243, 249, 259, 269–271
Farming, viii, 43, 172, 173, 183, 186, 187, 193, 251, 266
Fast-moving consumer goods (FMCG), 71, 76, 80, 92
Feeding, x, 23, 153, 265–282
Field, vi, 1, 2, 5, 12, 19, 21, 24, 35–37, 40, 41, 44, 46, 48, 49, 51, 53, 54, 65, 71, 79, 81, 82, 105, 108, 109, 114, 120, 122, 131, 133, 212, 272
Field-based, 34, 35, 38–40, 45–49, 52–54, 56–58
Financial, 11, 12, 16, 22, 23, 25, 34, 52, 77, 146, 161, 162, 173, 174, 179, 191, 202, 203, 214, 270, 271, 273
Flows, 2, 9, 17, 20–25, 27, 36, 38, 77, 78, 81, 85, 87, 102, 109, 114, 122, 123, 130, 133, 155, 174, 186, 209, 236, 238, 324
FMCG distribution, 71, 80
Food, 1, 71, 102, 148, 172, 201, 265
Food distribution, vii, viii, x, 10, 20, 173, 178, 194, 203, 237, 251, 257, 265–282
Food hub, 233–261

Food insecurity, viii, 1, 2, 4–25, 27, 266
Food logistics, vi, 120, 121, 133, 281
Food retail/food ratailing, vii, xi, 7, 10, 15–17, 23, 289–292, 294, 296–299, 301, 309, 321, 324, 325
Food security, vi, vii, xi, 1, 2, 4, 5, 7–9, 15–27, 148, 265, 270
Food supply chain, vii–ix, 1–27, 34–36, 103, 173, 174, 176, 178, 181, 201–204, 209, 211, 225, 226, 233, 235, 238, 239, 249, 251, 254, 259, 265, 267, 281, 289, 293, 297, 301, 325
Food waste, v, vi, viii–xi, 34, 102, 148, 149, 151–157, 162, 163, 201–207, 209–211, 213–221, 224, 225
Food waste solution, 203, 204, 206, 207, 211, 213, 219–225
Food waste value, 155, 162, 164, 202
Footprint, 34, 51, 52, 102, 108, 133
Formal, 13, 17, 24, 25, 52, 146, 147, 151, 152, 154, 162, 164, 205, 206, 209, 211, 212, 218, 224
Formats, 102, 289–292, 294, 296, 298, 301–321, 324
Forms, 15, 77–79, 83, 121, 151, 163, 178, 179, 181, 183, 193–195, 204, 209, 221, 224, 302, 317
Framework, 7, 14, 15, 17, 21, 24, 27, 35, 72, 84, 147, 189, 195, 206, 207
Freight, 73–75, 78, 81, 83–85, 235, 257
Fresh, v, viii, 17, 23, 33, 77, 80, 120, 152, 177–181, 193, 195, 203, 213, 219, 220, 226, 235, 237, 244, 257, 276, 292, 302, 309, 313, 316–319, 323, 325
Frozen, 302, 313, 317, 319, 323
Fruit, 27n1, 121, 155, 186, 217, 218, 220, 237, 242, 251, 268, 271, 302, 309, 313, 317–319, 323, 325
Function, x, 13, 14, 21–24, 26, 121, 124, 125, 188, 192, 205, 208, 235, 239, 243, 247–249, 251, 260
Future, vi, viii, ix, xi, 2, 26, 35, 52, 54, 82, 92, 114, 122, 133, 183, 192, 195, 204, 209, 226, 269, 274, 281, 325

G

Generate, xi, 53, 72, 83, 85, 87, 88, 129, 163, 182, 184, 203, 208, 216, 220, 235, 238, 295
Generated, x, 44, 84, 124, 128, 154, 157, 182, 260, 269
Geographic, 87, 89, 149, 176, 179, 254–257, 275, 282, 290, 292, 294, 298, 301, 302, 305–307, 309, 311, 313, 317–319, 321–323, 325, 341n7
Geographical, vi, xi, 14, 21, 27, 206, 243, 251, 254–261
Global, v, 2, 8, 11–15, 18, 25, 27, 71, 80, 102, 114, 133, 187, 226, 269, 272, 292
Goods, viii, 71, 74, 76, 77, 81–84, 87, 102, 122, 123, 128, 133, 174, 178, 209, 269, 276, 289,

302, 307, 309, 313, 316–319, 323, 324
Governance, 146, 147, 152, 154, 155, 164, 225, 244
Government, 12, 27, 82, 102, 149, 151, 152, 158, 161, 162, 164, 211, 213, 247, 269, 271, 273, 281
Green, ix, 35, 53, 54, 101, 102, 122, 155, 161, 171, 178, 180, 267–268, 276
Greenhouse (GH), v, viii, 34–37, 39–49, 51–55, 58–65, 70, 108, 180, 186
Green supply chain, ix, 115, 131, 172, 174–181
Grocery, v, 4, 24, 77, 78, 277, 289, 297–300
Growing, vii, 12, 34, 36, 39, 41, 45, 73, 78, 102, 115, 120, 150, 161, 177, 181, 183, 190, 192, 194, 233, 268, 324
Growth, 11, 13, 34, 54, 69, 70, 75, 102, 155, 172, 174, 186, 204, 236, 293, 295, 296, 299

H

Harvest, 214, 221
Health, v, 5, 20, 77, 114, 121, 133, 146, 147, 154, 155, 162, 236, 270
Healthy, 1, 8, 17, 251, 266, 268, 270, 293
Holistic, 2, 20, 21, 23, 24, 26, 27, 121
Hours, 19, 39, 40, 103, 128, 130, 220, 253, 317, 318
Household, 2, 9, 17–19, 24, 145, 149, 157, 159, 160, 192, 205, 289, 324
Human, vii, 3, 9, 10, 25, 121, 146, 147, 153, 163, 164, 202, 249

I

Identify, viii, 2, 53, 83–85, 91, 92, 114, 115, 133, 182, 187, 234, 235, 237, 239, 266, 267, 272, 274
Impact, viii, ix, 15, 35–39, 43, 44, 48–49, 53, 54, 71, 75, 76, 79, 81, 83, 84, 89, 91, 101, 102, 108, 122, 133, 146, 154, 155, 161–163, 178, 203, 204, 210, 213, 214, 216, 220, 226, 249, 250, 260, 266, 269, 275, 281, 289, 296, 326
Implementation, 34, 36, 71, 77, 83, 84, 127, 151, 206, 234, 271
Implemented, 73, 77, 88, 91, 92, 237, 240, 267
Inclusive, vii, 1–27, 151, 293
Income, 11, 15–18, 23, 149, 193, 225, 270, 281, 295, 309, 311
Indiana, 42, 44, 46
Indicator, vii, viii, 11, 17–20, 24, 37, 52, 103, 115, 121, 272, 281
Individual, 17, 18, 24, 43, 53, 74, 78, 81, 87, 146, 150, 188, 202, 205, 221
Industry, 10, 12, 15, 21, 35, 39, 43, 45, 102–115, 120–122, 131, 133, 149, 178, 204, 205, 207, 208, 210, 220, 221, 233, 295, 323

Influence, 9, 11, 17, 18, 22, 72, 73, 115, 260
Informal, 4, 11–13, 16, 24, 26, 145–149, 151, 152, 154, 162, 164, 205, 206, 209, 211, 212, 218, 224, 225, 227
Informality, 2, 11–13
Information, x, 19, 34, 35, 37, 40, 41, 43, 45, 52, 72, 76–77, 80, 84, 87, 90, 92, 114, 125, 181–184, 188, 191, 204, 206, 209, 212, 219, 236, 241, 242, 260, 297, 317, 325
Infrastructure, 9–12, 14, 15, 71, 78, 81, 84, 86, 146, 147, 176, 186, 202, 208, 269, 272
Initiatives, 13, 75, 83, 154, 158–163, 181, 189, 203, 209, 217, 226, 234, 248, 258
Innovation, vi, 12, 13, 21, 163, 208, 209, 292
Insecurity, viii, 1, 2, 4–25, 27, 147, 266
Institutional, 15, 21, 147, 154, 173, 209
Institutional catering, x, 234–239, 241, 244, 247, 257–259, 261
Integration, 14, 79, 147, 151, 162, 189, 224, 249, 290, 293
Interaction, 9, 22, 70, 105, 120, 155, 161, 205, 206, 208–210, 341n3
Interdisciplinary, 1, 3, 15, 22, 25, 27, 272
Interest, vi, vii, ix, 10, 15, 43, 52, 77, 79, 105, 109, 159, 171, 178, 179, 187, 194, 205, 233, 239, 244, 269, 297
Intermediaries, 146, 151, 174, 176, 208, 211, 214, 216, 234, 243, 257, 294, 341n4
Internal, 77, 203, 207, 208, 235, 291, 305
International, 8, 15, 36, 71, 80, 217, 237, 292, 295, 324
Interview, x, 85, 183, 187, 190, 191, 211–213, 240, 241, 247, 258, 260, 293
Inventory, v, ix, 71, 75, 80, 102, 104, 114, 115, 121, 206, 305
Investment, viii, 34, 36, 43, 52, 60, 75, 92, 147, 163, 187, 225, 266, 296, 324

K

Knowledge, 2, 4, 12, 53, 122, 133, 147, 171, 180–182, 195, 202, 206, 217, 219, 225, 292

L

Labour, 13, 15, 34, 35, 40, 41, 46, 48, 52, 121, 145, 146, 148, 152, 157, 162, 172, 181, 214, 305, 322
Landed cost, 35, 36, 40–43, 45–48, 51–54, 66n2
Large, 23, 26, 36, 70, 74, 75, 87, 128, 149, 163, 173, 193, 203, 214, 290, 296, 341n3, 341n4
Leading, 10, 21, 23, 114, 148, 164, 268
Learning, 21, 26, 127, 176, 178, 225, 273
Lebanese, 290, 294–298, 301, 305, 317, 321, 324, 341n6
Lebanon, x, xi, 289–326
Legislation, 13, 148, 149, 152, 154, 233, 241, 260, 261
Lettuce, viii, 33–54, 218, 276

Level, v, vi, 1, 14, 15, 18, 21, 37, 39, 43, 52, 54, 70, 71, 74, 77, 78, 80, 81, 92, 103, 120, 146, 157, 164, 173, 188, 202, 205, 207, 208, 243, 244, 292, 293, 300, 341n3, 341n5
Lighting, 35, 42–44, 46, 48, 49
Limited, x, 5, 15, 19, 37, 52, 53, 77, 87, 92, 120, 124, 125, 131, 147, 152, 158, 172, 174, 178, 184, 195, 257, 260, 269, 290, 297, 317, 322
Linear, 280, 281
Living, v, vii, 11, 149, 150, 162, 266, 294, 295, 341n4
Loading, 74, 83, 87, 89, 90, 257
Local, vi, ix, x, 15, 16, 34, 37, 52, 54, 74, 83, 145, 146, 149–152, 155, 157–162, 164, 171–195, 219–221, 224, 225, 233–261, 266, 268–273, 276, 277, 280, 281, 290, 292, 293, 301, 309, 322, 325
Local food, ix, 34, 150, 174, 176, 219, 256, 261, 266, 268–270, 290, 292, 293, 309
Local products, 219, 220, 237–239, 257–259
Located, 17, 37, 39, 42, 74, 76, 84, 85, 104, 122, 127, 155, 158, 211, 226, 235, 251, 253, 266, 271, 276, 278, 294, 309, 322
Location, viii, 4, 10, 14, 16, 17, 23, 24, 35–39, 42–44, 46, 48–51, 53, 54, 75, 79, 80, 82, 84–87, 90, 91, 145, 163, 164, 172, 175, 212, 235, 251–253, 255, 258, 267, 273, 275, 277, 291, 296, 300–302, 306, 310, 311, 317–319, 323, 325
Logistics, vi, ix, x, 5, 14, 23, 71–83, 91, 102–104, 109, 114, 115, 120–123, 133, 148, 150, 152–162, 164, 178, 202, 224, 234, 235, 237, 238, 244, 247, 249, 251, 256, 259, 260, 266, 267, 269, 272–277, 280–282, 318
Long-term, vi, xi, 71, 102, 161, 323, 325
Losses, ix, x, 102, 120, 151, 201–227, 276, 282
Lot-sizing, ix, 104

M

Major, v, 14, 16, 25, 35, 38, 39, 43, 54, 69, 70, 73, 75, 84, 92, 133, 201, 204, 205, 323
Majority, 2, 5, 15, 82, 92, 104, 174, 211, 251, 299–301, 309, 318, 320, 321, 341n3
Marginalized, 2, 13, 21–23, 25–27
Market, ix–xi, 1, 4, 5, 9, 10, 14–16, 21, 23–25, 35–39, 41–48, 51, 52, 54, 72, 74, 80, 102, 147, 150, 154, 157, 162, 172–174, 176–179, 187, 189, 194, 195, 202, 205, 207, 209, 214, 216–221, 224, 225, 234–238, 243, 244, 247, 249, 258, 268, 271, 290, 292, 293, 296–298, 300, 301, 313, 316–318, 321–325
Marketing, 35, 41, 48, 114, 172, 173, 179, 181, 188, 202, 214, 216, 218–221, 224, 249, 290, 291, 296–298, 301–313, 321, 324, 325

Material, 3–5, 10, 22, 23, 39, 108, 114, 115, 120–122, 145–149, 151, 155, 157, 162, 183, 290
Meals, v, 154, 155, 235, 236, 241, 247, 259, 266, 270–272
Means, viii, 10, 16, 26, 33, 52, 70, 73, 92, 103, 120, 123, 127, 149, 178, 179, 182, 193–195, 253, 258, 259, 298, 302–305, 308, 309, 314–315, 317, 341n7, 342n9
Measures, viii, 11, 19, 71–73, 76, 77, 82–84, 91, 115, 260, 278, 279, 282, 341n5
Mechanisms, 4, 26, 80, 203–207, 209–213, 224–227, 321
Members, ix, 11, 13, 80, 154, 179, 180, 186–188, 191, 201, 212
Membership, 178, 180, 182, 183, 186–191, 193, 194
Method, 2–6, 20, 34, 36, 38–40, 49, 54, 79–81, 125, 182–184, 195, 205, 211–213, 239–243, 247, 267, 269, 272–274, 277, 280, 281, 290, 296–298
Metropolitan, 33–36, 40, 42, 43, 51–54, 61–64, 66n2, 71, 72, 85
Micro, viii, 16, 24–26
Model/models, viii, ix, 36, 49, 72, 73, 79, 81–85, 87, 91, 104, 114, 120, 121, 124, 128, 133, 145–164, 174, 180, 181, 187, 191, 193, 205, 207–209, 226, 247, 281
Modeling/modelling, ix, 35–37, 41, 42, 44, 55, 71, 81, 84, 125–127, 269, 273
Modern, 4, 10, 121, 289–294, 296, 299–326
Modern-format, 296, 298, 299, 321–323
Mogi das Cruzes, x, 266–268, 271, 273, 274, 280, 282
Monthly, 155, 159, 221, 271, 277, 306, 307
Months, 19, 42, 49, 162, 218, 219, 320, 321
Multiple, viii, 1, 15, 79, 120, 146, 150, 152, 163, 182, 203, 205, 208, 233, 235, 237, 293, 307
Municipal, ix, 34, 148, 150–152, 154, 155, 162, 271, 281

N

National, 15, 157, 159, 164, 234, 244, 258, 268
Natural/nature, 4, 9, 10, 12, 16, 18, 20, 23, 36, 50, 52, 53, 101, 114, 145–147, 156, 182, 187, 202, 214, 221, 243, 249, 258, 268, 270, 272, 291
Negative, 15, 20, 27, 36, 53, 54, 70, 101, 133, 180, 203, 216, 218, 234, 266, 293, 295
Neighbourhood, 2, 9, 17, 22–24, 26, 149, 155, 160, 272
Network/networks, v, vi, viii, ix, 3, 14, 15, 77, 78, 81, 87, 90, 102, 122, 128, 145–164, 173, 177, 204, 206, 225, 244, 258, 271, 276
New York, vi, 39, 42, 44, 45, 48, 49, 51, 53
Non-perishable, 302, 313, 316–319, 323
Non-standard, 202, 203, 209, 210, 214, 217–219, 221, 225

Number, viii, 8, 11, 14, 21, 22, 35, 36, 42, 44, 49, 54, 74, 76, 81, 82, 85, 89, 90, 92, 102, 105, 106, 108, 124–128, 130, 131, 134, 149, 153, 158, 177, 181, 182, 188, 194, 234, 241, 243, 257, 259, 260, 265, 270, 271, 278, 280, 291, 298, 300, 302, 305, 307, 309–312, 316–319, 322, 323, 325, 341n4, 341n5
Nutritional, 5, 8, 9, 18, 164, 202, 214, 219, 251, 271, 294

O

Objective, 4, 35, 38, 81, 114, 131, 150, 211–213, 219, 235, 239, 267, 341n5
Observation/observed, 21, 87, 90, 103, 109, 125, 126, 128, 176, 179, 182–184, 190, 211, 212, 214, 220, 227, 293, 296, 309, 322
Offer/offering, 9, 16, 22, 27, 84, 157, 172, 174, 179, 181, 191, 193, 195, 208–210, 216, 219, 241–243, 247, 249, 251, 257–259, 271, 282, 291, 293, 307, 311–313, 322
Operate/operating, ix, x, 16, 26, 36, 39, 41, 43, 54, 75, 77, 85, 92, 122, 127, 131, 150, 151, 173, 179, 180, 187, 188, 208, 217, 220, 236, 297, 322
Operational, 25, 71, 77, 78, 81, 115, 120, 123, 131, 183, 188, 189, 280, 281, 291, 296, 297, 301–313, 322, 324, 325
Operation/operations, 14, 24, 34–37, 39–46, 48, 51–54, 60–65, 80, 81, 83–87, 91, 101–104, 108, 114, 115, 120, 121, 133, 183, 187, 195, 260, 267, 269, 271, 273, 275–277, 280–282, 290, 297, 298, 301, 303–305, 309, 322, 325
Opportunity/opportunities, 8, 13, 14, 16, 24, 26, 34, 43, 52, 74, 78, 80, 83, 91, 148, 149, 155, 157, 160, 161, 163, 164, 171–173, 176, 204, 217, 226, 266
Optimisation, 26, 76, 78, 87, 103, 121, 125, 205, 243, 259, 341n1
Order/orders, 12, 74, 80, 87–89, 92, 103, 108, 114, 120, 177, 179, 187, 189, 191, 192, 216, 219, 226, 235, 257, 267, 309, 312, 320, 323, 325
Organic, ix, x, 145, 151, 154, 171–195, 233–235, 237–240, 242–244, 247, 249, 251, 258–261, 268, 318
Organic local, ix, 171–195, 237–241, 243, 244, 247, 251, 257–259, 261
Organisations, 4, 5, 18, 24, 101, 114, 147, 151, 152, 164, 178, 187–189, 205, 206, 209, 211, 235, 244, 248, 249, 260, 266, 277, 291
Orientation, 290, 296, 297, 300–302, 305–307, 309, 311, 313, 316–319, 321–323, 325, 341n7

Outcomes, 4, 7, 34, 35, 49, 50, 52–54, 78
Owned, 208, 248, 260, 302

P

Packaging, 40, 41, 46, 102, 108, 123, 127, 202, 220, 249, 251, 301, 320, 323
Parameters, 35, 42, 87–89, 103, 124, 126–128, 134, 182, 195, 251, 252
Participation, xi, 15, 41, 76, 271, 272, 289, 311, 325
Patterns, x, 11, 81, 209, 218, 266, 272, 273, 290, 292–296, 322
Payment, 209, 249, 321, 324
People, v, vii, 1, 8, 9, 11–13, 22, 27, 69, 122, 148, 163, 173, 177, 182, 187, 192, 211, 213, 219, 221, 247, 259, 341n4
Perform, 76, 84, 85, 238, 241, 249, 251, 260, 274, 276, 277
Performance, 14, 35, 46, 52, 77–79, 114, 115, 121–123, 127, 134, 206, 218, 256, 273
Performed, 86, 241, 247, 251, 277, 298, 301, 302
Period, 49, 74, 80, 87, 130, 182, 218, 270, 272, 292, 294, 295, 298, 299, 320, 321, 324
Perishable, 120, 269, 272, 313, 323
Peri-urban, 33, 39, 152, 153, 162, 172, 226, 293, 295, 296, 341n3
Personal, 35, 147, 162, 163, 173, 179, 191, 227
Phase, 84, 174, 211–213, 217, 219–223, 240–243, 260
Physical, 1, 8–10, 12, 14–19, 22, 23, 80, 92, 145–147, 172, 212, 234, 249, 305, 322
Pickers, 145–147, 151, 155, 162, 164
Picking, 145–149, 151, 163
Pickup/pick-up, viii, 74, 79, 101–134
Pickup and delivery problem, 123
Place, 9, 10, 14, 17, 71, 78, 103, 115, 150, 159, 186, 187, 190, 193, 195, 202, 217, 219, 255
Planning, ix, 8, 25, 27, 71, 80, 81, 103, 104, 114, 120, 133, 172, 190, 192, 202, 216, 220, 224, 273
Plant, viii, 34, 58–64, 92, 146, 154, 220, 267
Plant Factory (PF), 34–37, 39–49, 51–54, 65
Platform, ix, x, 77, 174, 177, 181, 183, 191, 194, 201–227, 235, 237, 243, 244, 249, 273, 297
Point, 2, 4, 10, 16–19, 27, 74, 82, 83, 85, 103, 122–124, 127, 128, 130, 131, 135, 136, 149, 154, 157–159, 161, 217, 243, 258, 266, 267, 273–282
Policy, vi, xi, 8, 13, 14, 21, 25, 27, 35, 70, 73, 74, 121, 125–127, 145, 149, 151, 157, 164, 171, 234, 251, 258, 269, 271, 272, 293, 294, 325
Policy framing, 151

Political, vii, 6, 7, 9–10, 15, 17, 22, 83, 147, 205, 243, 257
Pollution, 70, 73, 114, 115, 148, 150, 161, 164, 273
Population, v, vii, ix, 8, 11, 14–16, 21, 25–27, 69, 71, 103, 145, 148, 149, 159, 161, 163, 164, 268, 270, 293–295, 297, 300, 341n3, 341n4
Positive, 16, 24, 54, 70, 75, 122, 162, 204, 218, 224, 225, 249, 260, 295
Possible, xi, 12, 23, 25, 37, 46, 82, 91, 109, 134, 146, 149, 164, 182, 202, 213, 214, 241, 242, 260, 267, 309
Potential, ix, 13, 15, 23, 34, 35, 37, 44, 52, 53, 66n2, 76, 77, 79, 80, 82, 83, 87, 115, 122, 125, 131, 133, 159, 164, 172, 181, 204, 234, 241, 258, 260, 273, 289
Poverty, v, vii, 2, 4, 6–8, 11–13, 21, 146, 295, 296
Power, 2, 9, 10, 23, 147, 149, 150, 159, 163, 180, 195, 205, 225, 269, 316, 317, 321, 324
Practical, 4, 12, 26, 83, 91, 122, 226, 282
Practice/practices, viii, xi, 12, 14, 21, 66n1, 71, 73–74, 77, 79, 83, 84, 92, 104–113, 131, 147–150, 153, 154, 157, 160, 178, 179, 182, 189, 194, 204–206, 224, 226, 235, 244, 249, 251, 277
Presence, 21, 149, 155, 176, 270, 293
Present, vii, 2, 5, 23, 52, 54, 66n1, 72, 104, 105, 108, 115, 122, 123, 127, 128, 130, 131, 149, 174, 203, 208, 213, 221, 226, 255, 270, 273, 280, 296, 312
Price, xi, 1, 16–18, 52–54, 66n2, 80, 129, 150, 158, 180, 190, 203, 214, 217, 225, 237, 249, 291, 294, 305, 307, 309, 310, 318, 320, 323
Pricing, 16, 307, 309
Primary, viii, 18, 26, 84, 87, 151, 183, 187, 207, 208, 211, 212, 236, 241, 247, 296, 297, 300–302
Private, 23, 72, 78, 80, 81, 102, 151, 157, 177, 178, 191, 226, 235, 236, 240, 248, 325
Problem, viii–x, 71, 79, 80, 82, 83, 86, 101–134, 146, 156, 159, 162, 179, 182, 183, 186, 189, 202, 203, 205, 206, 213–217, 219–221, 224, 272, 280
Process, x, 4, 5, 9, 10, 21–23, 25, 26, 36, 38, 45, 74, 76, 77, 81, 83, 85, 120, 121, 123, 125, 133, 146–148, 152, 155, 159–161, 163, 164, 183, 187, 188, 190, 192, 193, 207, 211, 212, 214, 216, 218–221, 225, 241, 249, 259, 272, 276, 280, 290
Processing, 35, 73, 146, 157, 163, 174, 191–193, 202, 249, 273, 274, 282
Produce, v, 2, 17, 22, 23, 25, 35, 37, 38, 47, 48, 121, 152, 159, 163, 173–181, 186, 187, 190–195, 218, 219, 238, 268, 323

Producer/producers, x, 121, 148, 154, 155, 162, 172–174, 176, 178, 179, 181, 194, 202, 203, 208, 209, 211, 213–221, 224–227, 234, 236–239, 242–244, 247–249, 251–261, 266–269, 275–278, 280, 281, 293, 313, 316, 321, 323, 324
Product/products, x, 16, 17, 23, 24, 34–38, 45, 49, 53, 76, 78, 81, 103, 115, 120–123, 159–161, 174, 178, 179, 182, 187, 188, 191–193, 202, 203, 207–212, 214, 216–221, 224–227, 234, 235, 237–240, 242–244, 247, 249, 251, 252, 257–260, 266, 269–272, 276, 277, 280, 290–294, 302, 309–311, 313, 316–325
Production, 8, 9, 54, 71, 102, 155, 173, 214, 268, 269
Profile, 120, 298, 301, 308, 309, 322, 325
Programme, 147, 152, 155, 157, 159, 265–282
Programming, 101–134, 280, 281
Promote, 173, 181, 206, 209, 210, 213, 219, 224–226, 247, 266, 270, 294
Proximity, 75, 150, 234, 243, 244, 251, 253–261
Public, 5, 15–17, 73, 74, 81, 114, 146, 149, 158, 159, 161, 194, 195, 233, 234, 236, 237, 243, 244, 248, 249, 267, 269–272, 282
Purchase, 51, 154, 181, 189, 191, 202, 217, 218, 221, 225, 235, 266, 270, 271, 313, 318
p-value, 301, 303–304, 308

Q
Qualitative, x, 102, 184, 195, 211, 239, 240
Quality, v, xi, 8, 54, 92, 101, 103, 121, 150, 152, 155, 157, 162, 172, 173, 186, 190, 192, 193, 195, 202, 203, 219, 220, 225, 234, 236, 241, 244, 247, 251, 257, 258, 266, 267, 269–271, 274, 275, 277, 281, 290, 293, 298, 309, 310, 318, 323–325
Quantitative, x, 102, 120, 122, 133, 184, 195, 239, 240, 242, 281, 298
Questions, 4, 19, 20, 26, 179, 183, 204, 212, 241, 267, 269, 290, 294, 296–298, 321

R
Range, v, 4, 52, 83, 173, 174, 177–179, 182, 191, 194, 195, 273, 277, 291–293, 305, 309, 311, 316, 320, 323
Rates, 15, 16, 42, 43, 46, 146, 147, 155, 163, 268, 306, 311, 318
Receive, 25, 54, 92, 109, 123, 173, 187, 191, 219, 220, 225, 266, 317
Recent, vi, 80, 82, 103, 115, 121, 149, 152, 176, 226, 265, 266, 290, 292, 296
Recycling, vi, 108, 122, 145–164
Reduce/reduced/reducing, viii–x, 34, 38, 53, 54, 71, 73, 74, 78, 79, 83, 87, 89, 92, 101, 102, 122, 123, 133, 146, 147, 152, 153, 161–164, 173, 176, 178, 179, 189, 191, 201–227, 234, 235,

242, 249, 251, 255, 257–259, 276, 281, 282
Reduction, viii, x, 71, 75, 76, 89, 155, 203, 206, 210, 224, 225, 256
Region/regional, x, 18, 37, 43, 87, 149, 157, 164, 214, 216, 219, 237, 240, 241, 243, 244, 247–250, 257–259, 261, 267, 268, 272, 277–279, 290, 292, 296, 324, 326
Regular, 87, 189, 191, 235, 244, 247, 249
Related, x, 6, 15, 41, 42, 73, 81, 113–115, 125, 133, 152, 202, 203, 205, 211–217, 219, 221, 224, 225, 233, 234, 239, 247, 273, 294, 297, 321
Relation, 9, 10, 92, 120, 204, 211, 214, 219, 220, 253, 259, 325
Relationship/relationships, ix, 4, 9, 10, 12, 114, 149, 152, 173, 174, 176, 178, 179, 188, 194, 205–208, 211, 212, 217–221, 224, 225, 239, 281, 290, 296–298, 323, 324, 326, 341n1
Relative/relatively, 16, 51–53, 75, 102, 105, 133, 150, 177, 180, 181, 190, 293, 294, 301, 310, 312, 317, 319–321
Relevant, 3, 5, 9, 10, 12–17, 19, 25, 27, 35, 39, 52, 53, 80, 81, 108, 202–204, 211, 214, 220, 225, 268–270, 277
Representative, 34, 35, 38, 133, 213, 272, 298
Require/required/requires, v, 4, 10, 14, 19, 35, 37, 39, 41–46, 48, 52, 70, 76–78, 82, 83, 87, 92, 102, 103, 115, 121, 127, 128, 147, 155, 161, 177, 188, 191, 192, 195, 203, 216, 220, 234, 269, 297, 319, 323, 325
Residents, ix, 11, 17, 23, 25, 26, 73, 160, 244, 266, 294, 341n3
Resource/resources, viii, 8, 9, 13–15, 25, 36, 39, 54, 78, 101, 102, 114, 121, 124, 145, 149, 157, 161, 164, 203, 206, 214, 267, 271, 273, 293
Respect, 11, 26, 102, 120, 121, 125, 133, 176, 194
Respondents, 19, 241, 297, 299, 301, 307, 309, 311, 312, 316, 320, 321
Response/responses, 42, 207, 270, 297–300, 307, 309–312, 316, 317, 319, 320
Responsible, 41, 121, 133, 154, 162, 178, 189, 190, 203, 208, 249, 274
Restaurants, 54, 151, 152, 154–159, 221, 235, 236, 291
Retail, vii, xi, 6, 7, 10, 14–17, 23, 24, 37, 187, 193, 205, 209, 214, 215, 224, 235, 289–294, 296–325
Retailer/retailers, viii, x, xi, 4, 10, 16, 24–26, 74, 84, 85, 123, 127, 157, 193, 202, 203, 214, 238, 247, 289–326
Retailing, v, 16, 176
Retail transformation, 289, 290, 324, 325
Return, 102, 124, 125, 128, 129, 162–164, 173, 243, 252

Reverse, vi, ix, 51, 103, 104, 109, 122, 123, 133, 145–164, 270
Reverse logistics, vi, ix, 103, 104, 109, 122, 123, 133, 145–164
Review, ix, 1–6, 14, 20, 26, 27, 36–38, 72–82, 103, 120, 131, 133, 174, 193, 247–251, 267–273, 290–296, 298
Route/routes, 14, 73, 80, 82, 102, 123–125, 130–132, 172, 243, 251–253, 256, 259, 276, 278, 279
Routing, vi, viii, ix, 86, 101–134
Rural, vii, xi, 6–9, 11, 18, 25, 78, 148, 152, 187, 195, 219, 234, 268, 271, 272, 274, 290, 292–296, 301, 302, 305, 306, 313, 321, 323, 341n3, 341n4
Rural-oriented, 305–307, 311, 313, 316–321, 323

S

Safety, xi, 72, 152, 154, 162, 172, 176, 194, 293, 310, 318, 319, 323–325
Sales, 16, 43, 102, 114, 161, 184, 186, 187, 189–193, 214, 224, 234, 249, 291, 296, 299, 300, 317
Sample, 76, 125, 155, 241, 301, 322, 325
Sao Paulo, vi, 203, 211, 214, 217, 221, 226, 267–268, 272
Savings, 72, 76, 78–80, 82, 92, 120, 203, 208, 292
Scale, 9, 16, 19, 24, 26, 53, 54, 70, 83, 182, 203, 226, 249, 257, 259, 269, 272, 291–294, 305
Scenario, 35, 43, 45–46, 50–51, 53, 54, 72, 82, 84–92, 104, 195, 251, 253, 260, 267, 280
Scheduling, ix, 71, 104, 120
School, x, 13, 24, 154, 155, 159, 187, 218, 236, 266–268, 270–282, 291
School feeding programme, x, xi, 265–282
Secondary, 85, 147, 211, 212, 235, 241, 297
Sector, ix, 4, 12, 13, 16, 24–26, 42, 70, 71, 77–80, 92, 146, 147, 149, 151, 152, 157, 181, 184, 204, 208, 213, 226, 235, 244, 249, 258, 259, 269, 271, 290, 292, 293, 295, 317, 324, 325
Security, 151, 164, 193, 226
Selecting, 3, 312, 318, 320, 323
Selling, 53, 163, 176, 187, 191, 192, 211, 214, 221, 289, 292, 307, 309, 313, 316, 318–320
Sense, vi, 21, 82, 102, 158, 190, 202–204, 218–220, 225, 239, 282
Served, x, 10, 124, 157, 176, 181, 220, 237, 247, 277, 281, 282
Serve/serves, x, 2, 75, 83, 123, 127, 221, 236, 237, 252, 276, 278, 299, 301, 309
Service/services, vi, x, 11, 12, 16, 41, 71, 76–78, 83, 89, 91, 102, 128, 129, 149, 155, 173, 179, 182, 186, 188, 189, 194, 195, 205, 208, 209, 216, 219, 235, 237, 238, 244, 247, 249, 269, 291, 295, 312, 313
Serving, 49, 53, 85–87, 123, 130, 131, 179, 234, 276, 302, 305

Setting/settings, vi, 8, 79, 83, 104, 123, 124, 127, 130, 134, 163, 172, 174, 209, 235, 244, 296, 307, 341n3
Share, 69, 70, 80, 88, 89, 163, 173, 205, 207, 226, 234, 237, 259, 260, 268, 292, 294, 297, 300, 310–312, 317, 325, 341n3
Shared, 75, 77, 78, 159, 187, 206, 208, 269
Shared Harvest, 186, 191–193
Sharing, 78, 80, 84, 204, 206, 225, 237, 248, 273, 281
Shelf, 270, 302, 305, 306
Short, 80, 164, 172, 176, 186, 193, 237, 243, 278
Short food supply chain (SFSC), 176, 234, 237, 254, 281
Short supply chain, 176, 179–181, 194, 195, 224, 226, 240, 241, 244–251, 259–261
Simulation, viii, 41, 42, 71, 72, 80–86, 89–92, 125, 127, 280, 281
Single, 18, 41, 181, 182, 188, 189, 195, 207, 208, 221
Small, 36, 38, 44, 88, 123, 128, 133, 146, 152, 154, 159, 176, 178, 181, 186, 187, 195, 202, 211, 214, 216, 217, 219, 225, 226, 271, 278, 290, 291, 297, 301, 305, 309, 317, 322
Small-scale, 9, 13, 76, 172–174, 182, 187, 191, 193
Snack, 302, 313, 317, 318, 323
Social, ix, 2, 8–10, 12, 13, 15, 16, 18–23, 26, 34, 83, 84, 89, 102, 114, 115, 122, 133, 146–148, 151, 152, 157, 161–163, 172–174, 176, 177, 179, 203, 204, 209, 225, 226, 234, 236, 249, 257, 260, 270, 281
Society, 2, 8, 10–13, 15, 21, 83, 91, 161, 234, 244, 272, 296
Socio-economic, xi, 10, 22, 146–148, 322
Soft-drink/ Soft-drinks, 103, 104, 120–122, 131
Solid, ix, 148, 150, 151, 154, 155, 161, 163, 164
Solution/solutions, x, 4, 13, 25, 73, 74, 78, 81–83, 124, 125, 149–151, 179, 182, 201, 203, 204, 206–208, 211, 213, 219–221, 224, 225, 234, 244, 257, 258, 269, 275, 277–280
Space, 8, 11, 39, 43, 73, 74, 127, 148, 164, 178, 219–221, 302, 305, 306, 322
Spatial, 23, 25, 267, 272–274, 277, 281, 290, 292–294, 297
Speed/speeds, viii, 89, 103, 115, 120, 122, 124, 125, 128, 130, 131, 163
Square, 276, 302, 305, 306
Stages, 26, 121, 149, 155, 161, 181, 182, 202, 210, 212, 213, 216, 224, 225, 271, 301
Stakeholder/stakeholders, ix, x, 12, 14, 20, 73, 86, 152, 204, 206, 211, 213, 226, 236, 238–242, 244, 247, 257, 259, 260
Standard/standards, 16, 36, 42, 44, 78, 150, 154, 155, 203, 209, 214–217, 224, 269, 293, 310, 319, 323–325, 341n8

Start, 21, 27, 124, 158, 174, 186, 192, 206, 218, 258, 282
Started, 18, 159, 217, 218, 221, 258
State, 13, 72, 86, 91, 103, 104, 125, 150, 152, 156, 157, 159, 172, 270, 281
Storage, 114, 120, 121, 202, 220, 221, 267, 269, 274, 275, 302, 305
Stores, 16, 146, 193, 219, 238, 289, 291, 297–299, 302, 305–307, 309–311, 323
Strategy/strategies, 20, 24–26, 34, 71, 76, 78, 79, 81, 150, 159, 172, 204, 206, 218, 234, 237, 240, 256–260, 290, 292–294, 307
Structures, vi, x, 11, 24, 26, 36, 39–41, 43–46, 49, 52, 57, 60, 84, 90, 146, 150, 164, 187, 189, 205, 224, 237, 241, 247–249, 259, 260, 272, 289–291, 297, 313–321, 324, 325, 341n1
Success, 14, 16, 54, 80, 176, 179, 181, 187, 195, 224
Suitable, 83, 86, 172, 176, 182, 274
Supermarketisation, 289, 290, 293, 294, 322
Supermarkets, x, 10, 15–17, 23, 26, 74, 158, 160, 211, 218, 289, 291, 293, 297–299, 305, 322, 324
Supplier/suppliers, xi, 54, 76, 101, 114, 159, 206, 210, 212, 213, 221, 225, 234, 236, 238, 240, 258, 270, 281, 290, 313, 316–321, 323–325
Supply chain, 1–27, 33–54, 104–113, 171–195, 205–207, 244–251, 313–321
Supply chain simulation, 69–92
Supply Chain Simulation, viii
Supply/supplies, 33–54, 69–92, 104–113, 171–195, 205–207, 244–251, 313–321
Survey, 18, 19, 79, 180, 270, 289–326
Surveyed, 301, 302, 305, 307, 309, 311, 313, 319, 322, 323
Sustainability, vi, ix, 12, 15, 20, 21, 25–27, 33, 102–104, 108, 109, 113–115, 121, 122, 131, 133, 162, 174, 176, 201, 203–206, 209, 217, 226, 233, 234, 253, 256
Sustainable, vii, xi, 4, 21, 26, 72, 73, 76, 102, 114, 115, 122, 150, 156, 163, 171, 174, 178, 194, 203, 204, 206, 209, 212, 224–226, 233, 234, 238, 249, 260, 266, 268, 270, 271, 293, 325
Sustainable food supply chains, xi, 225, 325
Systematic, 3, 133, 247, 290, 302, 305, 309, 313, 317, 321, 324
System/systems, vi, ix, xi, 10, 14, 16, 34, 36–39, 41, 42, 45, 49, 52–55, 70, 72, 73, 75–77, 80, 81, 83, 102, 108, 109, 121, 146, 147, 149–151, 154, 156, 157, 160, 161, 163, 164, 179, 181, 192, 202–205, 208, 209, 220, 224, 233, 234, 237–243, 249, 251–256, 259, 260, 266, 267, 269, 272, 273, 275, 279, 281, 282, 289–293, 305, 325, 326

T

Technologies/technology, ix, 37, 50, 52, 71, 74, 83, 114, 163, 182, 194, 202, 204, 208, 209, 219, 305
Theoretical, 7, 104, 122, 133, 204–210, 226
Time, vii, 1, 17, 25, 27n2, 37, 42, 43, 73, 74, 77–80, 82, 89, 92, 101–103, 115, 120, 121, 124, 125, 127–129, 131, 151, 154, 156, 158, 160, 177, 193, 206, 209, 214, 218, 220, 239, 241, 243, 253, 281, 291, 295, 297, 301, 317, 318, 320, 321, 323, 341n1
Time-dependent, 103, 125, 130
Time-dependent vehicle speed, viii, 103, 115, 120, 122
Tools, ix, 15, 18, 25, 34, 72, 76, 83–85, 115, 120, 121, 133, 134, 195, 208, 243
Total, viii, 5, 37, 39, 41, 43, 44, 46, 47, 51, 53, 60, 69, 70, 85, 87, 89, 102, 108, 115, 124, 130, 131, 211, 225, 241, 278, 281, 294, 296, 297, 299, 300, 302, 305–307, 309–312, 316, 341n4
Trade, 14–16, 41, 188, 244, 296, 324
Traditional, x, 16, 162, 172, 173, 193, 194, 204, 205, 208, 219, 238, 273, 289–291, 293, 294, 296, 299–325
Traffic, viii, 70, 72–74, 81, 82, 85, 89, 90, 103, 115, 120, 121, 123, 130, 146, 235
Transformation, 13, 161, 289, 290, 293, 324, 325
Transport, viii, 15–17, 43, 44, 70–74, 76–79, 81, 85–90, 92, 102, 125, 147, 151, 154, 160, 178, 217, 238, 243, 251, 255, 258, 266, 270, 317
Transportation, x, 34, 35, 37, 39–42, 44–46, 48, 49, 52, 53, 58, 61, 70, 78–80, 102, 123, 124, 127, 133, 148, 154, 214, 217, 220, 221, 224, 237, 238, 257, 266, 277, 281
Travel, 40, 103, 115, 123–126, 130, 132
Truck, 42, 73, 77, 78, 80, 83, 87, 88, 221, 317
Trust, 172, 173, 176, 180, 183, 191, 194, 195, 205, 206, 220, 225, 227, 233

U

United Arab Emirates, 37
United Kingdom (UK), 23, 37, 114, 298, 324
United Nations, vii, 5, 8, 11, 69
United States (US), 23, 33–54, 150, 270
Urban, v, 1–27, 69, 101–134, 145, 172, 174–181, 203, 234, 268, 290
Urban agriculture, 268, 275
Urban consolidation centre (UCC), viii, 72, 75–76, 83–87, 89–92
Urban/suburban-oriented, 306, 307, 313, 316–323

V

Valuable, 12, 24, 149, 204, 260, 280
Value/values, 35, 42, 43, 45, 46, 49, 51, 53, 78, 84, 85, 102, 114, 120, 121, 125–127, 146, 148, 149, 155, 157, 161, 162, 164, 173, 176, 177, 181, 182, 191–193, 202, 204–206, 214, 224, 225, 234, 239, 243, 249, 251, 258–260, 266, 269, 271, 272, 281, 289, 293, 296, 298–300, 303–304, 306, 314–315, 324–326, 341n1, 341n5
Variety, vi, 5, 18, 80, 115, 121, 149, 191, 267, 268, 289
Vegetable(s), 27n1, 34, 35, 180, 182, 183, 186, 187, 190, 191, 217, 218, 237, 242, 251, 267, 268, 271, 272, 277, 302, 309, 313, 317–319, 323, 325
Vehicle, viii, ix, 23, 70, 72, 75, 76, 79–82, 84, 85, 87–89, 92, 102, 103, 115, 120, 122–124, 126, 128, 130–132, 252, 253, 257, 273, 276, 278, 280, 282
Vehicle routing, viii, ix, 131
Volume, vii, xi, 48, 74, 76, 78, 87, 90, 155, 190, 249, 258, 291, 309

W

Waste, v, vi, viii–xi, 23, 34, 102, 122, 123, 145–164, 201–227
Water, 10, 11, 34, 37, 39, 41, 42, 44, 45, 53, 54, 82, 108, 127, 146, 156, 157, 159, 161, 164, 251, 268
Wholesale, 38, 39, 42–45, 47, 48, 51, 52, 177, 193, 238, 313, 317, 318, 323, 325

Printed by Printforce, the Netherlands